Regulating the Revolution

I0644208

Regulating the Revolution

Democratic Deliberation and Biotechnology

NAOMI SCHEINERMAN

The University of North Carolina Press Chapel Hill

© 2026 Naomi Scheinerman
All rights reserved
Set in Charis by Westchester Publishing Services

Library of Congress Cataloging-in-Publication Data
Names: Scheinerman, Naomi author
Title: Regulating the revolution : democratic deliberation and biotechnology / Naomi Scheinerman.
Description: Chapel Hill : The University of North Carolina Press, [2026] | Includes bibliographical references and index.
Identifiers: LCCN 2026018185 | ISBN 9781469697215 cloth | ISBN 9781469697222 paperback | ISBN 9781469691725 epub | ISBN 9781469697239 pdf
Subjects: LCSH: CRISPR-associated protein 9—Moral and ethical aspects | Gene editing—Political aspects—United States | Technological innovations—Law and legislation—United States | BISAC: MEDICAL / Ethics | POLITICAL SCIENCE / Public Policy / Health Care
Classification: LCC QH438.7 .S38 2026
LC record available at https://lccn.loc.gov/2026018185

Cover art: DNA strand © Adobe Stock / rookielion.

For product safety concerns under the European Union's General Product Safety Regulation (EU GPSR), please contact gpsr@mare-nostrum.co.uk or write to the University of North Carolina Press and Mare Nostrum Group B.V., Doelen 72, 4831 GR Breda, The Netherlands.

For Yoni, Echo, and Motek

Contents

Illustrations

Figures

Tables

Acknowledgments

It is a labor of love to read and offer feedback. I am extremely grateful to the many wonderful friends, colleagues, and mentors at different institutions over the lifetime of this project whose insights and wisdom were invaluable in making this book happen. They not only contributed their own significant time and energy to the ideas and their organization but also readily volunteered encouragement, affirming that the things I wanted to say were worth saying.

Much of the work on this book began while I was in graduate school at Yale University's Department of Political Science under the invaluable guidance of Hélène Landemore, Ian Shapiro, Susan Rose-Ackerman, and Steven Latham. I was also supported by wonderful friends who read the earliest versions of these chapters and provided consistent emotional labor: Vatsal Naresh, Mie Inouye, Anne Mishkind, Nica Siegal, Joy Wang, Isabelle Laurenzi, Alicia Steinmetz, Lisa Gilson, Lizzie Krontiris, and Alyssa Battistoni. While an AI Fellow-in-Residence at Harvard University's Edmond J. Safra Center for Ethics and the Berkman Klein Center for Internet and Society, I was greatly aided by my fellow Fellows-in-Residence Roni Hirsch, Anna Lewis, Attila Mráz, Jacob Sherkow, Allison Stanger, and Deva Woodly.

During my time as a T32 Postdoctoral Fellow in ELSI Genetics and Genomics from the National Institutes of Health, National Human Genome Research Institute, I received significant support from the University of Pennsylvania Department of Medical Ethics and Health Policy's incredible scholars. I wish to thank Matthew McCoy, Jonathan Moreno, Steve Joffe, and Holly Fernandez Lynch, and my superb co-fellows Kate Saylor, KP Callahan, Rebecca Mueller, and LaKisha David, who at times read multiple drafts of the chapters.

At The Ohio State University, I have found myself enmeshed in a fabulous group of bioethicists, political theorists, and legal scholars. These individuals contributed many hours of their time reading and giving considerable feedback at a manuscript conference: Adrian Calmettes, Bridget Dooling, Sahar Heydari Fard, Abraham Graber, Dana Howard, Benjamin McKean, Michael Neblo, Blaine G. Saito, Emma Saunders-Hastings, Winston Thomas, Shameka

Thomas, Pim Trommelen, Piers Turner, Jada Wiggleton-Little, and Patricia Zettler.

Thank you profoundly to my editor, María García, for her thorough reading and guidance during all parts of this process, and to the entire editorial staff and production team at UNC Press. My thanks as well to Payton Fetzner for her assistance with citation formatting. Thank you as well to The Ohio State Library whose Open Access Monograph Initiative provided subvention funds.

And finally, in the biggest way possible, I thank my friends and family for their love and support. Thank you to Yoni Naveh, my life partner and rock, for feeding me and proofreading this entire work.

Regulating the Revolution

Introduction

Creativity amid Crisis

The United States currently faces a cacophony of overlapping crises: the crisis of *deregulation* and the gutting of the US administrative state; the crisis of *democracy*, with violations of the rule of law and the integrity of elections; the crisis of *expertise*, which discredits expert authority and scientific findings; the crisis of *health*, given rising medical costs and the imminent loss of coverage for the poorest and sickest among us; and the crisis of *climate*, with rapid temperature rises contributing to extreme flooding, drought, wildfires, and disease outbreaks. Finally, there is the crisis of *trust*: The loss of trust in the veracity of our regulatory system, our democracy, and our experts is rapidly unwinding our social fabric and shredding any unified belief system.

First, the crisis of deregulation. Early in 2025, President Donald Trump appointed Elon Musk, one of the world's wealthiest individuals and CEO of Tesla, SpaceX, and X, to head the newly created Department of Government Efficiency (DOGE). There, in the name of stopping "waste, fraud, and abuse," Musk took a "chainsaw" to the federal workforce, cutting staff and programs already approved by Congress that did everything from distributing Social Security checks to establishing affordable housing for domestic violence survivors.[1] DOGE's plans include eliminating the Department of Education, which is tasked with providing aid to struggling schools, students with disabilities, and college students through loan programs.[2] Cuts severely impacted the United States Agency for International Development (USAID), whose work in international aid relief throughout the world, such as researching and combating HIV/AIDS, was responsible for saving an estimated 90 million lives since the early 2000s. USAID cuts could, as of this writing, impact as many as 14 million people.[3] Many of these cuts specifically targeted programs that benefited women, racial and sexual minorities, and non-Americans, such as a suicide hotline service for LGBTQ+ youth, a high-risk group.[4] And yet, as of this writing, no evidence of "waste, fraud, and abuse" has been produced by DOGE. Meanwhile, Musk—a CEO put in charge of a governance system designed to curb the ills of capitalism—left the administration in May 2025. Six weeks later, one of his companies announced a $200 million contract with the Pentagon.[5]

Far worse, perhaps, is that rampant deregulation will have dire costs for human lives and welfare, largely because of cuts into health science regulation. The United States Department of Health and Human Services, founded in 1979, is mandated with enhancing "the health and well-being of all Americans, by providing for effective health and human services and by fostering sound, sustained advances in the sciences underlying medicine, public health, and social services."[6] In 2025, Trump appointed Robert F. Kennedy Jr. (RFK Jr.), a known vaccine conspiracy theorist, as HHS secretary. In addition to drastically cutting the HSS workforce from 82,000 to 62,000 federal workers, RFK Jr. fired the advisory board at the Centers for Disease Control and Prevention (CDC) that was responsible for overseeing and approving the vaccine schedule and the annual flu shot.[7] He then convened a new advisory committee, appointing many members who are not vaccine experts, and instead peddled widely debunked claims linking the measles, mumps, and rubella (MMR) vaccine to autism and the COVID-19 vaccines to higher mortality rates.[8] Soon after, on July 7, 2025, it was announced that there were 1,281 recorded measles cases in 2025—the highest number since measles was declared eliminated in the United States in 2000.[9] The new advisory board then removed the COVID-19 vaccine recommendation for children and pregnant patients, despite the fact that COVID-19 vaccination has been proven safe and effective for both groups and that contracting COVID-19 during pregnancy carries a "high risk of severe disease and death."[10]

The National Institutes of Health (NIH)—HHS's juggernaut for funding medical research in this country—comprised 80 percent of grant investment in scientific research in 2024 and supported 99.4 percent of drug approvals from 2010 to 2019, based on 356 drugs studied.[11] And yet, as of May 2025, the NIH had canceled $1.9 billion in grant money (previously allocated by Congress) for research on cancer, environmental studies, health equity, LGBTQ health, vaccine hesitancy, COVID-19, and mental health.[12] Examples of canceled grants include "A Project to Improve Recruitment of Participants in Alzheimer's Clinical Trials," "A Study to Increase Vaccine Uptake in Underserved Populations," and "A Study Investigating in Utero Exposure to Contaminants in Public Drinking Water."[13]

Also under HHS, the Food and Drug Administration (FDA), whose role since 1938 has been to oversee food quality and pharmaceutical safety and efficacy, suffered significant staff cuts, leading many to worry about its capacity to monitor the nation's milk and produce quality.[14] The FDA has also declared a central focus on accelerating drug approval pathways, potentially shortening the timeline for pharmaceutical authorization. While the FDA already had pathways to speed up the approval process, this announcement has caused many to worry that it could sacrifice proof of drug effectiveness, leading to "glorified (and expensive) placebos."[15] The loss of experts and procedures established over decades to protect food and drug quality at the FDA has been so severe that former FDA commissioner Robert Califf lamented in April 2025 that the "FDA as we've known it is finished."[16]

The Environmental Protection Agency (EPA), founded in 1970 with a broad mandate to protect human health and the environment, has had its power almost completely stripped. As with the NIH, the EPA canceled millions of dollars of research funding for environmental projects, such as those that limited pollution from tailpipes and smokestacks, increased protections for wetlands, and reduced carbon dioxide and mercury emissions from power plants.[17] The EPA also terminated $15 million in funding for protections against perfluoroalkyl and polyfluoroalkyl substances (PFAS), also known as forever chemicals because they are "everywhere and forever," can accumulate in the body at high concentrations, and "will scarcely break down over time."[18] Even at low doses, PFAS can lead to immune system suppression and cancer.[19] As in many other cases of companies hiding evidence that their products are dangerous (e.g., Monsanto, Purdue, Johnson & Johnson, among others), PFAS were produced by companies that went to inordinate lengths to convince scientists and the public that they were harmless.[20] Then, on July 25, 2025, the EPA reversed its "Endangerment Finding," which showed that greenhouse-gas emissions are warming the planet and gave the EPA the power to place limits on industry carbon emissions.[21] This flew in the face of decades of mounting evidence that climate change is human-caused and existentially dangerous.[22]

The United States Department of Agriculture similarly experienced billions of dollars in cuts to such programs as Farm to School, which provides underserved children with nutritious food, and the Supplemental Nutrition Assistance Program (SNAP), which currently provides food financial assistance to 42 million people living in 22 million households.[23] Cuts to SNAP are projected to reduce federal spending by $287 billion over the next decade and cause several million Americans to lose food aid, leaving the poorest Americans worse off.[24]

Congress's passage of the "One Big Beautiful Bill Act" in July 2025 created requirements to prove employment in order to qualify for Medicaid, an income-based federal medical coverage program. Many experts fear this will create administrative barriers to receiving Medicaid, thus leading to an estimated 51,000 preventable deaths.[25] In addition, hospital emergency departments will be burdened with newly uninsured patients, and rural hospitals in particular will become imperiled and face higher rates of closure.[26] This comes on top of an already crumbling health care system that can barely keep up with the demand for medicine in this country and whose high costs are prohibitive for many people.[27] Compounding this situation, the United States has withdrawn from the World Health Organization, threatening the country's preparedness for disease outbreaks, alongside an over 50 percent cut in funding to the CDC that will severely impair its ability to collect information and monitor diseases.[28]

As these programmatic budget cuts and regulatory rollbacks threaten actual public health, the administration instead presented an alternative agenda under the guise of addressing the "true" health needs of the country. RFK Jr.'s "Make America Healthy Again (MAHA)" plan is composed of selective politically popular issues, such as cracking down on chronic disease and type 2 diabetes. It also favors certain industry players and discredits expertise. In mid-2025, MAHA announced various victories, including requiring companies to swap out seed oils and food dyes in favor of "natural" ones and cracking down on "ultra-processed foods."[29] However, the claims that many of these agenda items will improve health are nebulous at best; for example, many seed oils have been shown to be good for heart health "if used wisely."[30] At worst, they are rooted in conspiracy, such as linking the MMR vaccine to autism, and fomented by the backlash from the 2020 COVID-19 pandemic, which led to misinformation about the very existence of the disease as well as the efficacy and safety of the COVID-19 vaccines.[31]

Crises of regulation, climate, and health have also required a systematic project of discrediting expertise, such as accusing experts of being out of touch with the American people or motivated by "gender ideology."[32] RFK Jr.'s efforts, for example, to overhaul vaccine safety oversight were, according to him, intended to "restore public confidence in the FDA."[33] However, firing vaccine experts may instead lead to a further crisis of expertise as those who *do* have confidence in the safety and efficacy of vaccines, which by many estimates is the majority of Americans, will no longer trust the government's role in ensuring their safety.[34]

This crisis of expertise is also present in the crisis of democracy and democratic legitimacy.[35] In 2025, the majority of Americans (76 percent) felt that there was a "serious threat to the future of our democracy."[36] However, in a poll from 2022, Democrats and Republicans differed on their reasons.[37]

For Democrats, the majority of whom say there is a democratic crisis (89 percent), the threats include the weakening of representative institutions, assaults on electoral systems and voting rights, and the erosion of constitutional protections.[38] This can be seen in the Trump administration's violation of court orders and its attempts to overturn birthright citizenship as established in the Constitution.[39] Meanwhile, only a little over half of Republicans believe there is a crisis; for them, the fears are the deep state, "voter fraud causing stolen elections," and "woke culture imposing out of touch liberal values."[40] Distrust of experts and elected officials permeates these concerns. Many experts add to these worries that the United States may be better described as an oligarchy rather than a democracy because of the connections and capital required to get elected.[41]

Widespread loss of trust in experts and in the stability of democratic institutions culminates in a loss of perceived legitimacy. As a consequence, the government is seen as operating in the interests of the powerful and wealthy rather than the public. The systematic dismantling of the regulatory state and the public's confidence in experts is a devastating problem. It undermines the mechanisms that protect vulnerable people from the excesses of capital, specifically regarding environmental impact and pharmaceutical safety, and will affect all areas of technology, health, and medicine.

Amid all this, pressing questions of technological and biotechnological oversight loom: Can we limit the spread of medical and vaccine misinformation on social media? Will artificial intelligence (AI) replace my doctor? Should we edit the genes of future generations? These questions require robust political solutions, so finding the proper institutional arrangement during this moment of mounting crises can feel daunting. Ethical questions, then, also quickly become political conundrums: What should be done to counteract all these crises? What is the responsibility of our government to address them? And how can we best arrange our institutions to chart a path forward?

There may be a natural urge to return to the previous system, to reinstate regulatory pathways as we knew them, or to entrench expert rule even more in the face of its rapid erosion. But despite the difficult strain of the moment, I offer that we resist the temptation to revert to the status quo wherein many felt disaffected and estranged from administrative governance. Rather, this book reimagines the ways in which the public can be incorporated into the regulatory process while simultaneously upholding expert authority, scientific integrity, and the protection of vulnerable populations. I argue that to do so, we should convene deliberative minipublics—randomly selected groups of lay participants (like a jury)—within regulatory agencies because they are effective modes of creating considered judgments on values, risk, and knowledge of technology's impacts on society. To examine how and why these diverse bodies can bolster the substantive and justice-enhancing outcomes of regulation, I explore the case of CRISPR, a powerful and efficient gene-editing tool that promises great medical advances and environmental protections while also posing significant moral risks and societal anxieties. As we will see, regulating revolutionary technologies requires a revolution in democratic thinking itself.

Expertise, Democratic Deliberation, and Technology

Tensions between democracy and expertise, though certainly heightened and, in many ways, acutely dangerous in this moment, are neither new nor unanticipated. As Sheila Jasanoff writes, the "oscillation between deference and skepticism toward experts can be observed at almost every stage in the rise of the modern administrative state."[42] Liberal democracies, as systems that centralize rule by the people through individual rights and freedoms, have long struggled with how much authority to reallocate from the people to those with unique qualifications and training in particular matters of governance.[43] Indeed, many commentators over time, including as early as the third century BCE with Plato's *Republic*, have feared democracies not just because of the way that popular rule can have mob-like tendencies—as on display during the January 6, 2021, insurrection—but also because of the general inadequacy of the people to perform the job of self-rule, such as by voting for the "wrong" people or those whose interests do not actually align with their own.[44] Hence, expertise, as we will see in chapter 3, has been seen as necessary for modern governments to answer questions that require empirical knowledge rather than ones that reflect public preferences.

Correspondingly, often when democracy and its institutions seem to be failing us, there is a technocratic tendency to entrench expert rule or safeguard the halls of power from the populist demands of the people. For example, during emergent public health crises, like pandemics, the impulse is to call for a vaccine mandate. Because vaccine mandates can save numerous lives, particularly vulnerable lives, and pose far less risk than exposure, they are justifiable. However, requirements like these also do not help to decrease vaccine hesitancy and exemption requests, leaving us (still) with a pernicious public health problem, such as rising measles outbreaks, and a populace willing to support the authoritarian tendency to discredit experts. Instead, we should take our cues from evidence in medicine around how trust can be formed and lost: People who are vaccine-hesitant tend to vaccinate their children when their provider takes their fears and anxieties seriously, rather than dismissing them.[45] Conversely, vulnerable patients—including women, pregnant, LGBTQ+, obese, and minoritized patients—may avoid seeking medical care when their pain and experiences are discounted.[46]

Given their lived experiences, patients often offer important perspectives and contribute knowledge about how people access medical care, how doctors engage with them, and what risks they are willing to accept. This can, in turn, better inform our understanding of medicine writ large and the environment in which technological innovation is introduced. Take the case of intraoperative pain during a cesarean section: Only by listening to "non-experts" have doctors learned that anywhere between 11.9 to 22.7 percent of patients undergoing cesarean sections are able to feel the surgery.[47] Adding to this understanding, the historical dismissal of women's pain and gynecological experimentation on enslaved people further illuminate the significance of this statistic—this rate of interoperative pain would not be acceptable in *any* other surgical context.[48] Thus, the combination of expert and nonexpert perspectives culminates in a broader understanding of the worrisome patient experience and provides insights into ongoing injustices.

Thus, the political system should establish mechanisms that allow modes of personal interaction to impact policy more effectively, especially when reimagining regulatory procedures for quickly evolving biotechnologies that often affect different populations inequitably. Our political system should resist rather than protect the technocratic response to populist power because it threatens to further destabilize democracy. Democratic governance, whose very core is the egalitarian distribution of power and the creation of conditions that allow individuals direct input in governance, should figure strongly when reimagining regulatory procedures, especially for biotechnologies. *Deliberative* democracy, featuring discussion, listening, and narrative telling rather than just voting and petitioning, can allow individuals to realign the imperative for regulation, the authority of experts, and commitments to equality, evidence, and representation in "reclaiming democracy."[49]

Democratic deliberation is not a novel notion—it was an integral facet of the birthplace of democracy itself. Athenian democracy in the fifth and fourth centuries BCE featured both a smaller council that prepared laws and was elected by lot every year, as well as a much larger assembly that included thousands of citizens (non-enslaved men) who engaged in prolonged deliberation and decision-making.[50] As Jon Elster has written, "The idea of deliberative democracy and its practical implementation are as old as democracy itself."[51] Indeed, many contemporary deliberative democratic theorists remind us of its ancient origins, including Amy Gutmann and Dennis Thompson who cite Thucydides (a historian of the Peloponnesian War in the fifth century BCE) in detailing that Athenians saw deliberation "'not as a stumbling-block in the way of action' but as an 'indispensable preliminary to any wise action at all.'"[52]

American democracy has also embraced a rich deliberative tradition since its founding as a republic in the mid-eighteenth century CE.[53] This was emphasized in the writings of Alexis de Tocqueville in 1835, who praised local modes of deliberation and participation in civic life as essential to the democratic fabric of America.[54] Soon after, the pragmatists of the late nineteenth and early twentieth centuries depicted democracy as a concept that was constantly evolving, required innovation and intelligent problem-solving, and was always meant to be in service of social ends and progress.

Pragmatism arose during the Progressive Era of the United States, which roughly spanned from the 1890s to the 1920s and brought large waves of immigrants—primarily from Southern and Eastern Europe—as well as widespread outbreaks of communicable diseases, largely due to unsanitary living conditions.[55] This period, correspondingly, witnessed the rollout of numerous vaccines, including those for rabies, tetanus, typhoid, tuberculosis, diphtheria, scarlet fever, pertussis, yellow fever, and influenza.[56] The Progressive Era was also associated with significant economic growth that increased class inequalities, saw the rise of industrialization, and witnessed dangerous factory work that sparked a labor movement, with workers demanding safer conditions and the power to negotiate their employment terms. Threads of these themes—economic upheaval, growing economic inequalities, and fear of immigrants as sources of disease—are all present today. As we will explore further in chapter 3, due to the growth of the population, disease, and factories, the federal government's response was to build a robust regulatory framework. The emergence of the disciplines of economics, sociology, and statistics spurred the regulatory agencies' technocratic foundations aimed at using the tools of empiricism to establish safety requirements and protect the public from the prowess of industry.[57]

The Progressive Era led many public officials and philosophers to lean into the idea that the ills of democracy required *more* democracy, such as through the creation of schools, educational opportunities, and social networks to aid and uplift the poor and sick by giving them greater voice and agency over their lived conditions. Jane Addams, in particular, believed that action was required to improve the lives of all of us and that knowledge of this improvement necessarily came from the "mixing" of diverse members of society by giving them "a wider acquaintance with and participation in the life about them."[58] Addams argued that "the foundation and guarantee of Democracy" required this kind of "diversified human experience and resultant sympathy."[59]

American democracy has in many ways always been one giant experiment, requiring modifications and reinvention over time in response to diversity. John Dewey, a contemporary of Addams, saw democracy not just as a form of government but as an ethical and social ideal that required active participation and continuous social reconstruction. Dewey held that democracy could not be isolated to one particular mechanism but rather was a "wider and fuller idea."[60] This, he maintained, required experimentation with participation to maximize the best outcome. Dewey argued ultimately that a "society which makes provision for participation in its good of all its members on equal terms and which secures flexible readjustment of its institution through interaction of different forms of associated life is in so far democratic."[61]

Today we often think of specific mechanisms as democracy—calls to "strengthen" democracy or even just to "democratize" often evoke images of voter rights and liberal protections, such as enfranchising ex-felons and securing free and fair elections by declaring Election Day a holiday and expanding access to mail-in ballots. It may also include protecting the right to protest, unionizing workplaces, and fighting gerrymandering. While election integrity and social justice goals are crucial rights protections, democratic *engagement* can mean more than voting, protesting, demonstrating, and organizing. If we imagine democratic mechanisms as a kind of toolbox, wherein we use different processes for inclusion and entry points for participation, we can create better guidance, representation, and outcomes related to how the laws of our state impact our lives. As Archon Fung writes, "Citizens can be the shock troops of democracy. Properly deployed, their local knowledge, wisdom, commitment, authority, even rectitude can address wicked failures of legitimacy, justice, and effectiveness in representative and bureaucratic institutions."[62]

Beginning in the late twentieth century, scholars began experimenting with deliberative minipublics (DMPs) as promising modes of institutionalized deliberation that harness productive nonexpert contributions to policymaking. DMPs are bodies of randomly selected individuals acting as proxies for the people. Experimentation with DMPs has been an ongoing and expansive process for several decades, well into the twenty-first century, as DMPs have been organized throughout the world in a variety of different contexts, institutions, and topics. According to a 2020 report by the Organization for Economic Cooperation and Development, which analyzed hundreds of DMPs worldwide, the most popular topics for DMPs were health, environment, science, technology, and innovation.[63] DMPs rose in popularity during the 2020 COVID-19 pandemic to deliberate on resource allocation and vaccinations, as in the case of Oregon's online Citizen Assembly on COVID-19 Recovery.[64] More recently, there has been a wave of Climate Citizens' Assemblies throughout Europe and Australia, some supported by climate-action social movements.[65]

DMP scholarship aspires to institutionalize what Jürgen Habermas referred to as "informal deliberation," or talk that happens generally in the public sphere, directly into realms where it would have an impact on policy and law, such as legislatures.[66] A lottery system, the argument goes, could produce a deliberative body comprised of a more accurate representation of the people, creating a kind of "democratic mirror" that reflects what the population truly looks like.[67] This "age of sortition" reflects a deliberative process that can truly deepen our democratic commitments by "equaliz[ing] agency."[68] The "systemic turn" in deliberative democratic theory refers to deliberation becoming equated with DMPs themselves, as opposed to other institutional or noninstitutional modes of "talking politics."[69]

Most DMP theorizing has been applied to legislative lawmaking to disrupt power by reforming representative systems. For example, the 2019 "Real Utopias" volume *Legislature by Lot* explores the ways that sortition has or could herald the next wave of democratic governance.[70] The book's essays predominantly focus on the use of DMPs in either supplementing or replacing representative-based institutions. In tandem, researchers also analyze DMPs for opportunities to "scale up" or create a "spreading effect," wherein the DMP deliberation rejuvenates or strengthens a kind of civic culture in the broader public.[71] The impacts range from enhancing our knowledge on how and when to convene DMPs to producing tangible legal changes. One of the most prominent examples is the Irish Citizens' Assembly, which paved the way for legalizing marriage equality and abortion rights in 2016.[72] While there have been successful applications of DMPs in the legislative context, there has been less scholarship and application in other realms of governance, including regulation.

I build on this foundational work by transplanting the case for "unelected power" into the administrative state, arguing in favor of *regulation by lot*.[73] Echoing Robert Dahl's quest to "perfect and deepen" our understanding and commitments to democratic ideals, I contribute to the deliberative democracy discourse by carving a path forward that will bolster the democraticity of the regulatory process.[74] DMPs, I will show, not only offer ways to make legislatures more representative and reflective of the people but also reveal how the public can engage with experts and scientists within administrative agencies on questions of scientific knowledge within a democratic system. Namely, they intervene in debates regarding the extent to which those deemed experts versus nonexperts should be granted authority or a say in rulemaking.[75] This shows up strongly in bioethics debates on the moral and political implications of biotechnologies, such as AI in medicine, mitochondrial DNA donation, artificial wombs, and, in this book, gene editing in particular, as exemplified by the 2021 Australian Citizens' Jury on Genome Editing.[76]

The regulatory and administrative agencies in the United States are thus an important next frontier for exploring the tension between democracy and expertise due to their institutional function as sites where normative and political debates on new and emerging technologies take place.[77] Reconfiguring regulatory institutions with expertise alongside nonexperts can help decrease the overt influence of industry and the public perception of illegitimacy. And while DMP justification often emphasizes the intrinsic value of using smaller subsets of the public to act on behalf of the broader public, I argue we should concern ourselves with whether these bodies aid in creating good outcomes. Namely, we should evaluate the legitimacy and merits of DMPs by examining whether they better the agencies' considered judgments. The aim is to incorporate deliberative minipublics into the realm of administration to create regulations that are substantive, mitigate abuses of power, and cultivate trustworthy institutions.

This book, then, reimagines the relationship between democracy and expertise as one of cooperation rather than opposition. Its central argument is that deliberative minipublics should be included early in the decision-making process within regulatory agencies because they can help improve the collective judgments of their hosting institutions by articulating and applying societal values alongside empirical evidence. They can also examine the risks and uncertainties inherent in regulating industries and other sectors, such as biotechnology. The mechanism by which our collective judgments improve is through diversity: Random selection acts as a proxy for diversity and equalizes power, the ultimate foundation of democracy. Diversity delivers on three main instrumental goals by (1) decreasing the problem of power abuses (particularly arbitrary and disciplinary power), (2) increasing epistemic outcomes (better informed decision-making), and (3) cultivating a healthy yet skeptical trust in the regulatory process, which in turn fuels democratic stability and the institutional goals of the regulatory agencies overall. These instrumental outcomes create the foundations of legitimate governance and are connected through a feedback cycle: non-domination begets epistemic merits, which allow for trust, which in turn reduces domination, and so on.

Is a DMP's authority, however, legitimate? In other words, can we justify allocating authority to nonexperts through this procedure, and for what reasons? While many scholars over the years have offered treatises on the concept of legitimacy itself, my understanding aligns with Hélène Landemore's in which DMP legitimacy rests on "instrumental, consequentialist, and 'best outcomes' arguments for democracy."[78] DMPs are legitimate, as such, insofar as they are instrumentally valuable or produce good consequences.[79] I argue, then, that DMPs are legitimate democratic innovations within the regulatory process because they aid in substantive rulemaking, promote justice-enhancing outcomes, and cultivate institutional trustworthiness.

Others have offered additional consequential defenses of DMPs, noting that deliberation creates a more compassionate, informed, and thoughtful citizenry, or that deliberation is valuable because the practice of conversation and mediated interests is healthy and good for individual well-being.[80] Still others argue that it leads to "better politics" by increasing mutual civility, which, in turn, makes citizens "happier" by feeling respected and less isolated.[81] While these benefits are indeed desirable and certainly needed in today's polarized and agonistic environment, I hold that DMPs are legitimate sites for more equally distributing power and thus sources of political and moral authority because they produce non-dominating, epistemically enhancing, and trust-cultivating outcomes. Regulatory agencies, as institutions of expert-based power intended to mitigate risk and promote safety, should remain expert-based, but they can benefit substantively when their democraticity is deepened through DMP inclusion.

The Case of CRISPR Gene Editing

Just as our democratic thought has, over thousands of years, expanded its horizons toward institutional and deliberative analysis, so too has our long history of scientific inquiry led to the discovery of how to edit our genes. Gene editing and democracy are often described as the culmination of years of evolution and innovation and as the basic building blocks of what make us us and society organized. Often portrayed as utopian projects aimed at achieving human flourishing, they are, in many ways, traveling along a converging path. And just as CRISPR confronts us with significant moral inquiry and uncertainty, raising important questions around justice, knowledge, and trust, the quest to understand the relationship between democracy and experts likewise requires modification and even experimentation.

CRISPR is a helpful case for illustrating the moral, political, and social goals of enhancing the democraticity and legitimacy of the regulatory process because, unlike food dyes and vaccines, gene editing is not as prominent in the public eye and, as such, is not (yet) divisive in ways that may preclude productive deliberation. Social groups are not as well organized in support of or against specific policies in this area, meaning that stakeholder groups and their corresponding positions are less clear. This results in not only less polarization but also fewer spaces and opportunities for people to find information and voice their concerns. Finally, CRISPR applications are not topics featured in campaigning or electoral politics. Few bills on the subject are introduced in Congress, so opportunities for legislative engagement are scarce. Lastly, CRISPR development is a profitable industry that requires moral guardrails and public safeguards. Thus, it lends itself well to considering which public groups are invaluable in crafting policy and serves as a helpful case to explore institutional mechanisms that can mitigate crises in regulation, democracy, science, and expertise.

Gene editing generally refers to the "modification of the genome within a living cell through the insertion, deletion, or replacement of one or more segments of DNA."[82] CRISPR (**C**lustered, **R**egularly **I**nterspaced **S**hort **P**alindromic **R**epeats) was discovered as a unique sequence within bacteria that had at some point in the past fought off viral attacks and from that had learned to dismantle DNA sequences. By harnessing this defense mechanism, CRISPR can act as a "Swiss Army knife" for editing genes by slicing and recombining DNA, or "knock-in, knock-out, modify, or correct genes within the genomes of living cells."[83] Cas9 refers to an enzyme that is particularly effective in allowing the cell to identify virus DNA that matches with CRISPR.[84] And while there have been other gene-editing technologies, such as zinc finger nucleases (ZFNs) and transcription activator-like effector nucleases (TALENs), CRISPR has "revolutionized gene editing as it is a

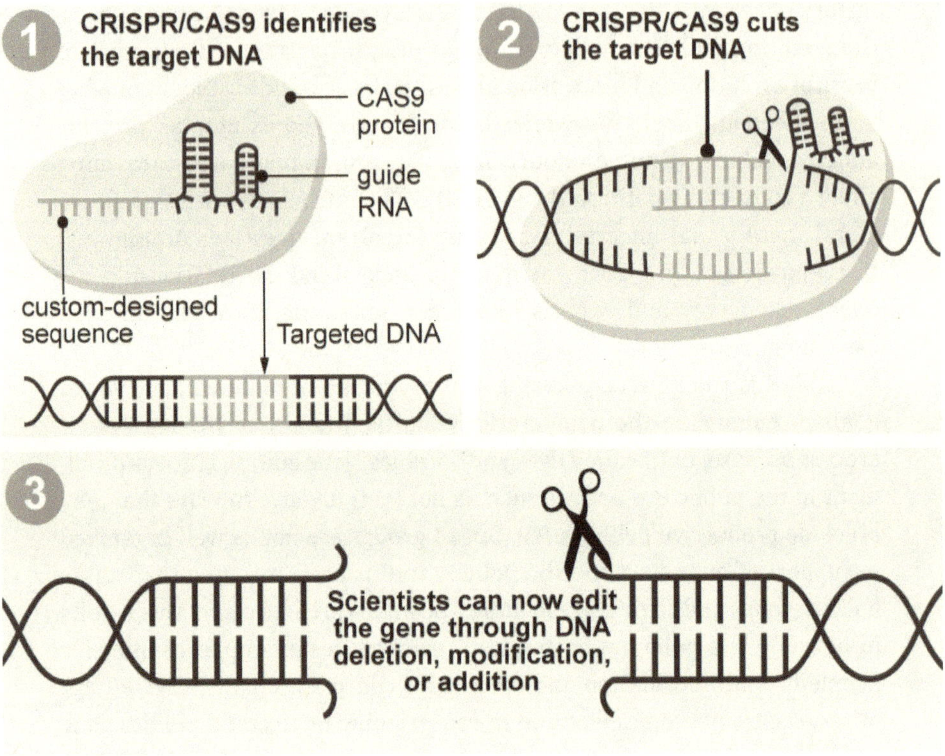

CRISPR is used to "cut" DNA at target sites, which allows it to remove sequences that code for certain diseases or traits. Source: US Government Accountability Office.

faster, more efficient, more accurate, and cost-effective technology" with the "potential to both cure disease and help feed the world."[85]

While research on CRISPR began as early as the 1990s, it was not until 2012 that Jennifer Doudna and Emmanuelle Charpentier, who would later be awarded the 2020 Nobel Prize in Chemistry, began publishing findings of its applications as an effective gene-editing tool.[86] CRISPR, unlike gene-editing tools in the past, quickly became affordable, leading many labs around the world to acquire it. Since then, research on CRISPR has grown exponentially. In 2018, it was estimated that there were over 5,000 publications with the word "CRISPR" in the title, and "from 2013 to 2022, the number of cancer-related publications involving CRISPR increased from 1 to 4,408."[87] Further, the CRISPR technologies industry has experienced "staggeringly high" job growth projections, with many observing that the technology could prove a "gold mine."[88] In 2023, the estimated CRISPR market size was $3.4 billion, with projections of $17.8 billion for 2034.[89] It even led to the availability of at-home CRISPR kits for do-it-yourself editing projects, causing many to applaud the democratization of science and others to fear rogue actors, such as "biohacker" Josie ("Jo") Zayner, who in 2017 "crispr-ed" herself onstage in front of a public audience.[90]

However, much of the historical controversy surrounding CRISPR can be dated to 2018. In November of that year, at the Second International Summit on Human Genome Editing in Hong Kong, Dr. He Jiankui "shocked the world" by announcing that he had used CRISPR to create HIV resistance in recently born twins.[91] Both those at the conference and in the wider scientific community vehemently and nearly unanimously condemned this step.[92] As *The Atlantic* journalist Ed Yong wrote, "The reaction was swift and negative. The crispr pioneer Jennifer Doudna says she was 'horrified,' NIH Director Francis Collins said the experiment was 'profoundly disturbing,' and even Julian Savulescu, an ethicist who has described gene-editing research as 'a moral necessity,' described He's work as 'monstrous.'"[93]

This largely reproving reaction took two main forms. First, many rebuked He for taking an "unreasonable risk" because "there are simpler ways of preventing H.I.V. infection that do not involve the risks of such a trial."[94] There were also later reports that the babies may have been "at risk of early death," that the intervention did not actually work, and that it was merely a "publicity stunt."[95] Second, others bemoaned that He "broke the rules" by violating the "global consensus" of "scientists all over the world" in creating humans whose genes had been edited using CRISPR—a hard boundary that most had tacitly accepted.[96]

Within this deontological critique, wherein the wrong was the violation of a rule, lay a consequentialist and political concern of public backlash: If the public caught wind of the perceived difficulty of restraining the "rogue scientist," as Hank Greely wrote, it could spark extreme "public opinions" that "scientists were pursuing their schemes with no regard for the law or for the opinions of their fellow citizens, citizens who were largely footing their bills."[97] Given that CRISPR generally was being described as a "magic bullet" to battle some of the worst genetic illnesses and to curb the dangerous and impending impacts of climate change, many scientists feared that Dr. He's actions could "slow" or "stifle others' research in the wake of public disapproval."[98]

Thus, the scientific consensus after the He debacle was to endorse "slow science" and "self-governance."[99] Perhaps in the most direct manifestation of these aspirations, in a letter published in *Nature*, a top scientific journal, on March 13, 2019, eighteen "specialists from seven countries," including Doudna and Charpentier as well as Feng Zhang (another important CRISPR researcher), called for a "moratorium on heritable genome editing."[100] This "global moratorium," they argued, should remain in place until an "establishment of an international framework in which nations, while retaining the right to make their own decisions, voluntarily commit to not approve any use of clinical germline editing unless certain conditions are met."[101] In their letter, these "CRISPR experts" and "leading geneticists" argued that the risk and efficacy of various medical interventions existed within the boundaries between enhancement and controlling the human population.[102] The letter also signaled support for achieving "broad societal consensus on the appropriateness of altering a fundamental aspect of humanity for a particular purpose" and that "unless a wide range of voices are equitably engaged from the outset, efforts will lack legitimacy and might backfire."[103]

Reflecting similar fears of public backlash, researchers pushed not just for individual restraint in certain kinds of research but also for broader institutional oversight.[104] Since 2018, numerous national and international bodies have followed suit by convening expert groups and releasing ethics reports, almost all recommending against heritable human genome editing (HHGE), in which edits are made to the genetic material of germline (reproductive) cells as opposed to somatic (non-inherited) cells, out of concern about the ripple effects of inheritance. To date, there have "been more than five dozen . . . other, similar reports about the ethics of human germline genome editing."[105] For example, the World Health Organization (WHO) met on March 18–19, 2019, and released an interim report warning that it would be "irresponsible at this time for anyone to proceed with clinical applications of human germline genome editing," followed by a final report in 2021 in which they called for establishing a registry that would include current somatic cells.[106] Similarly, the US National Academy of Medicine and National Academy of Sciences and the UK Royal Society convened the International Commission on the Clinical Use of Human Germline Genome Editing. In 2020, they released their report, "Heritable Human Genome Editing," in which they too declared that society was not ready to conduct HHGE while also naming some criteria for appropriate initial uses and trials in the quest to "establish safe and effective methodologies that could form the necessary steps in a translational pathway for any clinical uses of HHGE."[107]

Since these tumultuous years, CRISPR research has indeed produced medical breakthroughs, including in 2023 a gene therapy for sickle cell disease, a mutation of the blood cell shape causing extreme pain and organ dysfunction, and in 2025 a treatment for a ten-month-old baby with CPS-1 deficiency, a devastating genetic disorder that impairs the body's ability to remove nitrogen, leading to significant brain and liver damage and often fatal outcomes.[108] CRISPR has also been harnessed in various environmental applications, including to make crops more resistant to climate change and as an intervention to stop the spread of mosquito-borne diseases, which are among the leading causes of death worldwide. And yet concerns persist, particularly around the power of this technology to eliminate whole groups of people (primarily those with disabilities) by preventing them from even existing.

Thus, the discovery of CRISPR has been greeted with both an abundant enthusiasm to "snip away humanity's worst diseases" alongside cautious trepidation, sometimes from the same individual.[109] Among those expressing concern is Doudna herself, who famously recounts having nightmares in which Adolf Hitler wanted to learn more about CRISPR for his eugenics project.[110] This notion that the same technology has both beneficial and deleterious functions, or is "capable of elevating humanity and unleashing destruction upon it," is referred to as the "dual use of technology."[111] Nuclear technology often serves as the paradigmatic example: It can be used as a source of clean, cheap, and reliable energy, or it can be made into the world's most destructive weapon. Similarly, the internet and social media connect people in significant ways and give them more information than ever before, but these technologies have also exacerbated a crisis in teenage mental health and allowed for the dissemination of hateful and violent messages. Further, innovations are often met with fear of societal destabilization, as was the case with coffee (and coffeehouses) and electricity, and yet, they have become normalized and integrated into our daily lives.[112] AI technologies may follow this pattern as well.

Technology's deleterious and beneficial impacts are not random, but rather can often be anticipated because they are introduced into preexisting systems of power and inequality, social hierarchies and injustices, and institutional arrangements. Adding political layers to this, the question modulates to *who* should decide its treatment? For example, we do not often ask, "Should we have the internet?" but rather, "How and by whom should the internet be designed or limited?" Questions of politics—such as those of power, authority, disagreement, institutions, and justice—necessarily become language employed to answer pressing moral quandaries as we quickly realize that technology acts within preexisting societal conditions rather than abstracted from humanity's organization.

In answering these questions, another political pattern that has arisen in many of these expert-based groups and institutions is to call for "public engagement" or "democratic decision-making." The *Nature* moratorium letter, for example, called for "a wide range of voices" participating in these "public value" conversations.[113] The aforementioned reports similarly almost all express a political commitment to some form of public engagement or societal dialogue. The WHO, for example, "endorsed a strong committee to public engagement . . . that increases the voices of those outside the traditional science and policy circles and encourages the circulation of information, views and values."[114] The International Commission similarly noted throughout that "HHGE should not be attempted unless or until safety and efficacy were better understood and extensive public engagement and social decision making had taken place."[115]

These attestations of support for public engagement reflect a long history of similarly positioned institutions that publish reports on the ethics of science and technology and endorse a democratic process as a necessary part of solving moral problems, often with little or no explanation of why or how to do so.[116] Rather, public engagement is often treated as a necessary safeguard against widespread backlash, as scientists feared following the case of He and witnessed viscerally in the COVID-19 pandemic backlash against experts and regulatory institutions.[117]

In perhaps the most comprehensive treatment of the topic in the United States, the National Academies of Science, Engineering, and Medicine released a report in 2016 on gene drives (a potential application of gene editing wherein a particular trait can be quickly perpetuated throughout a population), devoting a chapter to "engaging communities, stakeholders, and publics."[118] For these committee members, public engagement operated as "a risk assessment" in the interest of informing government and future policymaking. And in the United Kingdom, the Nuffield Council on Bioethics has done perhaps the most to follow through on public engagement promises by incorporating the views of the public in its reports on mitochondrial DNA transfer and genome editing in farmed animals.[119] However, the Nuffield Council's rationale was still mainly to alleviate potential backlash. As the authors declared in their report "Genome Editing and Human Reproduction," public engagement among affected groups is important due to the threat of social media in propagating "misinformation and perhaps also scorn for evidence, intolerance of criticism and the rejection of expertise in favour of opinion."[120]

Thus, even as many expert-based institutions call for and endorse some form of democratic engagement or public participation, modulating the right kind of deliberation and determining where it should occur is a formidable task, particularly in our current era of rampant distrust and crisis of expertise. Again, the inclination here may be to favor *more* expert-based institutional control, but we should learn from the current moment of crisis that experimentation and modification are required to create trustworthy institutions that preserve expert guidance alongside nonexpert contributions.

Book Overview

I begin this book with a deeper dive into understanding CRISPR to explore the possibilities of regulation by lot. In the first two chapters, I provide an overview and analysis of the history, ethics, and regulatory framework of CRISPR. Next, I explore the insufficiencies of the administrative state in incorporating democratic practices into its governance structures. Following this, I advance an interconnected theory of justice and epistemology, explaining why diverse groups are important for non-dominating and knowledge-enhancing purposes, including in regulation. Then, I examine the possibility of using DMPs to incorporate these diverse groups into the regulatory process, paying particular attention to the design and precedent for using DMPs in CRISPR regulation. Finally, I show how deliberative minipublics (DMPs) can bolster institutional trustworthiness. The book concludes with global-level implications and proposes expanding regulators by lot to other domains, such as AI and artificial womb technologies.

Chapter 1 provides an overview of the current landscape of CRISPR applications and regulatory approaches. In so doing, I examine more deeply the duality of CRISPR: While CRISPR is a revolutionary technology that promises great human and environmental benefits, it also poses significant uncertainty and risk. As such, it pushes us to define human health and medical priorities and to confront existential problems of the climate crisis. To illustrate this, I examine three CRISPR applications: (1) the first approved gene therapy, Casgevy, to treat sickle cell disease (SCD), (2) interventions for hearing-loss reversal, and (3) gene drives to combat climate change. I then explore the burgeoning regulatory and legal framework by focusing on three agencies in particular—the FDA, the EPA, and the United States Department of Agriculture—that have jurisdiction over CRISPR applications developed for the market and examine the guidance they have issued.

Chapter 2 explores the historical, political, and moral context into which CRISPR is being introduced and analyzes, as such, the underlying moral concerns presented by CRISPR and its applications. In so doing, I distinguish these issues from the regulatory questions presented in the first chapter to show that the risks and benefits should be understood as part of systems of social inequality and disability rights. I also discuss the long history of genetic abuses and eugenics that affect racially and disability defined groups and explore lingering questions pertaining to reproductive freedoms, rights, and obligations.

Chapter 3 examines the philosophical roots and procedural details of the administrative agencies that are situated front and center in this debate, exploring the ways in which moral values and participatory input are, or are not, taken up. The regulatory agencies of the administrative state are often the first places to address these concerns, and their governance and mandates are based on risk mitigation and precaution, which are inevitably value-based. I argue that regulatory agencies suffer institutional drawbacks due to (a) insufficient opportunities for participation provided through notice-and-comment, (b) limited imagination resulting from expert and technocratic bias, and (c) corrupted decision-making caused by the problem of privately captured interests taking precedence over the public good. Captured interests by industry and advocacy groups distort the swath of people in the country who lack representation and are also limited in their ability to engage during the comment period. Systemic distrust is created in these moments, resulting in phenomena such as vaccine skepticism and potentially fear of SCD gene therapies or aversion to participation in clinical trials.

Running parallel to the perceived loss of trust in governance and experts is an important critique of technocracy and meritocracy, the philosophical underpinnings upon which regulatory agencies are built. Meritocratic logic falsely equates entitlement with authority, reinforcing patterns of power and dominance by a particular group of people due to deeply entrenched inequalities. Institutional spaces need transformations that include, rather than exclude, a greater diversity of people to hear from experts and share important experiences with others. This will help to decrease regulatory agencies' democratic deficit.

Chapter 4 analyzes how the exclusion of an engaged citizenry from the halls of power creates the conditions wherein the government and others exercise two primary illicit abuses of power. First, biopower concerns ableist conceptions of health that translate into disciplining, or eliminating, disability from the population. Second, domination is experienced when people face arbitrary interference, resulting in systemic injustices that create medical sites of exclusion and hinder access to needed public goods and services. These concerns are particularly pronounced in the case of CRISPR, which promises to entrench inequalities and thus perpetuate systems of domination through new eugenics-type results.

Randomly selected groups of people are capable, knowledge-producing bodies that can better inform policy through their engagement and interaction with experts and the subjective accounting of their lived experiences. In turn, by attesting to one's experiences of domination, as well as the potential for marginalization, one should also be treated as an "epistemic agent"—that is, as someone who is knowledgeable about their domination. When seeking collective judgment, institutions should not be singularly populated by those with one specific type of qualification; rather, they should include diverse groups that are better able to incorporate experiential wisdom than homogeneous ones. In showing this, I also connect ideas of power and knowledge production as mutually reinforced goals: Power begets knowledge, which reinforces those in power. Epistemic democracy, as non-dominating *and* knowledge-enhancing, is necessarily related to reducing domination because it increases knowledge.

Chapter 5 provides an overview of the latest "wave" of deliberation and sortition, examining the potential of DMPs as a promising path forward for bolstering the democraticity of regulatory agencies because they are indeed capable of reducing captured interests while facilitating learning and providing knowledge. In this chapter, I offer an overview of DMP models, including consensus conferences, Deliberative Polls, and citizens' assemblies and juries, followed by an analysis of (some of) the many design-based considerations that go into convening a DMP—for example, the selection mechanism for the participants, the role of experts and moderators, the merits of online versus in-person formats, and whether participants should be given decision-making authority (as opposed to simply providing guidance). I offer here a typology of decisions one makes when organizing a DMP, with examples of how similarly situated DMPs have operated in the past to serve as guides for how they could function in a regulatory agency.

Finally, chapter 6 ties this analysis together, showing that by creating knowledgeable policy and reducing domination, the institutions that house DMPs are perceived as more trustworthy in ways that positively reinforce the regulation or policy they produce. I understand trust here as an invaluable political and social resource, one that can bind groups together but can also produce fissure. When institutions take people's experiences of domination and their experiential knowledge contributions seriously, they are better able to cultivate public trust by becoming trustworthy.

However, trust exists as a paradox. Not unlike the dual use problem of technology, too much and too little trust is dangerous. Too little trust undermines a healthy, functioning society, while too much creates the conditions for domination. DMPs are effective strategies for approximating the right kind of trust in a democracy, one that encourages institutional transparency and trustworthiness alongside a more confident and discerning public. This, in turn, will reassure the cloistered science experts who crave self-regulation that the public will not fear their actions or lose trust in them. Similarly, this will benefit the regulatory agencies' mandates as well because they will be better able to protect the environment and public health.

In the conclusion, I show that these instrumental reasons create a feedback cycle, for when public trust increases, people are more inclined to engage with government and lawmaking institutions, further reducing domination and improving epistemic outcomes, even when trade-offs exist in some cases. Exclusion and isolation beget distrustful and anxious politics. Inclusivity of the right kind, of a well-designed institutional space, captures the important safeguards of our system that yield collective judgments free from special-interest capture and polarizing partisanship. I also explore the important role of political will and civil society, which are dual requirements for the successful incorporation of DMPs, and suggest future avenues of exploration in these domains, including the international context and other biotechnology cases.

The current crisis of deregulation leaves us with a dangerous vacuum in regulatory authority. However, it also creates an opportunity to build new regulatory processes that are more inclusive and democratic. The slashing of government programs is not going to reduce fraud or corruption; if anything, it is going to make our system *more* susceptible to industry capture, since industries now stand to benefit from fewer guardrails, making America sicker. So while there is great need for repair, we should not simply revert to the status quo. Instead, we should imagine different arrangements in which experts and nonexperts can come together in institutions to create policy and solve problems that bolster trustworthiness and institutional legitimacy while preserving substantive and knowledgeable outcomes, rights, justice, and freedom protections.

1 CRISPR Science, Benefits, Risks, and Governance

Medical endeavors are often driven by a moralized mission to heal, but the medical industry itself can create barriers to patient participation in research and access to its products. CRISPR embodies a duality frequently found in biotechnology and technology: hope, especially among those with rare diseases, coupled with fear, from those who have already experienced harm and wrongdoing. Similar patterns have emerged in the past with other advancements, such as electricity, and are present today in other domains, such as AI. It is not unusual for significant portions of the population to feel eager for change while others are apprehensive about societal undoing.

As an affordable, efficient, and effective gene-editing tool, CRISPR promises medical miracles and environmental solutions desperately needed in many corners of the world. It also threatens to mirror, or even exacerbate, preexisting fault lines in society, including the marginalization of vulnerable populations, the amplification of a profit-driven pharmaceutical industry, and the climate crisis. Biotechnological innovation has yielded many medical marvels, such as vaccines and antibiotics. It has also generated significant trauma and discrimination, such as insufficient studies on women and pregnant patients and the purposeful targeting of vulnerable populations in medical research, such as the dermatology experiments conducted on imprisoned individuals at Holmesburg Prison in Philadelphia and the hepatitis studies done on mentally disabled children at the Willowbrook State School in Staten Island.[1] Whether one views CRISPR as heralding the next medical golden age or a dystopian future often hinges on past experiences with medicine and genetics, as well as society and its systems.

I begin by tracing recent developments in CRISPR medical applications, highlighting two examples that demand patient and lay participant perspectives to more fully grasp the implications of gene editing: (1) sickle cell disease treatment and (2) hearing-loss reversal. I then do the same for environmental applications and discuss in greater depth the case of gene drives created to combat climate change–related problems, such as the proliferation of mosquitoes carrying malaria. Next, I offer an overview of the legal and regulatory framework governing these CRISPR applications. These cases help illuminate the moral concerns explored more thoroughly in chapter 2, including systemic inequities, historical injustices, medical distrust, and disability rights considerations, and following that, the regulatory considerations examined in chapter 3, including assessment of risk, precaution, and oversight.

CRISPR Medical Applications

Current human medical research involving CRISPR applications features a wide range of pathological endpoints and illnesses, including rare diseases. For example, a baby with "severe carbamoyl-phosphate synthetase 1 deficiency, a disease with an estimated 50% mortality in early infancy" was treated in May 2025 with a "first of its kind" personalized CRISPR therapy.[2] Scientists are also attempting to use CRISPR to treat transthyretin amyloidosis (ATTR), a progressive fatal disease featuring the accumulation of amyloid fibrils caused by a misfolded transthyretin (TTR) protein in nerves and organs that causes "nerve pain, numbness, and heart failure as adults."[3] An in vivo CRISPR study being conducted by Intellia Therapeutics and Regeneron that aims to knock out the gene that causes TTR entered its third phase of clinical trials in October 2023.[4] Intellia Therapeutics and Regeneron also announced plans at that time to use CRISPR-based gene therapy to treat hemophilia, a blood disorder that prevents clotting, causing patients to bleed out.[5]

Researchers are hoping to develop CRISPR therapies for neurological degenerative disorders including amyotrophic lateral sclerosis (ALS) and Alzheimer's, Parkinson's, and Huntington's diseases.[6] Huntington's disease, a particularly "progressive and devastating" disorder, serves as a paradigmatic example of CRISPR's potential because the gene is pathogenic, meaning it will cause disease expression, unlike many other conditions whose specific genetics may be less definitive.[7] Scientists are also using CRISPR to target HIV, reverse vision loss, fight hypertension, lower cholesterol, treat Duchenne muscular dystrophy, and aid the "genetic manipulation of the tumor and immune cells" leading to potential cancer treatments for glioblastoma (a common and dangerous brain cancer in adults), lung, and breast cancers.[8]

Other important developments include efforts to treat diabetes through a stem cell therapy that creates insulin-producing cells altered by CRISPR to better evade immune system attacks.[9] CRISPR may also help combat the growing dangers of antibiotic resistance and enhance vaccine effectiveness.[10] This could aid in preventing the transmission of viruses such as RSV, HIV, influenza, and EBV.[11] CRISPR was also leveraged during the COVID-19 pandemic as a diagnostic tool "that gives results in 23 minutes that match the longer laboratory-based tests—faster than other PCR tests on the market."[12] "It can [also] be adapted to test for a broad range of infectious diseases including not just Covid but also flu, strep, and other viruses that require fast diagnosis."[13] Scientists also want to use CRISPR to reduce rates of obesity, manage osteoarthritis pain, mitigate smoking addiction, and treat infertility.[14] And, in a truly out of this world application, space experimentation using CRISPR aims to "aid in the development of safeguards for future space explorers, who face the risk of genetic damage due to cosmic radiation."[15]

Finally, CRISPR could also help solve the organ transplant shortage that claims the lives of seven to eight thousand people a year by editing pigs to "shield donor organs from human immune responses and enhance the overall success of xenotransplantation."[16] The first of such edited hearts was transplanted in January 2022.[17] Unfortunately, the patient died roughly two months later.[18]

In addition to the general uncertainties and risks that accompany all medical research, and perhaps gene editing in particular, the primary medical risk of CRISPR interventions is "off-target effects."[19] Off-target effects refer to the problem that editing a gene could create "unintended" consequences, not unlike genetic mutations.[20] As Roberto H. Herai writes, "When the CRISPR/cas9 system does not cut only the target region on a genome, it can generate unforeseen off-target edits that could cause potentially harmful effects on a patient, for instance by activating an oncogene."[21] In a 2022 analysis, researchers showed that "unintended large gene modifications may persist" at the "cut sites," thus "altering the biological functions and reducing the available therapeutic alleles."[22] And because germline edits are inherited, unintended consequences in this domain may affect future generations.

Many researchers are working to mitigate the problem of off-target events, including by applying advances in machine learning to curtail these risks by better calculating their occurrence and modifying the DNA strand accordingly.[23] As Rui Lopes and Megana K. Prasad analyze in their 2024 review article, researchers are using various methods including anti-CRISPR proteins that serve as a kind of "failsafe mechanism to deactivate the

CRISPR-Cas complex or enhance precision by decreasing off-target effects."[24] Lopes and Prasad ultimately argue that despite the options available, the "relative novelty" of CRISPR therapies means that the risks are relatively "undefined," most likely requiring a "range of complementary methods as each method is limited by its inherent sensitivity."[25]

So far, I have highlighted CRISPR applications that may affect individuals. Next, I discuss two cases wherein CRISPR applications affect specific groups: first, applications for sickle cell disease (SCD), a heritable condition that predominantly affects Black populations, and then hereditary hearing loss (HHL) and the implications that CRISPR interventions have on the Deaf community. These cases highlight how the ways in which we treat certain groups and populations as in need of "correction" or "saving" have divergent societal impacts.

Sickle Cell Disease

Casgevy was the first FDA-approved treatment using CRISPR to "treat a genetic disorder in the U.S."[26] Casgevy involves taking cells from the patients' bone marrow and editing them with CRISPR to produce a protein that can compensate for the defective one that causes sickle cell disease (SCD).[27] It was introduced as a therapy for SCD in 2023 and for beta thalassemia, a related condition, a year later.

SCD is a hereditary blood disorder wherein the hemoglobin (the protein present in red blood cells that transports oxygen) is abnormally shaped due to a β-globin chain mutation. This causes the red blood cells to look C-shaped, hence the name "sickle" cell. When these blood cells travel through the body's small blood vessels, they can get stuck and clog the blood flow, causing tremendous pain, also known as "sickle cell crises," and leading to complications such as stroke and infection.[28] SCD affects 100,000 patients in the United States alone; as of 2024, about 90 percent of individuals diagnosed with SCD in the United States are Black or African American, and approximately one in every 365 Black or African American newborns in the United States has SCD.[29] And although this project largely focuses on US regulatory institutions, SCD is most prevalent in sub-Saharan Africa, where more than 66 percent of the 120 million people affected globally live.[30]

Discussions of SCD and Casgevy raise important throughlines throughout this book about the importance of navigating the duality of optimism and fear. They illustrate a particular case of a medically marginalized minority and chronic disease community that approaches treatment opportunities through this duality. Some of the fear stems from a long history of distrust due to medical exploitation and profit-driven motives, originating in the early days of the American republic, when enslaved people were used for medical experimentation, and continuing into the modern era with the Tuskegee Syphilis Study (1932–72), in which Black sharecroppers with syphilis were experimented on without being informed that they were part of a medical study and were denied treatment for their condition.[31] Finally, it highlights the importance of community engagement in research agendas and clinical trials as well as the nuances of language related to cures, illness, treatments, and social identity.

SCD is embedded in a rich social and cultural history that is intricately related to race and identity.[32] When asked, for example, about somatic versus heritable genome editing for SCD in infants, some parents expressed resistance to using CRISPR to alter their babies' genetics because of "genetic identity and connection to future generations."[33] Further, SCD is a historically neglected disease population. Though the disease has been documented for over a century, few treatments and interventions were available until Casgevy.[34]

Correspondingly, given that the primary diagnostic tool for identifying that a patient is in crisis is the patient's firsthand account, many patients report that medical staff ignore or underestimate the severity of their pain.[35] It is not uncommon for patients to describe experiences of being "mistreated," a "lack of confidence from providers," or being accused of "only coming in for drugs."[36] In a study from 2022, "participants reported providers often do not appear to have sufficient knowledge of SCD and do not respect the patient's own experience and expertise with their body, particularly among providers who are not SCD experts."[37] In another study comparing Black and white SCD cancer patients, researchers found that Black SCD cancer patients were ignored for their pain more often than white SCD cancer patients.[38] Treatment such as this, alongside poor ratings of provider communication, lack of resources, and difficulty accessing clinics and hospitals, all contribute to lower levels of trust in medicine.[39] This has further ripple effects, including leading some people to neglect or delay seeking necessary medical care or going home to dress in a manner that they believe will compel hospital staff to take them seriously.

There has also been a duality of hope and fear regarding the research on Casgevy itself. Individuals hoping to access Casgevy were frustrated with the strict and exclusionary eligibility requirements and with the high price of $2.2 million for the one-time treatment, which is most often not covered by insurance because gene therapies are not considered "drugs."[40] Given that most individuals with SCD live in sub-Saharan Africa, the disparities loom larger on the global level, where access and resources are already underprovided. On the other hand, the testing was met with apprehension about the safety of the therapy, such as fear of off-target events, and concerns about being used as a "guinea pig" by researchers with ulterior motives, such as profit.[41]

Lingering skepticism about the drug remains post-approval in part because it claims to be a "cure," and yet "curative endpoints" for SCD are not entirely clear. In 2019, the FDA, in partnership with the American Society of Hematology, convened a panel of clinicians, investigators, and patients to develop consensus recommendations for clinical trial endpoints. They proposed that the ideal endpoints should "reflect patient desires and integrate objective measurements to assess disease severity and progression."[42] They noted that pain was a defining feature and a quality-of-life limiting factor of SCD and recommended a comprehensive way to assess different pain thresholds clinically. In 2020, it was announced that one of the patients had indeed been "cured" of SCD.[43] Others received this news with some skepticism, arguing that a "fully effective" cure is "one that completely suppresses disease-related complications and costs and restores life expectancy and health-related quality of life (HRQol) to that of a comparable individual unaffected by the disease."[44] In 2024, patients who were receiving Casgevy reported significant alleviation of pain, though not the total loss of pain, thus raising the question of whether cure can be claimed in cases where one of the defining symptoms of SCD remains.[45]

The case of SCD illustrates key themes that will repeat across many of the case studies I explore. We see here the hope for a cure measured against resistance rooted in discrimination. It also challenges the bright line definitions that medicine and society so often reach for, such as "disease" (versus identity) and cure (versus treatment). It raises questions about what communities desire for themselves (versus individuals within them), as well as the government's approach to progress versus sincere optimism.

Hereditary Hearing Loss

Heritary hearing loss (HHL) CRISPR applications similarly challenge traditional dichotomies of cure and illness, health and suffering, and optimism and fear. But while SCD and HHL both share parallel themes, HHL offers us the opportunity to reflect on the medical and social models of disability to better understand how different communities perceive technological impacts.

HHL affects roughly three out of every thousand newborns in the United States and one out of five hundred newborns worldwide.[46] Approximately four out of five cases of newborn hearing loss are genetic (other causes include illness and injury). Genetic causes, however, are not attributable to a single gene but involve as many as 250 genes and operate in combination with many environmental factors.[47] While there is no current FDA-approved CRISPR gene therapy for HHL, CRISPR-based gene therapy research has thus far been demonstrated in mice to enhance "hair cell survival and improve auditory brainstem response thresholds in treated ears compared to controls."[48] Though initial studies focused mostly on inner hair cell function, further research using CRISPR has shown that it is effective in recovering hearing in mice when targeting outer hair cell function as well.[49]

Cochlear implants are the primary medical intervention for HHL since their approval by the FDA in 1984 for adults and in 1990 for children. Cochlear implants are surgically implanted electronic devices that operate by bypassing the parts of the ear that send signals to the brain, which then interprets the input as sound by directly stimulating the auditory nerve. While cochlear implants allow users to interpret sound better than they otherwise would, the sound produced through a cochlear implant is not experienced in the same way a hearing individual without a cochlear implant experiences it. This, alongside the environment and social conditions in which the cochlear implant is introduced, such as whether the child also receives American Sign Language (ASL) education, has contributed to a range of reactions and perspectives among the HHL population and cochlear implant users, many of whom do not see implants as a *cure* for hearing loss.[50]

Though cochlear implants are more widely accepted and used today, they are utilized by only 2 to 13 percent of the population that would qualify for them.[51] Part of the reason for such low numbers may be that they have played a contentious role in the Deaf community because they are seen, by some, as eroding a community built around shared language and identity, calling into question "the nature of disability and the definition of 'normal' bodies."[52] To these advocates, deafness is not viewed as a disability, nor as a deviation from normal human function, in part because a Deaf person's well-being is not necessarily less than that of a hearing person's.[53] Rather, deafness may be better understood as a "unique culture in which the disadvantages of deafness are socially constructed."[54] Hearing loss is only problematic insofar as schools do not teach hearing peers ASL, public events do not provide interpreters, phones do not include transcription services, and so on.

For some, deafness is *not* a medical deficiency but can be an asset, serving as an integral part of that person's identity and connection to others, providing psychological advantages that outweigh the disadvantages.[55] Still others have offered that deafness is a culture, contributing a rich form of body movement language to the tapestry of human connection. For some others, deafness is "mere-difference," meaning it is merely another way of being in the world and is not necessarily good or bad.[56] Both the positive and mere-difference approaches reject the idea that deafness needs "curing" as it is a social identity that should be preserved. The debate on cochlear implants extends to the reproductive realm: Should the ability of Deaf couples to purposefully create a Deaf child, such as through preimplantation genetic diagnosis (PGD), be limited or expanded?[57]

The association of deafness with culture and social identity parallels theories of ethnicity, religion, and race insofar as they often share similar rights-based claims to reproductive freedom and communal integrity.[58] Kathleen S. Arnos has compared deafness with SCD insofar as it has been "maintained at much higher frequencies in the population because of racial and ethnic homogamy, another form of nonrandom mating that is analogous in its effects to the assortative mating that exists among deaf people."[59] Even while significant differences exist between these groups, such as Deaf communities advocating for Deaf schools while Black communities have fought against segregation, both situations overlap around questions of genetic intervention and are illustrative of group-based experiences and attitudes toward CRISPR interventions.

In the case of attitudes of Deaf people toward CRISPR, the question of HHL as a medicalized condition versus a social identity similarly plays a large role, particularly in a future scenario where CRISPR is used in germline editing to prevent the existence of future Deaf people.[60] These views expand on concerns that emerged in response to cochlear implants and PGD, and they are felt more acutely due to genetic exceptionalism: the idea that genetics plays a more powerful and determinate role in one's life. It is rooted in profound experiences of alienation from the larger society and a pressure to be "fixed" or "corrected" rather than taught sign language and immersed in Deaf culture. Similar attitudes toward CRISPR are also found among individuals with retinal conditions and blindness who view their blindness positively and do not feel the need for medical intervention; instead, they advocate for expanded access to resources that make the world easier to navigate.[61] Deafness offers a perspective that understands hardship as resulting primarily from social conditions rather than from physical deficiencies themselves. As with SCD, the social identity of deafness is a critical context for understanding community-based attitudes toward CRISPR-based gene therapies.

Environmental Applications

CRISPR environmental applications primarily aim to course correct the climate crisis created by industry. The same duality present in medical applications reemerges here: hope for fixing the planet's problems alongside fear of further damage that overeager drivers of industry may unleash.[62]

CRISPR has been lauded as able to "revolutionize agriculture in a second green revolution to ensure food and nutritional security of the ever-increasing population of tropical countries."[63] The hope is that, like earlier genetic modifications, CRISPR will offer "new breeding technologies—including using molecular methods to edit or alter genes—promise higher yields, health benefits for consumers and climate change mitigation."[64] Designing crops to be more disease- and fungal-resistant could eliminate the need for pesticides and make them more durable to withstand extreme temperatures and weather.[65]

Crops identified as candidates for gene editing, with the goals of bolstering their climate resilience, nutrition, and taste, include apples, grapevines, kale, carrots, lettuce, and eggplant.[66] Citrus greening, a bacterial infection affecting oranges, grapefruit, and lemons, severely disrupts the citrus industry in Florida and is a prime target for CRISPR intervention.[67] Similarly, researchers are using CRISPR to improve disease resistance in bananas, which are susceptible to catastrophic crop loss due to their monoculture cultivation. Disease-resistant bananas would provide many countries with nutritional and economic benefits.[68] Many traits of tomatoes have also been edited to enhance features such as fruit ripening, quality and nutrition, disease resistance, abiotic stress tolerance, metabolism, and herbicide resistance.[69] For potatoes, the goal has been to create varieties that brown less easily, thereby increasing both their "quality and resistance."[70]

Scientists are also developing soybeans to increase yields in temperate regions where rain and flooding can limit crop survival.[71] Numerous kinds of rice have been created, including varieties that need less water, along with maize and cranberries.[72] Researchers are also working to alter yeast to make beer "taste better" and to enable the "growth of beneficial microorganisms . . . to prevent food poisoning and infections using CRISPR-Cas and, in turn [,] improve food safety."[73] This has promising benefits for dairy production and monitoring. In addition, scientists want to use CRISPR to harness algae for biofuel as part of the synthetic biology movement.[74]

Nonhuman animals are being genetically edited to make more resilient livestock for food production. The biotech company Acceligen developed short or "slick-hair coat" cattle; this modification, approved in 2022, makes cattle more comfortable in a warmer climate, which leads to reduced stress and "improved food production."[75] SciFi Foods is developing beef burgers using a "blended" approach to cultivating meat that combines plant-based meat with cultivated beef cells. These cells will be treated with CRISPR to grow outside the cow.[76] Another application has been editing Japan's red seabream, a fish with great cultural significance, to increase its size. As Jeremy Hsu wrote in 2019, "The next research steps involve studying those issues and seeing how efficiently the muscular fish converts food into flesh—major factors in determining whether the fish can prove commercially viable."[77]

CRISPR is also being tested to protect livestock by editing genes to make them less susceptible to illness, including swine fever, which has already killed thousands of pigs.[78] Other CRISPR environmental applications include conservation efforts for coral reefs (particularly the Great Barrier Reef), which have been damaged by overuse and climate change.[79] Scientists have also proposed a more "humane" way of farming through sex selection in vitro. Sex selection eliminates the industry practice of killing animals of a particular sex by avoiding the creation of animals that are killed at birth.[80]

Finally, in a perhaps more fantastical realm, some scientists are interested in using the technology for de-extinction in hopes of bringing back animals and plants that have gone extinct, such as the woolly mammoth, the passenger pigeon, and the American chestnut tree.[81] While chapter 2 focuses on different moral concerns that arise in environmental applications of gene editing, the ethics of de-extinction are multifaceted, raising questions of whether doing so is an act of retributive justice and conservationism, or whether de-extinction itself would be more destructive to the *current* ecosystem, even if it succeeds in reverting us to ecosystems of the past.[82] Yet again, we see dualities.

Gene Drives

A "gene drive" occurs when a particular allele is transmitted throughout an entire population within several generations. Gene drives perpetuate a particular trait throughout an animal or plant species such that the whole population contains that trait at the expense of a previous one. Gene drives naturally occur in the wild and can be produced using previous forms of genetic modification. Using CRISPR to create gene drives, however, can be more powerful than other techniques because the timeline for a gene drive to become fixed in a population decreases dramatically and can be far more targeted in its design.[83] CRISPR-created gene drives are potentially more stable and easier to engineer than those produced using contemporary modification strategies.[84] I introduce two examples of gene drives derived from genetic modification, not CRISPR, to illustrate how scientists would like to use CRISPR in the gene drive space.

One of the primary examples of how scientists hope to use gene drives is to eliminate certain mosquito populations that carry disease and whose migration is made easier by warming and changing climates.[85] Mosquitoes are harbingers of numerous diseases worldwide, including malaria, dengue fever, yellow fever, and chikungunya, which are responsible for about one million deaths per year.[86] According to World Health Organization data, in 2024 there were an estimated 263 million malaria cases and 597,000 malaria deaths in eighty-three countries, with Africa carrying a disproportionately high portion of the burden.[87] By December 2024, the Caribbean and the Americas experienced 12.6 million cases, resulting in 7,700 deaths, a more than 200 percent increase compared to the same time the previous year.[88] And in 2024, more than 13 million cases of dengue fever were reported worldwide, doubling the 6.5 million cases in 2023.[89] As Timothy C. Winegard writes, we are locked in a "life-or-death struggle for survival with the not-so-simple mosquito" that claims more lives each year and threatens even more in the future, as climate change favors the mosquito's survival and migration in warming environments.[90]

Researchers want to eliminate these diseases by wiping out certain mosquito populations using a gene drive that will perpetuate a modified gene that makes reproduction deadly.[91] Researchers at British biotech company Oxitec, in collaboration with the Florida Keys Mosquito Control District, modified the *Aedes aegypti* mosquito, which primarily spreads Zika, dengue, and yellow fever. They created the "OX5034 trait," a genetic modification that disrupts normal cellular function in the creation of a protein tTAV.

This causes the female offspring, who are the ones that bite and thus are responsible for disease transmission, to die out before reaching adulthood. The male offspring, who do not bite, survive to carry and spread the trait with the self-limiting gene, leading (researchers anticipate) to diminishing numbers over successive generations.[92] After conducting smaller tests in Brazil, where the company reported that the technology reduced populations by over 90 percent in some areas, on May 3, 2021, with approval from the Environmental Protection Agency (EPA), Oxitec released the first batch of 144,000 genetically modified mosquitoes in the Florida Keys.[93] The third and final wave of the project ended in late 2023 with the release of 5 million more mosquitoes, followed by post-release monitoring and preparation to seek EPA approval to sell its technology in other areas.

This genetic modification has been met with resistance, particularly from local Florida residents, who showed up at town hall meetings expressing concerns about being experimented on without their consent, unease about the possible negative unintended consequences of modified mosquitoes, and their overall distrust of Oxitec.[94] Many residents argued that the decision-making process did not sufficiently involve the affected communities.[95] This case highlights the uncertain environmental risks that gene drives pose and raises important research ethics considerations around consent when conducting large-scale experiments in the wild.

Advocacy organizations, such as Friends of the Earth, also expressed concern by emphasizing that the risk that modified genes could spread into wild populations, potentially creating hybrid mosquitoes that would be more difficult to control and might increase the spread of mosquito-borne diseases. Friends of the Earth also pointed out that the release of the mosquitoes occurred during the COVID-19 pandemic, raising the stakes of the risks in a time of great illness and uncertainty.[96] There were further concerns expressed for other species within the mosquito ecosystem, such as birds and insects, and whether their populations would be affected negatively either by the dwindling numbers of mosquitoes or the unforeseen consequences of genetic modification in the environment.

In a second major example, the Mice Against Ticks project was proposed by Kevin Esvelt in 2016 to stop the spread of Lyme disease by targeting the populations of white-footed mice, the carriers of deer ticks, that live in New England. In 2023, there were approximately 89,000 reported cases (compared with 63,000 reported cases in 2022) in the United States, with experts anticipating that the actual number is more likely 476,000 cases, with most Lyme disease cases concentrated in New England and the Northeast coastal area.[97] While many early detection cases can be cured, prolonged infection can lead to lifelong debilitation and complications.

The plan for the gene drive was to create a "lineage of rodents that would be naturally vaccinated against Lyme from birth. Any tick that bites them would fall off and die. Any Lyme bacterium that infiltrates their blood would be destroyed."[98] The project leads had hoped to begin releasing these mice on Nantucket and Martha's Vineyard, but the project was effectively blocked by the residents of these areas who expressed similar concerns as in the Florida mosquito case.[99] In 2024, the idea was reintroduced after a "scientific breakthrough" in which Esvelt, along with Joanna Buchthal and Sam Telford, "successfully created the first genetically-modified mouse resistant to Lyme disease . . . [whose] immunity was passed down to subsequent generations."[100] The researchers presented their findings to Nantucket's Select Board, beginning the process again of seeking local support. Federal regulators will need to approve the project before the mice can be introduced into the wild.

Similar to the mosquito project, the Mice Against Ticks project bears the risk of off-target mutations and unintended, potentially uncontainable environmental consequences.[101] Considering that the very definition of gene drives is the perpetuation of a trait throughout a whole species, off-target events may be rather costly, altering entire habitats and ecosystems.[102] As with human-based applications, Esvelt has warned that "you cannot simply sequester them and wall them off from the wider world. . . . They'll eventually spread throughout the full range of the species they target. And if that species is the brown rat, you're talking about the entire planet."[103]

One solution has been the promise of crowdsourcing data analysis and data transparency. If researchers release the complete genetic sequences of their CRISPR products, "people can look for unintentional mutations and other problems."[104] A similar mosquito-monitoring project is underway in Florida, where residents are being asked to record their mosquito observations.[105] Citizen Science refers to the widespread engagement of lay participants in scientific data collection and has proved instrumental in other cases, such as charting certain astronomical phenomena.[106]

These risks, like with any technology, must be weighed against potential benefits. As we turn to the topic of regulatory oversight, we must consider the risks of *not* developing CRISPR-based technologies that might eliminate diseases threatening millions of lives and mitigate the climate crisis affecting everyone.[107]

CRISPR Legal Landscape

US regulatory agencies have played a central role in developing the legal landscape of CRISPR governance. Preexisting legal structures meant to ban or restrict research on modification of embryos has affected the development of gene-editing oversight. In 2015, the National Institutes of Health (NIH) issued a statement declaring that it would not fund research in which human embryo genomes are edited.[108] NIH based this on the Dickey-Wicker Amendment, a rider Congress added to appropriations bills in 1996 that bars funds from being used in research in which human embryos are intentionally created or destroyed. Many states have their own version of bans on human embryonic research that could be expanded further to include germline editing.[109] In this next section, I address the patent disputes that are shaping the development of CRISPR and legislative actions before moving on to the primary agencies responsible for determining the trajectory of CRISPR technologies: FDA, EPA, and the US Department of Agriculture (USDA).

Patent Battle

Patent disputes serve as the underlying context and initial driving force directing biotechnological developments. The United States Patent and Trademark Office is often the government's first encounter with new technology, long before it reaches legislative or regulatory oversight. The patent office thus plays a significant role in shaping future policies and regulations. Patents have an inordinate impact on the market forces that encourage or discourage certain kinds of research and access.[110] The BRCA patent dispute is a prime example of the power of patents—Myriad Genetics was awarded the patent for a diagnostic test identifying a mutation highly correlated with breast cancer, which resulted in a prohibitively high cost for many to access the test and thus compromised research on BRCA.[111] Investors also prefer companies with strong intellectual property rights, which motivates biotech companies to spend significant resources seeking patent protection.[112]

In the case of CRISPR, the patent feud has followed a similar pattern. The intense amount of attention on research and industry investment has fueled protracted competition in the courts for various patent applications.[113] Though the Nobel Prize went to Jennifer Doudna and Emmanuelle Charpentier for their role in the discovery of CRISPR, Feng Zhang at the Broad

Institute was the one who was awarded a pivotal patent in 2014 and defended it in court in 2022, after multiple appeals by Jennifer Doudna and Emmanuelle Charpentier.[114] As a result, as Jacob Sherkow argues, "'The big-ticket CRISPR companies, the ones that are farthest along in clinical trials, are almost certainly going to need to write the Broad Institute a really big check.'"[115] However, even though Feng, and the Broad Institute, triumphed, there are many more patent questions to resolve given that there are "now more than 11,000 families of patents on CRISPR-related technologies."[116] In September 2024, Doudna and Charpentier withdrew their patent applications from the European Patent Office.[117]

Legislative Actions

The US executive branch regulatory agencies have played, and will continue to play, a significant role in shaping policy around CRISPR governance. By contrast, the US Congress has only occasionally taken up the issue of CRISPR governance. In the years since the CRISPR discovery, there have been only occasional speeches from the floor of Congress, such as during a 2022 speech on insulin coverage and after the FDA's approval of the SCD therapy.[118]

There have also been only a handful of occasions when Congress has introduced a bill pertaining to CRISPR. Most significantly, in 2019, after He Jiankui's announcement and the self-declared moratorium, the House Appropriations Committee "restored language to a 2020 spending bill that bars the U.S. Food and Drug Administration (FDA) from considering requests to approve any clinical trial 'in which a human embryo is intentionally created or modified to include a heritable genetic modification.'"[119] This effectively imposes a moratorium on germline editing in clinical research. Another Senate bill introduced during the 2019–20 congressional session called for "the development of international standards regarding the ethical clinical use of genome edited human embryos."[120]

Some of the legislative action on CRISPR has been to authorize or support executive action. On May 20, 2021, a bill introduced in the House Committee on Agriculture declared support for the Department of Agriculture to "lead efforts . . . to prepare for the disruptive transformation of the entire agriculture industry" and listed CRISPR, along with synthetic biology, as interventions to help crops survive a changing climate.[121] Finally, on August 9, 2022, Congress passed the CHIPS and Science Act, which details various science initiatives and goals, including designating that the "office of Science shall carry out research and development activities in genomic science," including "fungal, microbial, and plant biosystem design research to advance the understanding of how CRISPR tools and other gene editing tools and technologies work in nature, in the laboratory, and in practice."[122] The US Congress and state legislatures have yet to pass more substantial laws to govern the use of CRISPR, direct its agenda and priorities, or set moral boundaries around its uses.

Regulatory Framework

The regulatory approach to governing "emerging disruptive technologies" primarily features (a) weighing risks and uncertainty and (b) evaluating whether the new technology requires novel rulemaking or can be considered similar enough to technologies that came before, thus applying preexisting oversight.[123] Regulatory agencies are tasked with approving the aforementioned environmental applications (such as the mosquito release in Florida) and medical applications (such as the SCD therapy), which include clinical trial oversight and testing in the wild to assess safety, risks, and impacts on human, nonhuman animal, and environmental health and welfare.

The main source of regulatory governance that applies to CRISPR is the Coordinated Framework for the Regulation of Biotechnology, revised for the first time in thirty years in 2017 by the White House Office of Science and Technology Policy. This update outlined a path forward for the Food and Drug Administration (FDA), the US Department of Agriculture (USDA), and the Environmental Protection Agency (EPA) to "share the task to review the introduction of products of modern biotechnology into the US market."[124] Within this superstructure, Scott J. Schweikart argues, individual agencies are also meant to retain unique jurisdiction over certain gene-editing applications to "establish and monitor the safety, efficacy, and ethical nature of gene editing treatments and research."[125]

FOOD AND DRUG ADMINISTRATION

The FDA is responsible for oversight of the safety, quality, and effectiveness of food, cosmetics, and drugs in the United States.[126] As such, it has authority over "clinical trials for drug, biologic, and medical device products."[127] This entails reviewing evidence of efficacy and adverse effects, or risks, weighing pharmaceutical interactions with other drugs or alcohol, supervising advertising strategies, and requesting specialized trials in certain populations.[128] The FDA weighs safety and efficacy and requires companies to communicate a reasonable threshold of their products' risks, such as the most common and severe side effects.[129] For CRISPR, the FDA reviews requests for clinical trials and examines the evidence based on how distinct they are from current gene therapies and modifications, while also monitoring off-target effects.[130]

In January 2024, the FDA published a (non-enforceable) "guidance document" entitled "Human Gene Therapy Products Incorporating Human Genome Editing" that details "recommendations regarding information that should be provided in an Investigational New Drug (IND) application in order to assess the safety and quality of the investigational GE [genome editing] product."[131] The guidance further "includes information on product design, product manufacturing and testing, nonclinical safety assessment, and clinical trial design."[132] In May 2024, the FDA issued another draft guidance, "Heritable Intentional Genomic Alterations in Animals: Risk-Based Approach," and declared its intention to develop a risk-based voluntary approach for overseeing and approving these alterations using genetic technologies, which it notes will most likely continue to develop and improve.[133]

Guidance documents from federal agencies differ from rules or "regulatory actions" in that they are not legally binding. However, they are vitally important because they signal how regulations and laws are intended to be applied in practice. For CRISPR, guidance documents are intended to transparently communicate to both industry and the public how to interpret existing law and obligations, as well as agency procedures, in developing and submitting evidence for CRISPR products.

In addition to overseeing pharmaceuticals intended for human therapeutic consumption, the FDA has some responsibility for animals and plants as stipulated in its 2019 updated Plant and Animal Biotechnology Innovation Action Plan.[134] For example, in 2016, the FDA initially approved the "uncaged field trial of GE *Aedes aegypti* mosquito" discussed above.[135] However, jurisdiction for this project switched to the EPA, which approved further field trial releases in 2020. According to Zahra Meghani, much of this depended on how to classify the mosquitoes themselves: "The switch in jurisdiction of the OX513A GE *Aedes aegypti* mosquitoes from the FDA to the EPA occurred because it was decided by the US that regulatory jurisdiction depends on how applicants classify their products. If a bio-product is classified by an applicant as a bio-pesticide (rather than a means for preventing or reducing the incidence of mosquito-borne diseases in humans), it will be regulated by the EPA and not the FDA."[136] This procedure for classifying products, as in determining agency jurisdiction, was established by statute or by congressional design and is discussed further in chapter 3.

In another example of potential jurisdictional overlap, in 2020, the FDA approved the "intentional alteration" of "domestic pigs . . . for both human food consumption and as a source for potential therapeutic uses."[137] The alteration was to the alpha-gal sugar on the surface of pig cells that causes mild to severe allergic reactions in some people. However, in December 2020, the FDA proposed a rule change in compliance with the USDA to switch regulatory oversight entirely over to the USDA.[138] In March 2022, the FDA also "made a low-risk determination for the marketing of products, including food, from two gene-edited beef cattle and their offspring after determining that the intentional genomic alteration (IGA) does not raise any safety concerns."[139] As a result of the low-risk determination, farms can use the short-haired cattle without registering with the FDA.[140]

DEPARTMENT OF AGRICULTURE

The USDA's mandate includes oversight of "food, agriculture, natural resources, rural development, nutrition, and related issues based on public policy, the best available science, and effective management." This agency is also tasked with overseeing "economic opportunity through innovation" and "helping rural America to thrive; to promote agriculture production that better nourishes Americans while also helping feed others in the world; and to preserve our Nation's natural resources through conservation, restored forests, improved watersheds, and healthy private working lands."[141]

The primary regulatory question for the USDA pertaining to CRISPR has been whether gene edited crops differ substantially from the already existing genetically modified crops.[142] Jennifer Kuzma explains that part of this debate hinges on process versus product distinction: Is the product significantly different, or do we care that the process to create the product is different?[143] The European Union (EU), for example, answered this question in 2018 by stating that, in contrast to the USDA, gene-edited products would be regulated as though they are GMOs, in part because of the comparison of need for safety.[144]

In 2016, the first CRISPR-edited crops, "corn with extra starch," "drought-resistant maize," and "browning-resistant mushrooms" were exempted by the USDA from regulatory approval.[145] Other crops followed suit, including "green bristlegrass" in 2017 and "loblolly pine" and "drought tolerance and yield stability corn" in 2020.[146] Crops "cleared for commercial release" include "camelina," "drought-and-salt-tolerant soybeans," and "high-yield tomatos."[147] In departing from the EU's ruling, the USDA declared in 2018 that it would not regulate CRISPR gene-edited crops as it does GMOs if the end product was indistinguishable from those using more traditional modification or breeding techniques.[148] Toward this end, in May 2020, the USDA's Animal and Plant Health Inspection Service released its final rule, "Movement of Certain Genetically Engineered Organisms," governing products of genetic engineering and editing. More commonly referred to as the "Sustainable, Ecological, Consistent, Uniform, Responsible and Efficient (SECURE)" rule, the regulation does the following:[149]

(1) establishes exemptions for plants modified by genetic engineering techniques where the modification could otherwise be achieved through conventional breeding techniques, ensuring that such plants are treated similarly to conventionally bred plants from a regulatory perspective;
(2) uses a risk-based approach to determine whether an organism is regulated, rather than relying on whether the organism was developed using a plant pest;
(3) provides a mechanism for a rapid initial review to efficiently distinguish plants developed using genetic engineering that do not pose plausible pathways to increased plant pest risk from those that do and thus require further evaluation.[150]

As Theresa Selfa, Sonja Lindberg, and Carmen Bain report, "SECURE will exempt most gene-edited plants from regulation and will allow developers to decide whether their products qualify for these exemptions."[151] Thus, the USDA has largely answered the process versus procedure question by focusing on the product (a.k.a. "the traits themselves rather than the technology used to create them") and whether it can be distinguished from other forms of engineering and breeding techniques.[152]

ENVIRONMENTAL PROTECTION AGENCY

The EPA, which considers both the environmental and human health impacts of biotechnology products, will have significant overlap with the USDA in reviewing crops and pesticides. This includes "crops with certain traits such as insecticidal properties," or crops that have genetic modifications causing them to express plant-incorporated protectants.[153] These are often created, as noted, to be either resistant to disease or insects (thus reducing the need for pesticides and harsh chemicals that harm the environment). However, as Jochan Menz and coauthors argue, "As many genome edited products will not produce pesticides *per se* it is expected that EPA will only play a minor role in the evaluation and release of most genome edited crops in the United States."[154]

In May 2023, the EPA released a final rule entitled "Pesticides; Exemptions of Certain Plant-Incorporated Protectants (PIPs) Derived from Newer Technologies" that aimed to relax the regulation of PIPs created through CRISPR. It listed exemptions intended to "benefit the public by ensuring that human health and the environment are adequately protected, while also reducing the regulatory burden for the regulated community. These exemptions may also result in increased research and development activities, commercialization of new pest control options for farmers, particularly in minor crops, and increase the diversity of options for pest and disease management, which could provide environmental benefits."[155] This summarizes well the aim of this regulatory oversight conversation: striking a balance between precautionary risk analysis and enabling the market to develop technologies that may deliver societal benefit.

In the next chapter, I explore the moral concerns of gene-editing technologies, examining whether this risk-based approach and the procedures of the regulatory agencies adequately address them. And while patient advocacy groups have done significant work bringing regulators and industry experts closer to the patients themselves, the moral nuances of patient concerns deserve time and space to explore.

2 Gene Editing's Historical Context and Moral Considerations

While many praise CRISPR for its potential to enhance human well-being and bolster environmental protections, others hold deep and abiding moral and political concerns. To accompany the previous chapter's framing of the duality of fear and hope and the concerns over the profit-driven objectives of biotechnology companies, in this chapter, I offer a historically informed perspective on the ethics of gene editing. By telling a story of the past woven together with the current moment, I illustrate the deeply rooted moral and political considerations surrounding gene editing that go beyond the safety and risk governance approaches prevalent in regulation.

My intent is to show that gene editing reiterates preexisting moral concerns in the field of genetics and genetic technologies, but in a much more heightened and accelerated form, given the power, capital, and promise of CRISPR. The story of gene editing is impossible to divorce from the broader story of genetics' historical uses, the social conditions in which discoveries and advances are made, and the narratives of the people participating in or affected by them.[1] Advances in genomic technologies and their applications reflect a long quest to decipher, predict, and, at times, perfect the ways our bodies operate and perform. They also demarcate social identity and groups, thereby tying us to, as well as distinguishing us from, others by geographical regions and even historical experiences.[2]

First, I explore the history of genetic technologies and eugenics, emphasizing the ways in which genetics has been harnessed to perpetuate racial myths and conduct nonconsensual research—research that has in turn been wielded to stigmatize and marginalize certain populations. Gene editing also reiterates concerns about the overzealous ambition of the researcher and the lingering sting of gene therapy gone wrong. Second, I argue that gene editing unjustly discriminates against disabled individuals by perpetuating ableist constructs or medicalized models of health. The enhancement argument (that we have a positive duty to edit even beyond the baseline of health because it makes the individual better off) serves as a claim countered in this domain. Third, I discuss the problem of unequal access: Whereas we fear some populations will be too readily targeted in the editing process, we also fear that some will be excluded from accessing important health benefits because of systemic inequalities. Fourth, I take up the parental and maternal, fetal, environmental, and animal rights considerations, paying special attention to the often-gendered experiences of those who will carry, deliver, and most likely be the ones to raise genetically edited fetuses and, ultimately, children.[3] Fifth and finally, I examine the significance of the history of genetically modified organisms in the CRISPR story in signaling deep fear and skepticism about our regulatory approaches.

Eugenics and Race

In some ways, the discoveries of "the genetic age" have bolstered the authority of biological empiricism in defending racialized and ableist social orders as an important historical context for CRISPR.[4] This history illustrates that CRISPR's discovery needs to be understood within the complicated and nuanced story of genetics, morally worrisome uses of race, and the exploitation of vulnerable populations. By examining the origins of the eugenics movement, the use of nonconsensual research on Indigenous populations, and current research programs, I show that CRISPR, as a powerful technology, will be encountered by groups differently. As such, those who develop the technologies and regulate them may be unaware of these histories and the ways genetic technologies have affected diverse communities.

The field of genetics in many ways stems from harmful and incorrect methods of classifying individuals. And while these modes of understanding human relationships and disease expression are not necessarily ill-intentioned, they are also often complicit in constructions of racism and ableism. Findings in research journals on genetics differences have been taken up in policy and law, such as in welfare and immigration policies, and have exacerbated people's internal racial biases.[5] This in turn perpetuates systems of domination and marginalization. In extreme cases, these ideas have led to genocide, as in the Holocaust, and homicide, as in the 2022 grocery store shooting in Buffalo, New York, where the shooter cited genetics research as part of his basis for his belief in white supremacy.[6]

This becomes particularly relevant in the contemporary interest in using tools to analyze patterns from large datasets. In her 2019 book *Superior: The Return of Race Science*, Angela Saini argues that genetics research has long fueled explanations for the "success" or "failure" of particular groups, almost always within racialized and political contexts and to assert certain kinds of power structures.[7] As Saini argues, even well-intentioned and "liberal" geneticists aimed to genuinely discern differences between races, claims that have provided fuel for racism.[8] Saini further shows that many other geneticists have demonstrated that there is no genetic explanation for race given the similarity of genes across different socially constructed racial groups.[9]

As noted in the introduction, gene editing and democracy may share parallel revolutionary trajectories. So, too, the roots of genetics can be traced back to the same origins as democracy: the ancient Greeks. Hippocrates in the fourth century BCE and Aristotle in the third century BCE sought to identify the "building blocks of life" described as the key "humors" (blood, phlegm, black bile, and yellow bile).[10] They understood these as components of the human body that signal health when balanced and illness when imbalanced. These theories established the framework for medicine for several centuries thereafter.[11]

The contemporary context in which CRISPR emerges, however, begins in the early nineteenth century when Charles Darwin formulated a theory of evolution to explain how species develop distinct features and behaviors that have allowed them to perpetuate throughout time.[12] Several decades later (in the mid-nineteenth century), the Austrian friar Gregor Mendel conducted botany experiments (particularly on peas) that led to the discovery of the laws of inheritance to explain the passage of certain traits from two contributors to a new organism. Mendel distinguished the idea of inheritance as depending on "dominant" and "recessive" alleles (one of two or more DNA sequences occurring at a particular location in a gene).[13] He found that each organism requires the contribution of two genetic sources (an egg and a sperm in animals) and that the combination of alleles, in some instances, can determine certain traits.

Mendel observed that when the "dominant allele was present, the recessive allele seemed to disappear, but when a plant received two recessive alleles, the allele reiterated its character."[14] Eye color, for example, (mostly) follows this rule: Brown eyes are dominant alleles, so if someone has even one brown-eyed gene, then they will have brown eyes. Blue eyes are recessive, so they needed two blue eye genes. Sometimes, however, a mistake occurs when the genetic information gets transmitted. Dutch botanist Hugo Marie de Vries coined this a "mutation," the Latin for "change." Mutations occur as "mistakes" during this process of DNA replication, thus coding for new genetic sequences. Albinism, for example, is a mutation in the genes that code for eye (as well as skin and hair) pigmentation, expressed as pink eye color.

In 1905, English botanist William Bateson first proposed the word "genetics" to refer to this newborn science of distinguishing traits based on parental coded contributions. Bateson integrated Mendel's ideas on heredity into his studies of evolution, a critical step toward the modern synthesis, which combined heredity, evolution, natural selection, and morality into a cohesive theory about whose genes should be perpetuated and how public policy should reflect this. As Siddhartha Mukherjee writes in his 2016 book *The Gene*:

> [Bateson] grasped the idea that the discontinuous nature of genetic information carried vast implications for the future of human genetics. *If genes were indeed independent particles of information, then it should be possible to select, purify, and manipulate these particles independently from one another.* Genes for "desirable" attributes might be selected or augmented, while undesirable genes might be eliminated from the gene pool. In principle, a scientist should be able to change the "composition of individuals," and of nations and leave a permanent mark on human identity.[15]

Mukherjee details further that in heralding in the age of genetic power, Bateson "preempted the century of the gene" by writing that "when power is discovered, man always turns to it. . . . The science of heredity will soon provide power on a stupendous scale; and in some country, at some time not, perhaps, far distant, that power will be applied to control the composition of a nation. Whether the institution of such control will ultimately be good or bad for that nation, or for humanity at large, is a separate question."[16]

The trajectory of genetics proved largely to be as dark as Bateson anticipated.[17] Darwin's cousin, Francis Galton, in his 1869 book *Hereditary Genius*, advanced the concept of "social Darwinism" to refer to the idea that humans, and not just animals, evolve to help them survive and pass on their genes in the physical world. Certain groups of humans, the theory held, have *social* traits that similarly allow them to not just survive but thrive. He then proposed in his 1883 book *Inquiries into Human Faculty and Its Development* a social program called "*eugenics*," adopting the Greek word for "well born," which he likened to a "new religion," as a process of creating the "best families" by combining the "best traits."[18] It was also during this time that the philosopher Herbert Spencer "coined the phrase *survival of the fittest*."[19] Also referred to as "positive" eugenics, this project sought to create better humans through the alignment of desirable traits. "Negative" eugenics, in contrast, refers to reducing the "*breeding* masses . . . unwashed inferior breed," such as through forced sterilization, which, according to Mukherjee, "haunted" Galton later in life.[20]

By the early twentieth century in the United States, the Progressive Era's state programs and political tide advanced the twin goals of perpetuating the wealthy, healthy stock and eliminating the downtrodden, "unfit" ones.[21] In particular, the "eugenics" movement was intensely interested in uncovering racial differences. Race in the modern era had been invented as a social category for justifying the expansion of empires and their colonial efforts to subordinate local populations and situate themselves as the dominant rulers.[22] For example, in 1777, the famed philosopher Immanuel Kant, known for his deontological or rule-based moral theories, classified people based on his own ideas of race to justify certain social hierarchies.[23] Despite the fact that many, such as W. E. B. Du Bois, had shown by the early 1900s that the "concept of race was not a scientific category," the eugenics movement in the United States facilitated research and programs that aimed to unearth the genetic and biological differences between races to protect its established racial hierarchy.[24] Indeed, it was the US eugenics program that in many ways inspired the Nazis' own brand of eugenics in the 1930s, which justified the forced sterilization and genocide of Jews, ethnic minorities, the physically disabled (including thousands of Deaf people), the mentally infirm, political dissidents, and others, all in the service of purifying the gene pool and perpetuating the dominance of the Aryan race.[25]

Pseudoscientific methodologies for studying "racial hygiene" included measuring head sizes and other anatomical features and testing intelligence to demonstrate white supremacy. The Tuskegee Syphilis Study, a project funded by the US Public Health Service that began in 1932 and lasted until 1972, is known for its nonconsensual treatment of Black sharecroppers in Alabama who had syphilis. These individuals were enrolled in a study exploring the long-term effects of syphilis and were denied treatment when it became available. The study's inception was rooted in the fundamental belief that syphilis affected Black and white bodies differently, and as a result, Black men's "bad blood" (a term for those with syphilis) presented a significant threat to white populations.[26] According to Melbourne Tapper, "sickle cell anemia"—now known as sickle cell disease (SCD), as discussed in chapter 1—was originally identified as "a supposedly black-related disease," and "medical science and anthropology have used sickling to claim the racial distinctiveness of 'blacks' and 'whites' and to represent the distinctiveness as the product of a specific genetic structuration—as indisputable fact."[27] The subsequent maltreatment of SCD patients in hospitals, and the long history of ignoring their pain, stems from associating the disease with Black inferiority.

Eugenics programs used compulsory sterilization and institutionalization of the "criminally diseased" and "unfit" in service of "cleansing" the population.[28] Sadly, this implicated the early birth control movement. While it did much to advance some women's reproductive freedoms, early birth control experimentation was on non-consented Puerto Rican women and targeted Black and immigrant populations to discourage procreation. In particular, Margaret Sanger, an early and famous promoter of birth control access, weaponized contraceptives to aid in diminishing Black population numbers.[29] This political aim of controlling populations, though not novel, began to adopt the language of the emerging field of genetics. Paired with the arrival of the fields of sociology and statistics, which provided further methodological authority to explain divergent behaviors of groups of people, geneticists gained renewed confidence in genetics' deterministic moral authority.[30]

The eugenics movement also targeted disabled populations through forced sterilization and institutionalization. The US Supreme Court in *Buck v. Bell* (1927) approved the sterilization of Carrie Buck after officials at the Virginia State Colony for Epileptics and Feebleminded in Lynchburg had declared her "feebleminded" and her family as one with "bad heredity."[31] In the court's decision, Justice Oliver Wendell Holmes Jr. reasoned that it would be "better for the world, if instead of waiting to execute degenerate offspring for crime, or to let them starve for their imbecility, society can prevent those who are manifestly unfit from continuing their kind . . . three generations of imbeciles is enough."[32] However, according to Adam Cohen, as well as others, Carrie was "not in fact feeble-minded, despite what the state's unreliable intelligence testing reported," but rather the state used her to advance their interest in curbing the "threat" of immigrants and racial minorities.[33] The narrative of the diseased immigrant has reemerged throughout American history but with special fervor in 2025 with the Trump administration's labeling of immigrants as harbingers of "very contagious disease" who are "poisoning the blood of our country."[34]

Despite these atrocities, the allure of explaining the natural order of social groups and their hierarchies based on the supposed inevitability of different human behaviors continued to appeal to those in power, even after the catastrophes and presumed lessons of World War II. During the 1950s, not only was the world grappling with rebuilding itself after the horrors of war and the Holocaust, but it was also continuing its quest to understand *how* traits were inherited. In 1953, British scientists James Watson and Francis Crick, with the evidence that they took from Rosalind Franklin, published that they had used X-ray imaging and other photographic techniques to discover the structure of inheritance: deoxyribonucleic acid (DNA). DNA is composed of base pairs of nucleotides (a unique type of biomolecule) zipped in a double strand that then twists into a double helix.[35] The revelation of the structure of DNA showed geneticists how cells were able to "transmit copies of information from cell to cell, from organism to organism."[36]

DNA "informs" the cell of what proteins to build and when, and thus largely determines how the body develops. Two Stanford scientists, George Beadle and Edward Tatum, identified the "gene" as a unit of heredity that would "*carry the code* to build a metabolic or cellular function specified by a protein."[37] Soon after, French biologist Jacques Monod discovered that DNA transcribes RNA (ribonucleic acid), which codes for proteins in a process called "translation." In the 1960s, geneticists discovered that DNA also codes for its own "gene regulation—the turning on and off of genes by proteins."[38] This ability to "turn on" or "off" a gene has proved as important as what the gene codes for because it serves as the initial triggering event. Subsequently, in the late twentieth century, the field of genetics pursued a full-force agenda to discover "how genes build, maintain, repair, and reproduce humans—and how variations in the anatomy and physiology of genes might contribute to the observed variations in human identity, fate, health, and disease."[39] This included attempts to explain human behavior according to genetics, the environment or social conditions, personal experiences, or some combination of factors.

As during the Progressive Era, genetics research during this time coincided with other attempts to use biology to support social and political projects of the state. The United States instituted intelligence quotient (IQ) tests starting in the early twentieth century initially to aid in identifying which students were likely to experience difficulty in school.[40] By the mid-1950s, adult IQ and literacy tests were used in numerous public arenas, such as voting, housing allocation, military recruitment, and Social Security claims.[41] Again, the justification largely rested on the assumption that intelligence was fixed and that genetics played a strong role.[42] James Watson himself claimed a connection between genetics and "intelligence" and concluded (unsurprisingly) that white Europeans had higher levels of intelligence.[43] IQ tests were then used to show that different races had natural tendencies toward different strengths and weaknesses, as recorded in Richard Herrnstein and Charles Murray's controversial 1994 book *The Bell Curve: Intelligence and Class Structure in American Life*.[44] Herrnstein and Murray's work has long been used to defend systems of meritocracy because of supposed intransigent genetic differences that dictate intellectual disparities.[45]

This line of inquiry, however, is dangerously erroneous. As Dorothy Roberts has argued, the emergence of genetics as a biological tool to understand human populations fueled attempts to categorize people by race, even though race cannot possibly be reduced to biological or genetic explanations.[46] As mentioned previously, the eugenics movement implemented intelligence tests to perpetuate the myth of a genetic relationship between race and intelligence. However, this claim has been widely and repeatedly debunked by showing that measures such as IQ tests are largely indicators of social awareness or knowledge of particular times and places, rather than actual cognitive capacity, and their results are largely not explained by genetics or inherent abilities.[47] Instead, IQ tests are predominantly indicators of access to certain kinds of information and language associated with the dominant cultural group that built and designed them.[48]

The themes of discrimination based on presumed genetic conditions of racial groups and the fear of discrimination against disabled individuals impact our perceptions of the current uses of genetic technologies and genetic testing. For example, in the 1950s and 1960s, the United States began a newborn screening program after the 1934 discovery of phenylketonuria (PKU), a disorder in which the body is unable to process phenylalanine, an amino acid essential for development, resulting in varying intellectual disabilities.[49] Until it was "scrapped" in 2025 as part of Trump's program of deregulation (as discussed in the introduction), the Department of Health and Human Services' Advisory Panel on Heritable Disorders in Newborns and Children recommended that states screen for thirty-four core and twenty-seven secondary conditions for which specific tests and treatments exist.[50] Even as the program has been widely praised for giving parents resources and access to care early in a child's life, others have worried that it could also lead to further stigmatization of disabilities.[51]

In addition to newborn screening, the US government has also made significant investments in understanding the role of genetics in diseases by mapping the whole human genome. Completed in 2003, the Human Genome Project was a government-funded initiative to map the 3.2 billion base pair long sequence of nucleic acids.[52] The many breakthroughs in medical applications of genetics afforded by this research include the discovery of the strong role of *BRCA1* and *BRCA2* mutations in breast cancer (whose patent dispute was mentioned in chapter 1); the mutations causing Tay-Sachs disease (a fatal and painful early childhood illness), which have led to its significant reduction; and the gene for Huntington's disease (a neurodegenerative disorder), among many others.[53] The *gene revolution* refers to the period beginning with the Human Genome Project in the 1990s and continuing through today, and it often expresses great optimism and hope.[54]

During this time, scientists also began exploring gene therapies as novel treatments for genetic conditions. However, the narrative of genetics overreach reemerged in 1999 when Jesse Gelsinger, an eighteen-year-old with a liver disorder, died during a gene therapy trial.[55] His death was (and still is) widely viewed as an unnecessary cost because his condition was manageable and he was overall young and healthy.[56] With the passage of time, though, gene therapies reemerged in the spotlight and began to be heralded as a new age in medicine, including the 2023 approval of treatments for hemophilia and SCD.[57] Today, gene therapies are also the most expensive medications on the market: Zolgensma, a one-time gene therapy for pediatric patients under two years old with spinal muscular atrophy, costs $2.1 million per patient; annual spending on gene therapies in 2023 in the United States was estimated at $20.4 billion.[58] As noted in chapter 1, Casgevy, the gene therapy for SCD, costs $2.2 million.[59]

In the history of genetics, there have also been more widespread, morally dubious research practices.[60] The paramount example is the Havasupai case: In 1989, researchers at Arizona State University (ASU) approached members of the 650-person American Indian Havasupai Tribe residing in the Grand Canyon to collect blood samples for a study on diabetes, a condition that had plagued both them and neighboring tribal populations. The tribe consented to the diabetes study even though participation involved giving blood, which for them has "deep spiritual meaning."[61] Though it turned out that the tribe did not have the genetic variant the researchers were looking for, ASU researchers continued to use the stored blood in subsequent years for numerous other research purposes, including to identify that the Havasupai Tribe's origins were not in the canyon but in Asia and published papers on "inbreeding and types of mental illnesses that are taboo to the Havasupai," such as schizophrenia.[62] This nonconsensual research was discovered in 2003 when a member of the tribe, Carletta Tilousi, attended a research seminar on one of these studies conducted without tribal knowledge or permission.[63]

In addition to the lack of consent, the experience damaged already low levels of trust and highlighted the Indigenous populations' historical, and ongoing, experiences of injustice and exploitation.[64] Other examples of similar abuses in genetics research involving Indigenous populations include the Nuu-chah-nulth First Nation in Canada, the San people in South Africa, and the Maori in New Zealand.[65] Today, Indigenous geneticists are rebuilding relationships with other geneticists through community-based processes, resulting in studies designed in conjunction with, rather than conducted on, the Navajo Tribe, and creating more opportunities for Indigenous tribe members to become geneticists themselves.[66]

Genome-Wide Association Studies (GWAS), which use extremely large sets of genetic data to find patterns of variants and traits across populations, have also stirred hope as well as controversy. This method of analysis is helpful for identifying genes implicated in certain illnesses, facilitating the development of novel interventions.[67] GWAS are also used to trace the ancestry and origins of populations. One of the largest studies of this kind, using data on Philippine populations, was published in 2021. Directed by researchers at Uppsala University in Sweden, the study described "the Philippines as a crucial gateway, with a multilayered history, that ultimately changed the genetic landscape of the Asia-Pacific region."[68] Yet again, while some see great promise in this method of understanding disease expression, others view it as a sign that "genetics has returned to its own controversial past, using gene-level differences across groups, enormous similarities, and a relative handful of variations across genomes as a template for reconfiguring groups and for building new notions of racial difference for a new era."[69]

Another worry with GWAS is methodological. First, correlations in medical data are often inevitable, leading to inflated conclusions.[70] Second, many of the conclusions that may emerge about social behavior, such as when identifying certain groups as having educational promise or capacity, are better understood as environmental or institutional problems rather than those of genetic determinism.[71] Finally, the majority of genomic studies (86 percent) have been conducted on individuals of European descent, "which represents an increase from 81% in 2016."[72] This underrepresentation of genetic variation is to the detriment of everybody. As Segun Fatumo and colleagues show, "as a consequence of underrepresentation, opportunities such as identification of new associations with population-enriched variants, pinpointing causal variants for functional follow-up, improving genetic risk prediction accuracy for all populations (particularly underrepresented populations) and understanding shared versus unique genetic and environmental population risk factors that influence health outcome" become stifled.[73] Underrepresentation, in other words, erodes the scientific quality of findings and thus the therapeutic benefits for the vast majority of the population.[74]

However, in a world where Indigenous groups and their data have been abused, the lack of diversity may be due to distrust in participation.[75] "All of Us," the National Institutes of Health's biobank program created in 2015 with $130 million aimed at collecting genetic and health data from one million volunteers, has strongly emphasized diversity and the recruitment of underrepresented groups.[76] This, however, evokes a moral tension that also pervades CRISPR conversations around balancing the value of diversity while learning from the past's exploitation of these communities by overtargeting them.

Alongside the development of our understanding of genetics and genomics, reproductive technologies have served as essential tools to further the goals of geneticists. Even before the gene revolution, advances in reproductive technologies offered parents new choices of genetic selection. In 1981, the first "test-tube baby" was born via in vitro fertilization (IVF), wherein an egg and a sperm are combined outside of the womb, and the resulting embryo is then inserted into the womb. Preimplantation genetic diagnosis (PGD) gives parents further options for testing and embryo selection during IVF.[77] Parents (who can afford to) can now create multiple embryos, sequence their DNA, and select for certain desired (or against undesired) traits, such as cystic fibrosis, a protein disorder characterized by the overproduction of mucus that affects the lungs and digestive tract.[78]

In the realm of reproductive technologies, prenatal genetic testing has also allowed parents who are already pregnant to test for chromosomal abnormalities in utero, such as Trisomy 21, or Down syndrome. Today there are two main tests. The "noninvasive" prenatal screening test (NIPT) requires a blood draw, and sometimes an ultrasound, and takes place around weeks 10 to 13. The other is an "invasive" prenatal diagnostic test using amniocentesis, which requires withdrawal of amniotic fluid, and takes place during weeks 15 to 18. The ableism and disability arguments in the next section largely arose from discussions on prenatal testing and PGD because they center on when and how to select for certain phenotypes, and thus when and how to describe something (or someone) as "desirable or undesirable."[79]

Disability Rights Versus Enhancement

Following a dark eugenics history that targeted disabled populations through forced sterilization and institutionalization, contemporary disability critiques of CRISPR note the potential for further discrimination and population elimination. Before examining the specific arguments, it is important to note the rich diversity in "disability rights views" and what critical disability theorists and advocacy groups think about gene editing. In this section, as in this whole chapter, I intend only to present ideas from the discourse as important considerations for why we should include these voices in our regulatory approaches, rather than to argue for *the* disability rights view.

In many ways, disability-based critiques of CRISPR stem from prenatal testing concerns that decry ableist approaches to describing pregnancies and children as "healthy" and "desirable" due to the ability to select against certain traits and thus certain groups of people.[80] In her 2017 book *The New Eugenics*, Judith Daar argues that access to resources and technologies, such as PGD and IVF, constitutes a "new eugenics" because they feature subversive eugenics-like results even when institutions are not aiming to control populations per se.[81] This concern is bolstered by a systemic problem of discrimination against those with disabilities by the medical community, which leads practitioners to not fully explain or appreciate what living with a disability or parenting a child with a disability is like.[82] These arguments become more pronounced with the arrival of CRISPR because it is seen as having the potential to achieve what modification and other interventions have not been able to.

Disability arguments that stem from prenatal testing, and PGD, have three main formulations. First, editing embryos is wrong because it stigmatizes and marginalizes the existing group of individuals with the condition being targeted.[83] Second, by editing the embryo of a disabled future individual, it diminishes the inherent value—and, for some, the identity—of that future individual. And third, editing embryos creates problematic conceptions of both parenthood and childhood expectations.[84]

Each of these formulations rests on another foundational critique in disability literature: In contrast to the "medical model," which identifies disability as something in need of fixing due to deviation from normal body functioning, the "strong social model" of disability holds that the environment, rather than some inherent characteristic or feature of the person, is the source or explanation of the difficulties disabled people face.[85] For example, the social model holds that individuals with a disability have a hard time transporting themselves through a city because of the lack of wheelchair lifts on public buses and curb cuts on sidewalks. Thus, the solution to the problem of exclusion from society should be to make the world more accessible rather than to focus on surgical interventions solely or perhaps even at all.[86] Given that the world is created for "normal" functioning human bodies and minds, modifications to our infrastructure should be made to accommodate diverse needs. Thus, the goal of medicine should not be to harness gene editing to modify embryos, but rather to alter the environment that perpetuates inaccessibility and discrimination or to focus therapies on improving the lives of existing individuals, rather than embryos, with genetic disorders.[87]

Each formulation also seeks to debunk the association of disability with poor well-being and simultaneously replace it with the understanding that people with disabilities can flourish and lead fulfilling lives. The "disability paradox" refers to the phenomenon in which individuals who experience traumatic events or chronic illnesses that limit their mobility or alter their lives report high levels of contentment and happiness, and a quality of life as high as those who have not gone through such life changes.[88] For example, in a study of individuals with traumatic brain injuries, the severely injured group did not indicate lower satisfaction scores than the moderately disabled group.[89] This, of course, is not truly a paradox; rather, it is just surprising to able-bodied individuals because most people assume disabled individuals have a lower quality of life.[90] The disability critique would caution us not to fold our assumptions about disability into specific targeting of certain traits we assume are undesirable for our children.

Disability theory on CRISPR often confronts the "enhancement" view, or the idea that it is permissible or even at times obligatory to genetically modify (or edit) embryos.[91] By maximizing the future child's chances of having "superior-to-normal performance" of "normal human species functioning," we make them better off.[92] Julian Savulescu, an infamous proponent of enhancement, argues that we have a moral obligation to enhance our children's welfare as much as possible. In so doing, he aims to distance himself from eugenics, writing, "If one is committed to the moral obligation to treat and prevent disease, one is also committed to genetic and other enhancement in so far as this promotes human well-being. . . . This is not eugenic but expresses our fundamental human nature: to make rational decisions and to try to improve ourselves. To be human is to strive to be better."[93] For Savulescu, to fail to do everything within one's means to improve the lives and welfare of our children, such as by bolstering their intelligence and attractiveness, is to fail in our obligation to them when we chose to create them in the first place.[94] Other enhancement enthusiasts argue that enhancements need not only be physical but also moral.[95] John Harris defines "moral enhancement" as "those forms of cognitive enhancement that operate across a wide range of cognitive abilities and do not target specifically 'ethical' capacities."[96] Such "cognitive abilities" include being "more empathetic" and "less violently aggressive" with a "higher potential for complex moral reflection."[97]

For many, enhancement is distinguished from "therapeutic" applications and medically/nonmedically necessary interventions.[98] Nicholas Agar offers a middle ground approach, arguing that enhancement should "moderate" the realm of general improvements but not significantly assist those who already "*greatly exceed* what is currently possible for human beings."[99] Thus, Agar identifies enhancement as a kind of "improvement" within "*human norms*."[100]

A core foundation of enhancement reflects a naturalist or evolutionary argument—namely, that humanity is on an upward trajectory of constant innovation and progress in medicine and morality. Technology allows us to attain better health and welfare, so we should embrace it. As Françoise Baylis and Jason Scott Robert argue, "The inevitability of the technologies results from a particular guiding worldview of humans as masters of the human evolutionary future."[101] This sentiment is also reflected in the "transhumanist" or "posthumanist" movements that aim to push beyond normal species functioning by harnessing qualities or attributes of other species—such as the sonic hearing of bats or the night vision of owls—or by incorporating cybertechnology as part of their own embodiment.[102]

Disability theorists often situate their interventions around the question of what should be considered a human *norm* and, as such, what an enhancement even is. Arguments responding to these pro-enhancement positions, then, either question the very existence or possibility of enhancement or offer a contrasting positive depiction of the world with disability as one we want to live in rather than one to avoid.[103] Opposition to enhancement arguments sometimes contains an essentialist construct of humanness, a kind of moral foundation beyond which we should not push ourselves, otherwise we risk losing our humanity. George Annas, for example, argues against genetic enhancements because it threatens our species integrity.[104]

Disability arguments do not exclusively counter negative portrayals of disability; rather, they offer positivist views on the value of human diversity.

Learning from and appreciating a range of different kinds of bodies and ways of thinking is inherently valuable and can better the world by enriching the vast tapestry of human life forms.[105] Isabel Karpin refers, for example, to the idea of "negative enhancement" as selecting *in favor* of a disability because it is in itself desirable or because it is desirable for the flourishing of the community, as in the case of the Deaf community.[106] From this, another concern is that gene editing could lead to the loss of communities and identities. As we have seen in chapter 1, in the case of heritary hearing loss, Deaf advocates have long argued that Deafness is a culture. Concerned with the erosion or even loss of their community, Deaf advocates argue for access to reproductive technologies to ensure that Deaf children not only exist but thrive.[107]

Others articulate similar worldviews with a deep appreciation for the intrinsic value of humans as they are. Erik Parens offers an "authenticity" critique of the false prospect of perfectionism.[108] The "designer babies" concern refers to this anti-enhancement authenticity approach to genetic interventions because it denies the essential value of the human being or, at times, even the very right to exist for certain groups of people.[109] Another consequence of this worry is echoed in lessons learned from the eugenics movement. If our moral mandate is to maximize our child's welfare, then, the logic follows, we should pick traits that are most favored in society, such as whiteness, maleness, heterosexuality, and so on. This, many may note, would not be morally preferable due in large part to the discriminatory prioritization of certain people existing.

Michael Sandel, in his 2002 book *The Case Against Perfection*, argues that the emphasis on "perfecting" children creates untenable expectations. Enhancement, Sandel contends, invites problematic expectations of our fellow humans and creates toxic approaches to valuing human life. It does so, in part, by eroding our solidarity with parents who "fail" to enhance, for those parents will be seen as not doing enough for their children. Josephine Johnston and Eric Juengst further argue that enhancement creates an incessant quest for perfection, through the demands of sequencing genomes and selecting embryos with "better" genetic profiles, merely achieves placing a "mental and emotional burden on parents" due to the incredible uncertainty and lack of knowledge about what sequencing can accomplish or predict.[110]

Gene editing, on this account, then, is morally threatening because it suggests that certain kinds of people are "less than human."[111] Critics say that enhancement enthusiasts necessarily seek to eliminate their populations, perhaps not (anymore) through "positive eugenics" (the purposeful sterilization of undesirables) but through "negative" efforts, such as by making it impermissible (or at least very difficult) to perpetuate genes deemed unhealthy or unfit.[112] Similar to the aforementioned "new eugenics" critique, the argument is that systems could incentivize (such as through insurance coverage or reimbursement policies) the bodies that are valued above others.[113]

Unequal Access

The negative or "liberal eugenics" consideration is that if gene editing were to be deemed safe and worthwhile, then it would require great means and fortune to do so and would thus only be accessible to the privileged few. This would lead to further genetic inequalities.[114] In contrast to Sandel's claim that "the fundamental question is not how to assure equal access to enhancement but whether we should aspire to it," many argue that already unequal levels of privilege, power, resources, wealth, and access to technologies have created a world in which there are many more healthy individuals than others.[115] Introducing gene editing into that context ignites worries about perpetuating or even entrenching injustices, which is far more worrisome, for many, than the threat of enhancement. As Collier Meyerson writes, "The danger [of CRISPR and IVF] isn't in the procedure itself, but who has access to this type of medicine—and right now that group is limited to those who can pay."[116]

Thus, while the previous section's primary disability-based concern was the *over*use of editing, resulting in certain groups being disproportionately targeted or eliminated, in this section, the worry is about *under*use by other populations owing to barriers to access. Myriad social problems result from inequality, such as the lack of access to lifesaving therapies and divergent experiences of the climate crisis.[117] Introducing new technologies, such as facial recognition, into an already stratified social system can exacerbate these problems, widening the gap between those who have been surveilled and those who have not.[118] Just as in the genetics history section, new technologies often bring out accelerated versions of previously rooted social conditions. For example, during the COVID-19 pandemic, early access to personal protective equipment, testing, vaccinations, and educational technologies was, and continues to be, harder for those who live in impoverished areas and for minorities, leading to unequal infection and death rates across populations.[119] Before increased competition drives costs down, such as in the case of the BRCA test, the costs of new and powerful technologies can be high, thus limiting access.[120]

In the case of CRISPR, we are already seeing high costs for the two gene therapies for sickle cell disease. Casgevy is listed at $2.2 million, and a non-CRISPR-based therapy called lovotibeglogene autotemcel (a.k.a., lovo-cel) costs $3.1 million. As Edward R. Scheffer and Frazer A. Tessema write, "These higher out-of-pocket costs might . . . actually penalize patients for the historical neglect of such conditions and may exacerbate existing health disparities, especially for low-income patients."[121] The common response, as in the case of BRCA, is that companies need to charge a significant amount at the beginning to recuperate the costs of creating the therapy in the first place, and it is better to benefit some than none at all. Further, companies need the incentive of high payouts to make the medicine at all.

Scheffer and Tessema respond to this line of critique by writing, "The paradox is that while high prices may incentivize health care and biotechnology ecosystems to increasingly develop drugs for neglected diseases, high prices will conversely limit access to these treatments, lead to higher out-of-pocket costs for those perhaps least able to afford them, and generate reluctance from insurers to cover multimillion dollar products, compounding access problems."[122] In response to this, the US Center for Medicare and Medicaid Innovation launched a pilot program, the Cell and Gene Therapy (CGT) Access Model in January 2024, to help alleviate gene therapy costs. CGT includes negotiating gene therapy prices, asking manufacturers to offer rebates when the gene therapy does not work on the patient, and offering state-based assistance to different programs.[123] Fortunately, the Trump administration's 2025 funding cuts to the Center for Medicaid and Medicaid, which threatened the viability of this program, were saved in July 2025 when it expanded Medicaid access to SCD gene therapy.[124]

Others, however, have pushed back against the idea that inequalities, such as those of genetics themselves, are a social justice problem. Kathryn Paige Harden, in situating herself as nonpartisan, offers a vision of using genetic information to help those in need, as opposed to discriminating against them, as history has shown it more often does.[125] Basing her argument on twentieth-century political philosopher John Rawls, who argued that society should *not* be based on genetic inequalities, Harden argues that knowing individuals' genetic predisposition toward cognitive struggles will help us design education systems that better identify and help struggling students so they can learn, grow, and succeed.[126] She writes, "The eventual development of a polygenic score that statistically predicts educational outcomes will allow researchers to control for genetic differences between people, so that the causal effects of the environment are thrown into sharper focus. Understanding which environments cause improvements in children's ability to think and learn is necessary if we want to invest wisely in interventions that can truly make a difference."[127] Because of genetic differences and their relationship with criteria such as weight, fertility, health, lifespan, and, sure enough, "our intelligence and success in school," she argues, there is benefit in knowing one's genetic predisposition toward educational success.[128]

Harden's claims have been strongly rebuked for both empirical and ethical reasons. Empirically, the relationship between intelligence and genetics is tenuous at best, as seen in the discussion of IQ. Ethically, critics such as Dorothy Roberts worry that this line of inquiry is problematic because "as soon as their findings are translated into knowledge about heritable intelligence, their research will take on the social implications that inevitably result from ranking human beings by cognitive capacity and attributing their rank in significant part to their genes."[129] Thus, knowing genetic differences will only entrench existing inequalities and find purchase in justifying hierarchy, rather than eliminating it. Daphne Martschenko similarly argues that such information of genetic inequality would be far more likely to be used by those with nefarious purposes to reify their racist and bigoted platforms, as has been occurring under the rise of Donald Trump.[130]

While most agree that vastly unequal systems are morally problematic, scholars of distributive justice have long debated what kind of social equality *is* desirable and, simultaneously, what kinds of inequalities are permissible. A core unresolved question, then, is how such a powerful, new, and innovative technology should be distributed across society and individuals, particularly when it heralds making lasting genetic changes to populations and groups of people that could have significant and long-term effects. What, ultimately, are our egalitarian commitments? Relatedly, which applications (such as those covered in chapter 1) should be pursued first and for what reason? The role of prioritization and the distribution of scarce resources have inordinate consequences for social justice that call for an examination of how CRISPR's emergence burdens different people differently. And as we will see, these matters require public inclusion and deliberation.

Maternal, Fetal, Environmental, and Animal Rights

Thus far, I have mostly focused on the moral concerns of specific populations that are poised to be directly affected, or even targeted, by CRISPR. Now, I examine the claims concerning mothers, fetuses, the environment, and nonhuman animals, also affected by CRISPR but often those whose specific interests going unheeded. These worries are neither unique to gene editing nor defining of the debate as they feature prominently in conversations generally within reproductive and environmental ethics and nonhuman animal rights literatures. But as with the rest of our moral quandaries, new technologies speed up already existing moral concerns and political patterns.

First, reproductive autonomy is the idea that those with uteruses seeking rights-based claims to their body and their decisions should have control over their reproductive decision-making. Early work on reproductive technologies in the 1980s and 1990s, during the advent of IVF and PGD, distinguished between what would constitute a "personal choice" and a "public health" consideration and how to balance "procreative liberty" with "reproductive conflict."[131] These considerations also later featured strongly in the aforementioned debates about non-invasive prenatal testing, particularly for Down syndrome. For CRISPR, the balance then becomes: How can we allow people to access reproductive technologies we deem a right while balancing the group-based claims of disability groups?

Alongside this paradigm of social choice and justice in liberal societies, materialist accounts of pregnant people's experiences have emerged. Contemporary feminist work on the topic also now explores *how* pregnant-capable people experience maternal (or parental) questions of reproduction and the technologies that transform their reproductive capacity. For example, Carla Lam's theory of "disembodiment" describes alienation from childbirth that women may experience when using reproductive technologies.[132] Others have voiced similarly Marxist approaches to the question of women's roles in the era of genetic technologies, with emphasis on labor power, domestic exploitation, and using pregnant individuals as a kind of laboring machine.[133] These perspectives play a central role in other important bioethics applications, such as ectogenesis (artificial wombs) and gestational surrogacy (carrying a child intended to be parented by others).

In centering the conversation on the role of the mother who became pregnant with the twins edited by He Jiankui, Rachel Adams penned a poignant perspective calling attention to the fact that the name of the mother who birthed the twin girls was never even mentioned in the flurry of outrage that ensued. Adams wrote that ignoring the mother "matters for reasons beyond this individual story" because it signals how we view the role of women in these decisions: as a side thought, if a thought at all.[134] Adams wrote further that we take a moral misstep in viewing the incident of Dr. He as though he is the creator of the twins. The *mother* was the true creator, and "regardless of how her twins were conceived, she is responsible for their well-being and development."[135]

Alongside women's "invisible labor"—the hours of emotional, logistic, practical support and work they perform without pay and often without recognition—the arrival of CRISPR will have a disproportionate effect on the *moral* labor of women.[136] Adams writes, "Like other new reproductive technologies, from ultrasound to amniocentesis and IVF, CRISPR is likely to present pregnant women with more responsibilities and decisions, not remove them from the process."[137] Adams goes on to argue that the "invisibility of the mother's body in images of the CRISPR babies" reveals a failure to understand that germline gene editing is still ultimately about reproduction and has inordinate implications for women's bodies, lives, and autonomy.[138] Similarly, in his 2020 book *Editing Humanity: The CRISPR Revolution and the New Era of Genome Editing*, Kevin Davies helps situate this narrative by reminding us that most nonconsensual experimentation has historically been primarily on women and young girls, leading to concerns about consent to carry an experimental CRISPR-edited fetus.[139]

Another important rights-based consideration is that of the edited baby and "future generations."[140] The concern also pertains to the consent of future generations to decide their own genetic modifications.[141] This conversation comes up in debates around cochlear implants, circumcision, and other non-lifesaving and potentially invasive procedures that children may not be old enough to consent to but may benefit from. These decisions may best be placed within the zone of parental discretion.

Finally, environmental and nonhuman animal considerations arise when we consider the needs of those with no voice.[142] What are the moral considerations and best interests we assume for them, and how do we make those determinations?[143] In the case of gene drives in Florida and Nantucket that aim to eliminate diseases affecting human populations, to what extent should we consider the rights and interest-based claims of rats or mosquitoes, or those of the larger ecosystem that may be affected? As Natalie Kofler and Colleen M. Grogan argue, "the inclusion of the voice of nature and the voice of future generations is required to ensure safe and just decisions about genetic engineering in the wild."[144]

GMO Legacy

In addition to the legacy of eugenics, medical experimentation on vulnerable and disabled populations, and the experience of uterus-bearing individuals in the reproductive autonomy discourse, the ghost of genetically modified organisms (GMOs) and the public reaction to them looms large in the moral and governance discourse on CRISPR. GMOs are plants, animals, and other organisms whose "genetic material has been changed by humans to a form that does not generally occur on its own in nature, particularly by transferring genes between species."[145] "Genetic engineering" constitutes this overall practice of selecting for and replacing DNA sequences in crops and livestock.[146]

GMOs, as we saw with CRISPR crops, were lauded as lifesaving, nutritious, and durable produce.[147] GMOs were first discovered in 1973 when Stanley Cohen and Herbert Boyer transferred a "gene that provided antibiotic resistance from one strain of bacteria to another that lacked this resistance," creating antibiotic resistance in the second bacteria.[148] By the 1990s, GMO products started to hit the market and were heralded as bringing in highly needed nutritious and stable produce—for example, tomatoes that could withstand miles of travel without getting smooshed, vitamin B–enriched rice ("Golden Rice") that would help to feed malnourished areas of the world that depend on rice production, and enlarged salmon that could be grown in short periods of time and could then curb wild salmon fishing.[149]

However, GMOs, still today, face vocal opposition from a particularly passionate yet small sector of society that decried the safety of new organisms and the devastating repercussions their introduction to the market would have on small (and Indigenous) farmers.[150] Especially in the early years of GMOs, many expressed a general distrust of powerful industry and the corrupted regulators who were meant to regulate that industry.[151] Advocates of GMOs responded that genetic engineering itself is not new, noting early domestication practices and standard contemporary farming and agricultural techniques that make it possible to choose and propagate some kinds of seeds and plants over others. Much of this public debate contributed to fraught disagreement on regulatory approaches, such as whether GMO products require labeling.[152] Finally, on January 1, 2022, the US Department of Agriculture promulgated the National Bioengineered Food Disclosure Standard that requires that all items containing GMO products be labeled as "bioengineered" or "derived from bioengineering."[153]

The fraught experience of the GMO legacy looms large in debates over CRISPR regulation for several reasons. For the public that opposes GMOs, there is a throughline of distrust in genetic technologies and concern over safety and potential medical interventions, as well as distrust of the regulators who are perceived as bought out by the pockets of these large companies. Conversely, many CRISPR experts similarly distrust the people who oppose GMOs, contributing to their worries about public reactions to CRISPR and calls for moratoriums and "slow science."[154] As discussed in the introduction, we often see calls for public inclusion in national and international reports on the topic in an effort to decrease the likelihood of backlash, rather than because the public may actually have important perspectives on being Deaf or having a family member with SCD.

From this conversation emerges certain "stakeholders": the people (or groups) with concerned interests, rights, and perspectives affected by CRISPR. These stakeholders may sometimes be obvious, such as those diagnosed with SCD or those making gene therapies, and at other times they may be more subtle, such as the residents of Alabama who live near Florida, where the modified mosquitoes were introduced. Institutions can also be considered stakeholders, and when these institutions accumulate enough power and influence, they become special interest groups, such as lobbyists or advocacy groups that push for a particular policy outcome.

From this, we can also start to see that the regulations detailed in chapter 1, which primarily focus on questions of risk and product novelty, do not address the more abstract or deep and abiding societal considerations surrounding claims of ableism, inequality, and reproductive agency. This is not per se a critique of regulatory agencies because, as we will see in chapter 3, they are *not* tasked with solving society's moral ills present in new biotechnology cases. However, regulatory and moral frameworks are importantly linked: Asking what constitutes acceptable risk evokes the enhancement debate; what is considered novel dictates the availability of products in markets; and what we consider marketable is implicated in reproductive welfare. This, as opposed to public backlash, forms a stronger positivist basis for imagining public inclusion in regulatory agencies.

3 Regulatory Agencies and Democratic Participation

Thus far, we have seen that CRISPR promises significant medical and environmental benefits while also entailing important moral and political considerations. Here, I continue the discussion by examining the origins, goals, and functions of regulatory agencies to better understand the importance of democratic deliberation on biotechnology regulation, particularly regarding CRISPR. I argue that regulatory agencies are well positioned to carry out their mandate but may also be too insulated from or unaccountable to the very public they are meant to protect. Thus, even though their authority and mandate come from Congress, they suffer from a "democratic deficit," insufficient accountability mechanisms that tie governmental decisions to the needs and interests of the people.[1] Regulatory agencies are technocratic (expert-based) institutions that enact risk-based governance in the name of protecting the public interest. This challenge of balancing expert-based rules with nonexpert inclusion is not unique to regulatory agencies, but it does provide an opportunity, particularly given their early involvement in CRISPR applications, to examine democratic institutional design, risk analysis, and public accountability.

Federal agencies provide opportunities for participation, primarily through the notice-and-comment period, when the public is invited to weigh in on drafts of certain rules and guidance. In this chapter, I argue that these opportunities require modification to increase the democraticity for some realms of CRISPR rulemaking because of its historical complexities and value-based risk analyses. Notice-and-comment alone are insufficient for garnering the kind of participation required in this domain. This is for three main reasons: (a) While agencies are obligated to respond to all comments, they are not required to substantively alter their rules based on the comments; (b) many populations we care about hearing from are underrepresented; and (c) in parallel, the ones we are seeking to regulate (industry) are overrepresented. This overrepresentation contributes to problems of regulatory capture wherein private interests begin to supersede public ones, resulting in at least the perception of illegitimacy. As such, some modification is required to address the democratic deficit by cultivating greater accountability to the people, such as by strengthening public input and the participation of dominated groups, as explored further in chapter 4.

Technocracy and the Democratic Deficit

US regulatory agencies are built on the idea that executive agencies should be insulated from politics to make rational, rule-driven judgments, particularly when it comes to scientific and technical matters. The authority of executive regulatory agencies comes directly from Congress, the legislative branch. The US Constitution created the power to regulate by giving the executive branch, with the approval of Congress, the authority to establish regulatory agencies. In *Federalist* 81, Alexander Hamilton describes how dividing power among separate departments to create independent, administrative agencies protects the republic from the changing tides of representative politics: "In a single republic, all the power surrendered by the people, is submitted to the administration of a single government; and usurpations are guarded against by a division of the government into distinct and separate departments. In the compound republic of America, the power surrendered by the people, is first divided between two distinct governments, and then the portion allotted to each, subdivided among distinct and separate departments."[2]

Congress, thus, has the power to create and provide oversight of regulatory agencies, such as by passing the federal budget and funding their programs, defining agency jurisdiction and areas of oversight, and delineating agencies' procedures and structures. Indeed, it was Congress that created the FDA, USDA, and EPA in the early- to mid-twentieth century, as well as the framework for how federal agencies conduct rulemaking and adjudication. Regulatory agencies were also created to perform tasks that Congress may be too burdened to handle or lack the specialized knowledge or qualifications to carry out on its own, such as analyzing gene drive technology and its projected environmental impact. And while regulators are accountable to Congress insofar as Congress has power over their appropriations and many aspects of oversight, they also have a different relationship to the public than legislators do because regulators are neither chosen by nor directly report to the people. However, they still have an important and intertwined relationship with the wider public insofar as the protection of the public and concern for the public's reaction shape their responsibilities.

To create a more specialized space where federal workers could focus more time, energy, and resources on specific areas of governance, executive agencies were also designed, in many ways, to be insulated from and independent of electoral politics. Of note, regulatory agencies are not alone in this logic—James Madison in *Federalist* 10 describes Congress as designed to be insulated insofar as it relies on electoral representation rather than direct participation; that is, it is designed to filter the views of the people and shield the government from the "passions" of the masses.[3] Creating space for agential independence, the logic went, would allow agencies and their civil servants to cultivate institutional wisdom by not subjecting them to significant turnover with partisan presidential changes. This was largely true until the 2025 Trump administration, which as of May 12, 2025, had cut 12 percent of the federal workforce, including 20,000 personnel at Health and Human Services where the FDA is housed.[4]

The founding of the United States was strongly influenced by the Enlightenment Era's theory of knowledge. From the late seventeenth to early nineteenth centuries, the Enlightenment promoted scientific (and philosophical) inquiry as a means to discovering justified true belief, realized through rational thought.[5] In the early twentieth century, the Progressive Era adopted this approach within the bureaucracy by centralizing objective, rational, and quantitative inquiry. Thus, since the founding of the United States, philosophers and theorists of bureaucracy have articulated their ideal vision of this system, almost all agreeing on the foundational role of expertise. Experts, as qualified conveyors of specialized skills such as scientific inquiry and rational thought, play a crucial role in interpreting data, and their objectivity makes them ideal regulators. Max Weber, the late nineteenth- and early twentieth-century German social theorist and philosopher, identified bureaucracy as an ideal location of "formal rationality," wherein procedures (the rule of law) were impartially applied with the aim of maximizing the most efficient means toward an end. Weber's account emphasized specialized function, arguing that employees ought to be divided into the kinds of work they are most capable of doing, often based on technical qualifications.[6] This construct of bureaucracy was seen as more efficient by streamlining information to limited cases, thus reducing the amount of information processing required for each person.

The "managerial" model of administration emphasizes the "managerial function as separate from and more important than any other job in the bureaucracy, as a difficult art that can be performed only by certain types of individuals who have 'rare qualities . . . and unique competence.'"[7]

According to Beverly Burris, this expert-driven model of a bureaucracy that requires qualifications morphs quite seamlessly into a meritocratic system. As a result, "in technocratic organizations, demonstrated expertise and credential certification tend to become more important than rank position as the basic source of legitimate authority. In contrast to the more personalized managerial authority characteristic of bureaucracy, technocratic authority rests on allegedly neutral decision making derived from expertise and systems maintenance."[8] Burris further shows that increasing complexity in power structures, technological innovation, and socioeconomic unrest solidified organizations' impulse to retain technocratic power, relying on the "polarization into 'expert' and 'non-expert' sectors."[9]

The "career bureaucrat" is someone who cultivates good political judgment because they possess a particular kind of expertise as a bureaucrat, with institutional wisdom and tacit knowledge developed through repeated affirmed judgments and responsibilities within the position.[10] For Peter Steinberger, political judgment is rooted in "intelligent performance," requiring both intuition and rational reconstruction.[11] Judgment here is defined as the range of inquiry of human mental activity meant to arrive at decisions on pressing questions. Political judgments incorporate the varied aspects of decision-making, such as evidence, values, and the distribution of rights and resources to decide how a technology should or should not be used. This is often presumed to be "value neutral," such as the idea popularized at the beginning of COVID that we should "just follow the science."[12] In other words, science would bring about clearly defined answers for public policy and in determining social risk, and epidemiologists were uniquely positioned to provide this invaluable insight.

And because regulators are expected to be more value neutral and qualified due to their "merit," this allows them to hold power quite often in the name of science and authority. Recall that in a technocracy, Ronald Beiner writes, the "monopoly of political intelligence is handed over to experts, administrators, and political technicians who coordinate the rules of administration and decision-making that accord with the reigning canons of method, rational procedure, and expertise."[13] This monopoly, Beiner continues, "goes unquestioned because the exercise of political rationality is assumed to be beyond the competence of the ordinary individual, whose proper sphere of competence is the choice of his own moral and social 'values.' Total political responsibility is ceded to the expert or administrator, provided that the individual's private spheres of values is not invaded."[14]

As such, the early days of regulation featured a technocratic structure within executive agencies. This served, as James Wilson writes, a kind of "administrative separation of powers," whose mandate of regulating the economy and the products of industry and capitalism hinged on specialized skills and knowledge.[15] Frank Fischer shows that specialized skill then gets equated with competency in policymaking, and "knowledge" becomes "a critical resource in the politics of class struggle, both inside and outside the workplace," by creating hierarchies of merit.[16]

This became particularly important during the early twentieth century when the development of the fields of statistics, biology, economics, and sociology as tools for measuring and quantifying populations coincided with a surge of immigration to the United States and labor upheaval. As Burris writes, "The ideology of technocratic management . . . veils political motivation and the range of choices as to the nature of the system, through recourse to the ideology of system preservation, which has more appeal in an era of clear socioeconomic crisis tendencies."[17] Technocracy helped to preserve the status quo by explaining why certain groups held certain stations in life. This featured heavily in the eugenics movement, which embraced the emerging methodologies of the time to further quantify, count, and justify differences between dominant and subordinate groups of people and subsequently impose policies that reinforced this hierarchy.[18] As discussed in chapter 2, this led to sterilization programs and the overall marginalization of nonwhite, non-American, and non-able-bodied individuals.[19] Fischer continues that expertise in a variety of fields was seen as needed to bring order to the chaos of the moment: "Given that it was a period of considerable social chaos—resulting from vast immigration, and unregulated economic activity, the rapid emergence of the new and more monopolistic business corporation, and various social groups looking for protection against unregulated economic competition—the problematic societal consequences of the time led many to see the revitalization of technical expertise as a source of hope."[20]

The Progressive Era, as such, fully embraced expertise as the backbone of its regulatory oversight of banking, transportation, environment, and other sites of federal authority, particularly with rising industrialization and mass production of commercial items and advances in medicine (such as the discovery of penicillin, X-rays, and blood transfusions). Particularly under the presidency of Woodrow Wilson (1913 to 1921), the United States witnessed a great expansion "of the administrative capacity of the federal government" to meet the needs of regulating the markets, industry, and movement of people.[21] According to James Miller, Wilson argued that the "proper execution of these policies will require the support of professional civil servants, trained like Wilson in the modern social sciences and with specialized knowledge about various topics (such as health care, the administration of justice, foreign affairs, etc.)."[22] Wilson further wrote that the regulatory agencies of this administration would require wisdom, knowledge, and experience because "'such things must be studied in order to be done well.'"[23] We can see echoes of this in contemporary calls for "evidence-based policymaking," the view that "policymakers should base their decisions primarily on scientific evidence."[24]

The Food and Drug Administration (FDA), for example, was founded during this period to meet the vastly unmet need to check products of consumption for ingredient integrity and safety. While the Department of Agriculture, founded in 1848, had been tasked with monitoring the chemical components of agricultural products, the Pure Food and Drug Act and the Meat Inspection Act of 1906 established much stricter requirements, including prohibiting the sale of misbranded or adulterated food and drugs in interstate commerce. That act created the framework for the FDA, which was formally renamed in 1930. In 1938, the passage of the Federal Food, Drug, and Cosmetic Act required that companies prove drug safety before marketing. Because of the significant regulatory oversight that these laws required, as Marc T. Law and Gary D. Libecap write, the role of experts within the FDA was presumed from the beginning: "Here was a productive role for product quality regulation by scientific experts, who had a comparative advantage in judging the quality of food and drug items."[25]

During the thalidomide "catastrophe" of 1957 to 1961, the drug thalidomide that had been sold worldwide to treat pregnancy nausea was found to cause significant birth defects.[26] However, thalidomide was successfully blocked in the United States due to the vigilance of Dr. Francis Kelsey, an FDA scientist who found there was insufficient testing data.[27] This experience spurred the passage of the Kefauver-Harris Amendments (also known as the Drug Amendments of 1962), which required drug companies to demonstrate a drug's effectiveness for its intended use, in addition to proving its safety. Prior to these laws and the regulatory framework to enforce them, there were few (or no) guarantees that the drugs companies sold did what the pharmaceutical companies claimed they did or that the ingredients were safe for human consumption.[28]

While having insulated experts in charge of decisions can lead to many good outcomes, such as the blockage of thalidomide, the approval of an annual flu vaccine, and the 2023 approval of Casgevy (the sickle cell disease gene therapy), technocracy itself is insufficient to meet the democratic threshold of governance. The Progressive response to the administration's "less than democratic feature of governance" was to "uplift the great unwashed by imbuing them with education."[29] Democracy would thrive by turning them into technically proficient and skilled workers.[30] This was famously supported by John Dewey, whose dream for democracy was to create a "new relationship between experts and citizens" wherein citizens are equipped to understand the issues and engage in the public debate surrounding technical matters.[31] But, as Fischer writes, this faced great institutional and societal impediments:

> As the role of the citizen grew in the political sphere—resulting in significant part from the expansion of basic political and legal rights in the nineteenth and twentieth centuries—there was at the same time a growth in the size and power of large corporate enterprises and government bureaucracies directed by managerial and technical expertise. In the very period when the political potential for citizen influence was emerging, it was being undermined by the expansion of large organizations and a more instrumental orientation to politics, depending on what Max Weber described as the rationalization of modern economic and social life. This elevated concerns about an informed citizenry and the future prospects of democratic governance.[32]

In other words, it may not matter how *educated* the people are considering the expansion of corporate power that still holds the reins of influence in society and life in a capitalist system. James Wilson identifies this as a kind of paradox: The combination of creating a bureaucracy designed to be "aloof from the people, distant for their concerns, and preoccupied with the power and privileges of the bureaucracy—an elaborate, grinding machine that can crush the spirit of any who dare oppose it" contrasted with a "system that is suffused with participation . . . this popular involvement would be taken as evidence that the administrative system is no system at all, but a bungling, jerry-built contraption wallowing in inefficiency and shot through with corruption and favoritism."[33] For Fischer, this paradox rests on the notion that the American people are too informal and spontaneous to be "other-directed . . . to be either neutral arbiters or passionless Gradginds," but also that the bureaucracy itself is not as passionless as we had hoped, nor as rule-bound as Weber desired. Another formulation of the paradox (or tension at least) is that educating citizens to be technocrats and engaging with them does not erode the anti-democratic elements of democracy; rather, it develops their ability to reiterate the status quo, not to challenge it.[34]

A further way that technocratic institutions suffer a democratic deficit is that the workers themselves tend not to be representative of the public. Representation, as opposed to expertise, has not historically been the goal by which regulatory agencies choose workers. However, because meritocratic systems assume that success is due to individual attainment rather than being built on the backbone of societal structures of privileges and disadvantages, the more powerful jobs are occupied by the more powerful social groups.[35] This in turn creates more homogenization. In 2021, only 43 percent of the federal workforce was female, with the proportion dropping significantly in upper management and leadership positions.[36] Similarly, racial minorities represented 46.07 percent of entry-level positions in the federal workforce, 38.4 percent of mid-level positions, 33.1 percent of senior-level positions, and only 22.7 percent of career senior executive service roles.[37] Patterns and trends in other industries, such as science, technology, engineering, and mathematics (STEM), as well as philosophy, also show significant gender and racial disparities.[38]

To be clear, regulatory agencies do indeed require the work of experts—they require scientists, like Dr. Kelsey, who have specialized skill in analyzing evidence and interpreting data, and they should have institutional longevity to cultivate the wisdom and experience required to keep our medications safe and effective. For CRISPR, they require geneticists who have training in identifying potential off-target events to evaluate risks of a particular gene therapy. However, what would be a historical mistake is to assume that the more privileged groups are the only ones capable of producing such skilled bureaucrats.

Further, the kind of work regulators do is rarely as value neutral as the Progressives envisioned, nor is their expertise the only kind needed in many cases. Just as it would be misguided to equate expertise with neutrality, so too, risk governance is not free from the social context in which it takes place. As Sheila Jasanoff argues, "risk assessment" is often "treated as a largely scientific exercise, insulated from the subsequent process of making value-based decisions."[39] However, "risk characterization estimates the amount of distribution harm at varying levels of exposure throughout a population. This may include the measurement of harm to especially vulnerable groups. . . . This determination in turn helps define reasonable policy options for reducing overall levels of risk to society."[40] In other words, the work of regulation is not merely objective and empirical but requires thoughtful judgment formation from the broader public. Thus, as Burris writes, "in actuality, technocracy is far from apolitical; technology and scientific knowledge have certainly been shaped by political interests, even as they also influence the shape of politics."[41] Jeffrey Friedman similarly argued that the uncertainty of social science makes it a problematic basis upon which to design a technocratic institution, precisely because it leaves significant room for interpretation into policy.[42]

The reason that "follow the science," then, is an insufficient policy directive is (a) empirical evidence always requires interpretation and (b) that kind of interpretation requires value-based analysis: ideas of what is good and bad, and how to prioritize competing interests. Indeed, regulatory oversight of new biotechnologies often involves evaluating risks, using precaution, and weighing uncertainty. Ideally, this should be in service of protecting the public against the private interests of industry, which may be inclined to care less about safety and more about profit, as was seen in the case of drug development prior to the FDA's enforcement of safe ingredients and efficacy.[43]

Applied to emerging technologies, particularly biotechnologies, the "public interest theory" of regulation holds that the government must curb the power of specialized interests by mandating how to create "safe" consumer products and through public health directives, such as mask-wearing requirements and vaccine mandates during a pandemic.[44] In the realm of economic regulation, for example, this may mean creating compliance mechanisms for enforcing antitrust laws or Dodd-Frank reforms or intervening in competition, such as by directing credit.[45]

During the HIV/AIDS crisis of the 1990s, the FDA played a role in stimulating and encouraging drug development by creating the "accelerated approval" pathway, which was then codified by Congress in statute.[46] Accelerated approval was a direct response to the demands of the public amid an emergent crisis, recognizing that at times the status quo of contracting and dying from AIDS was far riskier than the drugs that might be developed. It also recognized that the social needs of different communities require different regulatory responses in different moments, rather than being subject to the same oversight mechanisms at any given time. Further, because the HIV/AIDS population was (at least initially) largely comprised of gay men, the social stigma and marginalization of this group stood in the way of pharmaceutical and government action for years, showing the need for greater responsiveness and robust channels of engagement.[47]

Because risk governance is a robustly social enterprise, citizen engagement and the public interest matter. The framing or communication of risk, as Jasanoff writes, is a "widely recognized social process through which individuals and groups make sense of previously unorganized experience."[48] Evaluating risk changes as the ways in which humans live together and their environments change, whether due to the introduction of a new technology or a new policy, such as a tax cut or the closing of a hospital.[49] Media and changing political climates, or personal circumstances, for example, can create different conditions wherein the same statistical likelihood of risk is evaluated differently, especially when weighed against other pressing risks such as climate change or illness. Thus, the "traditional" understanding of risk governance as "positivist, rationalist, and information deficit model-driven attempts at defining, understanding, and communicating risks" has morphed over time to increasingly embrace "social constructivism and psychometric and cultural perspectives on risk."[50]

In a related example, the "precautionary principle," found throughout the bioethics literature to signal a restrained approach to new technology and medical practices, became situated at the "forefront of the debate" of how and when to curb the use of emerging technologies.[51] As Giandomenico Majone writes, the precautionary principle offers a way forward in moments of uncertainty while weighing competing risks and benefits: "Basically the principle says that scientific uncertainty should not be an excuse for regulatory inaction. Since regulatory science entails weighing uncertainty, any responsible risk regulator may be said to use a precautionary approach."[52] For David Resnik, the principle offers the "informal and qualitative" side of the coin to complement risk's "formal and quantitative" components.[53]

The precautionary principle is a guiding rule that suggests that regulations should take "effective action to prevent significant harms from occurring, even when scientific certainty is lacking."[54] As Elke U. Weber and Jessica S. Ancker write, "Conclusive evidence of adverse impacts is lacking for many potential risks, especially those that are novel, uncertain, or require extrapolating from known to unknown situations."[55] Thus, the precautionary principle has largely been used for environmental and medical regulation because of the ways in which innovation works: constantly reintroducing technology into ever-changing societal conditions and ideas about governance.

Notably, some reject the use of precaution and argue what agencies should be doing is "anticipatory guidance": Biotechnology's uncertainty leads us to *anticipate* future adverse events rather than use caution in slowing the tide of technology.[56] This aligns with the moral analysis in the previous chapter: that inaction on climate change may be more risky than introducing new technology meant to fight it (such as gene drives).

Thus, we can start to see the ways in which affected communities would want to weigh in on drug approval and the interpretation of risk. Returning to our cases from chapter 2: In the case of deafness, one could imagine a patient group's interest in weighing the implementation of interventions that may destabilize their community. In the case of sickle cell disease, one could imagine a community in high demand for intervention that is willing to push the FDA to approve certain therapies using different pathways, as seen during the AIDS crisis. And in the case of gene drives, one could imagine different communities approaching the intervention of gene drives within their environment differently depending on how much they are impacted by the climate crisis and the needs of small farmers.

Opportunities for Public Participation and the Problem of Capture

The Progressive Era's vision of an insulated administrative state is not quite what it is today. Over time, more avenues for public participation and representation of stakeholder interests have been incorporated. In many ways, also due to the massive social movement that arose in response to the AIDS epidemic, the FDA now has opportunities for patient and patient advocacy group representatives to serve on their advisory committees and is more responsive to the needs and concerns of patient populations.[57] The role of these patient advocacy groups and their own potential proclivity to capture will be discussed more in chapter 5. For the current conversation, the predominant institutional mechanism for participation and public outreach is through the comment process on draft rules as required by the 1946 Administrative Procedure Act (APA).[58]

The APA provides the framework for agencies' process of writing regulations and promulgating different kinds of rules. As Susan Yackee writes, "Before that time, select agencies were issuing rule-like standards as a means to implement congressional passed policy; however, the APA regularized and made transparent these agency policymaking efforts."[59] Section 553 of the APA requires that during "informal rulemaking," agencies

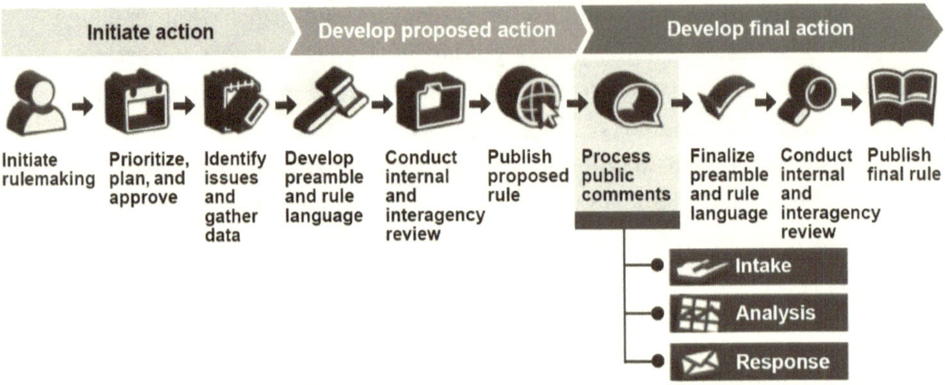

The Administrative Procedure Act (APA)'s formal rulemaking process. Source: US Government Accountability Office.

publicly announce draft versions (Notices of Proposed Rulemaking or NPRMs) in the Federal Register.[60] The APA requires that NPRMs include the time, place, and nature of the public rulemaking proceedings, the legal authority under which the rule is proposed, and the terms or substance of the proposed rule or a description of the subjects and issues involved.

This is followed by a period of collecting public feedback and collating the responses, constituting the "notice-and-comment period." Though the APA does not specify a minimum comment period, agencies have set sixty days as the default comment period length in accordance with Executive Order 12866 from 1993.[61] During this timeframe, "any concerned individual, group, or political entity may provide written comments regarding the proposed rule . . . that provides information regarding a commenter's opinions, preferences, concerns, and/or suggested changes."[62] While in past decades, this generally included a letter or an email to the agency, rules are now posted online at regulations.gov, and comments are submitted there as well. After the comment period, the agency then promulgates a "Final Rule," which acts as legally binding as the legislature's laws. And while there is no specific requirement that an agency alter the NPRM to reflect the concerns raised by commenters, agencies are required by law, and by the "arbitrary and capricious" statutory standard, to provide reasoned explanations for why they have or have not acted on certain comments.[63]

The comment period serves two primary purposes. First, it creates the condition wherein the agencies are more responsive and accountable to the public. The aspiration, as Daniel Lyons writes, is that "through the comment function, interested parties can make relevant information available to agency decisionmakers trying to divine the optimal solution to a public policy dilemma."[64] Second, this in turn helps to assuage concerns that the agency is not legitimate because of its independence or lack of accountability to the people. As Sidney Shapiro writes, "Despite the lack of guidance in the agency's mandate, the rulemaking process therefore legitimated a rule by inviting public involvement and ensuring that input could not be ignored. Or so the narrative claimed."[65] Shapiro goes on to lament that the potential trade-off of requiring comment periods is that the administrative process became less about excelling in the expertise that Congress had originally asked it to develop and instead moved it "towards the political process to legitimize rulemaking outcomes."[66] In other words, for Shapiro, there is a tension between these two comment period goals: Is the goal to solicit information, or is the goal to create the appearance of responsiveness to the people in order to aid in competing or pluralist interests?

However, specialized knowledge and legitimacy may not be terribly at odds with one another. Rather, there is indeed an important relationship between public input, legitimacy, and knowledge acquisition as they combine the important parts of rational inquiry and intuition that inform political judgments. There are indeed many rules for which public input may be less necessary, such as when determining toxic chemical levels in drinking water or sterility requirements for manufacturing drugs. There may be moments when the public interest wanes, such as when rulemakers are determining whether Swiss cheese must have "eyes" (the holes) to be considered Swiss cheese.[67] But, for more contentious or morally troubling rules that may affect specific populations differently, I offer three ways in which the comment period itself is a necessary but insufficient mode of democratic inclusion, primarily to achieve knowledge production and perceived legitimacy: The comments often underrepresent the broader population and are overrepresented by industry, contributing to the problem of captured interests. The combination raises concerns that the agencies' job of protecting public interests can become more skewed toward reflecting private ones, increasing the democratic deficit.

The first main worry, then, is that there are inequitable barriers to participation.[68] Commenters are often those who have the time, organization, resources, and thus power to comment, which often translates to invested stakeholders, such as businesses that are well organized and industries that are well funded and can disseminate tailored comments that serve their interests.[69] In the 1990s, Marissa Martino Golden examined eleven rules at the EPA, the National Highway Traffic Safety Administration, and the Department of Housing and Urban Development and found "a dearth of citizen commenters, the predominance of participation by business interests, and the presence of issue networks."[70] Golden argues that even though agencies often "fail to hear from all affected parties," they are nonetheless put in the precarious position of arbitrating among competing interests."[71] And finally, "when there was conflict, conflict rather than consensus among the commenters . . . the agency tends to hear most clearly the voices that support the agency's position."[72] This suggests the agency still has power to maintain its own status quo.

In a 2005 study, Susan Webb Yackee and Jason Webb Yackee found evidence that there is a "bias toward business" during notice-and-comment rulemaking because business commentators have more sway.[73] This occurs even in cases of prominent rules (regulations that are better known or more widely discussed), where there is greater participation from nonbusiness groups, such as environmental groups and other stakeholder organizations.[74] By analyzing thirty rules and 1,700 comments from 1994 to 2001, Yackee and Yackee concluded that "business commenters . . . hold important influence over the content of final rules . . . [and] as the proportion of business commenters increases, so too does the influence of business interests."[75] In a 2006 study, Susan Webb Yackee also found evidence from a dataset composed of 1,444 interest group comments in reaction to forty federal agency proposed rules that "formal participation of interest groups during rulemaking can, and often does, alter the content of policy."[76] More recently, Susan Webb Yackee's scholarship has revealed that not much has changed. Writing in 2019, she showed that the FDA can be swayed particularly by groups that "appear to lobby harder, expend more effort, and attempt to change a greater proportion of the rule's content during scientific conflict."[77] Yackee also found that in moments of scientific conflict, "politically oriented decisions are made."[78]

Yackee readily acknowledges that the overrepresentation of special interests may, and often does, benefit the resulting rule because the groups at the FDA are scientifically oriented and have done significant research to present their findings with data-driven exactness. However, the reverse side of the problem of overrepresentation of special interests is the underrepresentation of others.[79] While this is largely due to barriers to entry, such as access to the internet and networks that inform people of posted rules, it is also the result of collective action problems, such as the ways in which certain topics and problems tend to capture people's attention. John Coffee observed this phenomenon in the case of the Dodd-Frank Act, enacted in 2010 following the financial collapse of the US housing market, which implemented programs and measures that better constrained the risk-taking behavior of banks. Coffee showed that voter attention is episodic—that is, "easily distracted by other important issues, investors' and shareholders' attention span is short. In contrast the financial services industry is well organized, focuses on the issues that most affect it, and has an obvious incentive to maintain a powerful lobbying presence that will give the industry disproportionate influence."[80] Coffee argues that because of this, regulatory oversight increases after a market crash or other kind of crises but "wanes as society and the market return to normalcy."[81] He writes further that the "public's passion for reform is short-lived and the support it gives to political entrepreneurs who oppose powerful interest groups on behalf of the public also quickly wanes. This same pattern may characterize other forms of regulation—for example, environmental regulation may wax and wane with highly publicized, vivid environmental disasters."[82] Because financial regulation is "inherently opaque," this makes it even more of a problem.[83] This problem becomes particularly pronounced in the case of the financial industry, when special interests are well organized and ready to fight against regulation that seeks to curtail their market freedoms.

The overrepresentation of stakeholder groups in the comments, and corresponding underrepresentation of others, can be seen in the case of gene editing. Consider the guidance document "Human Gene Therapy Products Incorporating Genome Editing," the draft of which was published in March 2022, with final guidance issued in January 2024. As detailed in chapter 1, this guidance provides "recommendations to sponsors developing human gene therapy products incorporating genome editing (GE) of human somatic cells."[84] Specifically, the recommendations concern "information that should be provided in an Investigational New Drug (IND) application in order to assess the safety and quality of the investigational GE product, including information on product design, product manufacturing, product testing, preclinical safety assessment, and clinical trial design."[85] All of the twenty-six responses, except for one, were from major stakeholder groups, including pharmaceutical companies, academic research centers, professional societies, consortia of academic and industry members, alliances between industry, advocacy organizations, consulting groups, academic centers for research, and disease foundations. Only two comments were from non-stakeholder groups, and one of those was a public health professional.[86]

The USDA and EPA also posted drafted rules several years earlier, for which they received many more comments, likely because of the more popular GMO content of these rules. As also discussed in chapter 1, the Animal and Plant Health Inspection Service (APHIS) of the USDA published "Movement of Certain Genetically Engineered Organisms" on May 18, 2020, but prior to this version, it posted an initial draft in January 2017 that received 203 comments. Many of these were highly critical of the draft, stating "that the proposed requirements would be too burdensome and had the potential to stifle innovation," echoing the European Union's ruling on the topic. The USDA withdrew the rule and took time to discuss next steps with "organizations both supportive and skeptical of agricultural biotechnology."[87] In total, "APHIS met with more than 80 organizations, including 17 universities, State departments of agriculture, and farmer organizations."[88] The feedback fueled their efforts in the SECURE rule, leading to exempting CRISPR in cases where the final product is similar to other genetic modifications. The draft rule for this updated version was then published in 2019 and received 6,150 comments.[89] The SECURE rule maintained that the risk of off-target mutation did not exceed previous techniques and therefore did not require special oversight. Finally, the EPA published "Exemptions of Certain Plant-Incorporated Protectants (PIPs) Derived from Newer Technologies" on October 9, 2020, which received 8,120 public comments, many of which also decried the threat of PIPs, not CRISPR directly.

In addition to the correlating underrepresentation of non-interest group voices, the overrepresentation of special interest group voices during the comment period is a part of another body of concerns called "capture." Daniel Carpenter and David A. Moss define regulatory capture as the "result or process by which regulation, in law or application, is consistently or repeatedly directed away from the public interest and toward the interest of the regulated industry, by the intent and action of the industry itself."[90] Capture is related to more general corruption concerns and thrives in nontransparent conditions, manifesting when regulation reflects the demands of special interest groups and organized moneyed interests, rather than the broader goals of protecting the public good.[91] Capture then highlights, as Gary Becker once wrote, "pervasive imperfections in government behavior" due to powerful influences.[92] And, as Richard Posner writes, "Capture implies conflict, and regulatory capture implies that regulated firms have, as it were, made war on the regulatory agency and won the war, turning the agency into their vassal."[93] Capture can also manifest as firms creating conditions that limit market competition, which is less about deregulation per se and more about infusing the system with their interests in more subtle ways. Capture of personnel with industry ties can be seen throughout the 2025 Trump administration, especially in the EPA with the hiring of industry executives.[94]

The main concern with capture, then, is that the public interest, which regulatory agencies should be protecting, becomes subordinated to private interests. Private interests are too often prioritized based on power and wealth rather than on their value.[95] Hence, the corruptive force of capture is that the agency fails to provide oversight of an industry the way it should because it lacks the proper mechanisms or situatedness to do so.[96] One of the most stark and deadly examples of capture can be seen in the case of the Federal Aviation Administration's oversight of Boeing aircraft. Following the crash of two 737 Max airplanes, *The New York Times* reported that many agency officials "shuffle between the government and industry."[97] As Drew Nunn further writes, "Managers within the FAA's oversight program over the Boeing ODA [Organization Designation Authorization] were reportedly pressured to make sure Boeing met deadlines to deliver the 737 Max to its customers."[98] Some have called this form of capture a kind of "revolving door" wherein "often, at the end of their industry tenure, key industry personnel seek employment in government regulatory entities and vice versa."[99]

The social ramifications of capture, in addition to death and other kinds of harm stemming from failure to rein in the ambition of industry, include reduced accountability and reliability and, ultimately, less public trust and confidence.[100] As Carpenter and Moss maintain: "The widespread belief that special interests capture regulation, and that neither the government nor the public can prevent this, understandably weakens public trust in government and contributes to a sense that our political system is not capable of meeting the challenges it faces."[101] When the public perceives the agency as overly influenced by the interests of the corporate and wealthy, they often do not trust the resulting regulation, also known as a breakdown of "credibility markets."[102] Susan Webb Yackee notes that this is a phenomenon largely experienced "more readily" by the "policy losers than by policy winners."[103]

This is seen in the case of genetically modified organisms (GMOs) (see chapter 2), wherein the public did not trust their safety because the regulators were perceived as being in the pockets of the industry, and it has become a particularly important political touchstone for vaccine skepticism.[104] Health and Human Services Secretary Robert F. Kennedy Jr.

claimed on June 30, 2025, that doctors peddle vaccines because they create "perverse incentives . . . 50% of revenues for most pediatricians come from vaccines."[105] Investigative journalism from *The New York Times* revealed that the opposite is true: Doctors often lose money on routine childhood vaccinations.[106] Much of this anxiety played a role in attitudes toward the COVID vaccines as well, which were also seen as either produced "too fast" or "tested too quickly."[107] As Carpenter notes, the decline of trust in the FDA "has corresponded with a decline in trust in the pharmaceutical industry and its new products, and this relationship underscores the critically valuable confidence benefits that approval regulation of new drug treatments brings."[108]

Capture, then, is a concern not just because capture itself is antithetical to the public interest but also because *perceptions* of capture (such as in the case of vaccines) can fuel widespread distrust that contributes to behavior antithetical to the public interest (such as failing to vaccinate and creating widespread measles outbreaks).[109] Capture in the case of CRISPR could impede the trust of patients to take medications approved by the FDA if it is seen as corrupted by pharmaceutical interest rather than adhering to procedures that mitigate risk and uncertainty. Chapter 6 explores at greater length the relationship between trust, democracy, and perceived legitimacy. Importantly, problems such as vaccine skepticism and science "denialism" are not unique to regulatory agencies, nor are the agencies themselves the source of these problems. Similarly, CRISPR concerns, as seen in the last chapter, stem from multiple sources. However, they are certainly one part of the puzzle, and identifying modes of citizen engagement toward addressing this phenomenon will benefit public health.

Mitigating the problem of capture, however, should not necessarily mean expunging all opportunities for industry-regulator relationships. Regulatory agencies need at least some degree of coordination and relationship between the regulated and the regulators to create productive and practical regulation. This is often characterized as "ambiguous capture" because the government (quite often) needs to collaborate with industry to create advancements, such as in the case of the meningitis vaccine, which involved a rich array of coordinated efforts between regulators, scientists, and activists to meet a pressing global need.[110] Sara Singleton refers to this as the "paradox of co-management and community participation," noting that the agency is tasked with the "difficult balance that must be maintained by states if they are to encourage cooperation while at the same time resisting the 'capture' of state resource management agencies by the communities (or some subset of interests within those communities) with which they seek to form collaborative partnerships."[111] Dorit Reiss similarly shows that while aiming to form "collaborative" relationships, the space between cultural capture and good relationship building with governance can become blurred.[112]

The main worry, then, is when relationships cross the line into undue influence that creates problematic regulation, or regulation that fails to mitigate public risk because it is advancing industry or capital interests. Where to draw that line is something that is hard to see from within the institution and requires outside checks, such as through public scrutiny and greater transparency. Daniel Schwarcz, for example, advocates for preventing capture "through consumer empowerment programs" that may "provide 'countervailing power,' given the comparative ease with which organized powerful interest can participate in, and ultimately influence, certain new governance projects."[113]

Indeed, scholars have been more likely to offer solutions to the problem of capture that stem from further regulation, court interventions, or presidential power expansion, all aiming ultimately to curb the influence of interest groups and lobbyists. M. Elizabeth Magill, for example, envisions a role for the courts to use "judicial review" to "nudge" regulatory agency "back to the proper path."[114] Richard L. Revesz and Michael A. Livermore argue in favor of giving "greater presidential control over agencies" through "centralized review, which helps to ensure that administrative action is responsive to the broad national constituency represented by the president rather than a parochial special interest."[115] The last approach features the "nondelegation" doctrine, which states that Congress, the body traditionally thought of as the source of democratic governance and the creator of laws, has given too little guidance to agencies, and thus the transfer of power from Congress to them is unconstitutional.[116]

However, these alternatives may expose our regulatory agencies more to the changing tides of presidential or legislative authority, as witnessed during the Trump administration's program of deregulation and the erosion of evidence-based regulatory policies (see introduction).[117] Executive control can itself be considered another kind of capture in the corruptive sense of standing in the way of public interest when scientific integrity and our valued interpretation of it are undermined by the political goals of a business mogul.[118]

Another alternative solution is to streamline the notice-and-comment process, such as by bolstering mechanisms for "e-democracy" (an online interface for engagement and voting) and advertising and marketing to particular areas and populations.[119] There are also proposed attempts to diversify participation by "mandating that agencies give the public different types of opportunities to express its views during rulemaking."[120] This includes the most common method, the legislative hearing, "during which witnesses present testimony orally to those responsible for the rulemaking in much the same manner as congressional committees do their work. This allows a give-and-take not possible through written comments."[121] This sort of "negotiated rulemaking" creates more opportunities for participation and hearing from more people.[122]

Calls for transparency is a common refrain in the quest to fight capture.[123] There are often two main reasons for this: accountability and public engagement. First, increased transparency creates greater accountability.[124] This helps to mitigate capture because the government's acts of transparency make it more accountable to the needs of the public rather than private interests. It orients the administration more toward the wishes of the people and simultaneously reduces capture by requiring the disclosure of information about donations, relationships, and so on. The second reason, public engagement, suggests that when citizens have greater knowledge of government activity (and spending and relationships, etc.), then they can be better informed about when and where to protest or push for change. Without transparency, knowledge is diminished, and ignorance prevents individual and collective action, facilitating corruption.[125]

However, the goal of transparency may be insufficient to fight corruption. Archon Fung, Mary Graham, and David Weil argue that the goals of transparency are often thwarted by the public misinterpreting the information disclosed. Simultaneously, they maintain, transparency can increase corruption because private parties, which are (again) better organized and resourced, can use this information to advance their own causes.[126] In an empirical test of both accountability and public engagement, Brigham Daniels, Mark Buntaine, and Tanner Bangerter found that "once corruption and mismanagement are identified, it takes effective government institutions and action from civil society to successfully act as a disinfectant," which requires more resources and institutional mechanisms for inclusion than merely complying with transparency.[127] Alas, they concluded that "there is no certainty that if transparency exposed problems in the government, the government would work to address such problems."[128] Mark Fenster argues that much of this is because transparency of information often emerges as "noise," random disturbances introduced by something other than the communicator that inhibit the perfect transmission of information," thus standing in the way of democracy to properly do its job and produce "the consequences anticipated by transparency advocates."[129] Thus, transparency may yet again be another buzzword, not unlike "public engagement," as mentioned in the reports from the introduction—something that is often called for but with little institutional follow-up in ways that impact actual policy.

Reducing the democratic deficit in administrative agencies and during the rulemaking process, however, is indeed important in fighting capture. As Caelesta Braun and Madalina Busuioc argue, although not all stakeholder engagement practices are by default legitimate, some forms can "allay democratic deficit concerns, bolster accountability and serve as [a] key resource of informal regulatory authority—as well as a potential source of capture and bias, and hence de-legitimation, in regulation."[130] Democratic *flourishing*, then, requires cultivating active and diverse thinkers who can deliberate together to create new or modified judgments and engage with rulemakers as trustworthy sources of confidence rather than as social disciplinerss. It is through the iterative process of narrative sharing, contemplating evidence, interrogating experts, and hearing from stakeholders that a diverse array of individuals can weigh in on important regulatory questions about biotechnology. However, this requires giving people the means to translate knowledge into power and recognizing certain groups' knowledge as a form of power. This relates back to this chapter's beginning comments on political judgment: Often to create effective regulation you need engagement, participation, and good relationships with industry. Thus, much of the political solution to this perception requires bringing people closer to observe its true extent, rather than shielding them from observation.

4 Power, Knowledge, and the Value of Diversity

In the previous chapter, I argued that we should aim to increase the democraticity of regulatory agencies, considering their technocratic design, risk-based governance, overrepresentation of industry, and underrepresentation of affected groups. In this chapter, I scrutinize more closely why this is necessary. I argue that the democratic deficit, characterized by institutional isolation from public accountability and compromised protection of the public good, results in illicit uses of power and knowledge. Much of the justification for this hinges on the tight connection between power and knowledge: As we saw with IQ, those with power often determine what knowledge is, and those with this form of knowledge tend to acquire power. This can also lead to injustices, such as the marginalization and domination of those with different knowledge-based accounts. Incorporating the experiential wisdom of diverse groups can mitigate these abuses of justice and simultaneously enhance substantive policy outcomes.

To begin, I provide an account of power and, in particular, the important connection between power and knowledge: namely, that those in power create the conditions in which we declare something to be "known." I then offer several illicit uses of power in gene editing: disciplinary power (the correction of abnormal bodies, often targeting disability), biopolitical power (the organization of society using tools to control people's bodies), and domination (arbitrary impositions of power, often manifesting in inequities). Democratic systems are better able to correct these ills because diverse groups are better poised to testify to the conditions of domination as specialized knowers of their domination. Diverse groups of lay participants also have epistemic advantages over homogeneous ones. "Epistemic" refers to knowledge seeking, or understanding greater approximations of truth, even under fallibilist or nonideal conditions. When dominated and disciplined communities enter the halls of justice and power, their intellectual and experiential knowledge contributes to the considered judgment of institutions through the "co-production" of knowledge.[1]

Problems of Power

In keeping with Thomas E. Wartenberg's approach, identifying sources of power helps us to understand the ways in which people live together in more productive and robust ways. As Wartenberg writes, analyses of power offer us a "set of distinct social phenomena that play essentially different but nonetheless related roles in the constitution both of individual social agents and of society as a whole. . . . As a result, it provides social theory with a comprehensive view of social power, one that accounts for the negative as well as the positive role that power plays in the constitution of human life."[2] Analyzing power relations allows us to identify positive or instrumental power as serving societal good and noninstrumental or destructive kinds of power. For understanding governing gene editing, and other biotechnologies, generally, I want to offer a positive vision of power and three kinds of illicit power: disciplinary power, biopower, and domination.

"Positive" power, an instrumental power, is the ability or potentiality to create conditions that rule or act on others through curtailing freedom.[3] As Hanna Pitkin argued, power is the "capacity, potential, ability, or wherewithal" of some persons to produce both intended and unforeseen impacts on others' freedoms.[4] In *The Human Condition*, Hannah Arendt understood power as a "plurality of actors" who "act in concert" to "create new realities" for some common purpose.[5] Similarly, Clarissa Rile Hayward argues against power as a kind of un-freedom, defining power "not as an instrument some agents use to alter the independent action of others, but rather as a network of boundaries that delimit, for all, the field of what is socially possible."[6] In the case of gene editing, positive power can be patient advocacy groups binding together, as in the AIDS case, to apply pressure on the regulatory system to modify or expedite clinical research and treatment risk assessments.

Illicit power, then, is a kind of noninstrumental power, for it occurs when freedoms are problematic, lacking justification, or taken away. For example, in his famous articulation of permissible (or non-illicit) liberty restriction, nineteenth-century philosopher John Stuart Mill argued that "the only purpose for which power can be rightfully exercised over any member of a civilized community, against his will, is to prevent harm to others."[7] For Mill, power can only be wielded to prevent harm, as opposed to imposing one's will on others.

I offer three forms of illicit power helpful for clarifying moral concerns (presented in chapter 2) combined with regulatory procedural concerns (analyzed in chapter 3). Discipline, or the correction of "deviants" from "normal" body functioning, manifests in CRISPR's targeting of disability. Biopower, as a population-level form of this control, can be seen in genomic surveillance, which is complicit in race-based concerns about genetic disease associations. Finally, domination, as arbitrary interference, involves the exclusion of vulnerable populations from needed medical intervention and thus mirrors the underserving of certain groups. Bringing in the critique of regulatory agencies' democratic deficit and technocratic design (chapter 3), I analyze these manifestations of illicit power in the context of neoliberal policies as intentional state protections of market interests. These illicit forms of power undermine the good that gene editing can do and the importance of experts (such as those creating clinical interventions and evaluating their safety), and emphasize the role of democratic participation in regulation.

Biopower and Discipline

These two related but distinct concepts, developed in the work of twentieth-century historian of philosophy Michel Foucault, are important for analyzing illicit uses of power in the case of CRISPR. Discipline refers to the production of "docile bodies," or the use of a construct of normality to impose conformity on individuals.[8] Biopower refers to population-level attempts to "administer" or organize groups of people.[9] Thus, biopower (and biopolitics) uses the tools of discipline, but discipline acts on individuals within institutions whereas biopower operates on systems.[10] In the case of gene editing, we are concerned with both: individual encounters with disciplinary medicine and systemic features of medicine that create biopolitical effects on certain populations, particularly those defined by disability and race. Further, the development of these forms of power, for Foucault, is in service of neoliberal policies that protect markets rather than vulnerable groups. Definitions of health are then created by biotech companies' profit-driven pursuits, rather than by what we define as "normal functioning" for ourselves to identify the true needs of communities.

In Foucault's telling, biopower emerged during commercial society to replace violent power; governments became less interested in determining how and when subjects would die and entered the business of controlling the ways they lived.[11] With the widespread popularity of Adam Smith's legal and moral defense of the free market and its "invisible hand" in his 1776 account *The Wealth of Nations*, governments began the search for nonmilitarized tools to organize society.[12] In his 1979 lecture *Birth of Biopolitics*, Foucault cites the example of the Panopticon, a prison design proposed by philosopher of utilitarianism Jeremy Bentham, who lived in the late eighteenth and early nineteenth centuries. In the Panopticon, the inmates and the guards would be observed by a "single watchman" without them knowing when they were being observed. This created "a new mode of obtaining power of the mind over mind."[13] Further, for Foucault, the false promise of seventeenth- and eighteenth-century liberalism was that it would create a robust system of rights and freedoms, such as those supported by Mill (who was also a utilitarian, as well as Bentham's godchild) and by social contract theorists John Locke, Thomas Hobbes, and Jean-Jacques Rousseau.[14] Instead, liberalism bolstered governments' capacity to more efficiently discipline citizens by creating the illusion that the people's consent to be governed would protect them from overt interference and that, simultaneously, the government would respect their sovereignty.[15]

"Biopower," then, is practiced through the numerous and diverse techniques for "optimizing" populations through administrative control—methods that proved more effective than violence, could be easily feared, and faced less resistance through revolt or disobedience. Thus, biopower could be more subversive and produce even more docile citizens. This manifestation of power, Foucault notes, became widespread and diffuse, operating in every interaction, and not just between king and subject.[16]

The key to this system, Foucault argues, is the production of knowledge "legitimated by the sciences."[17] As we saw in the previous chapter, the late nineteenth and early twentieth centuries introduced new ways of quantifying populations, such as recording population birth and death rates, tracking the prevalence of disease, and measuring head shapes and sizes.[18] Indeed, the very development of the social sciences as disciplines, such as sociology and psychology, required the use of statistical, data-driven methods to define and categorize people's functioning for the purpose of understanding them.[19] Foucault's writings on sexuality, mental health, and criminal justice, in particular, detail long histories of the use of knowledge to protect expert authority and impose normalized expectations of human behavior, such as through imprisonment, conditioning, and forced therapy.

Disciplinary power becomes not just about conformity to a norm but also bears a strong moralizing component. Though Foucault did not situate his work as primarily normative, he shows in *History of Sexuality* that discipline erodes feelings of self-worth by marginalizing people's very existence.[20] Medicine, on this view, is not just about creating a healthier person but about shaping a person the government could more easily control by modifying the body and its behavior toward an idealized construct. "Abnormal" or "unhealthy" bodies are then *wrong* and in need of fixing.[21] As Majia Holmer Nadesan writes, "Health, purity, and American virtue are cast in contiguous terms by conservative religious authorities seeking to demoralize the nation. Sexuality is a particularly salient object of conservative health discourse, requiring endless examination and exhortation."[22] Nadesan shows that, for example, to give heterosexual relationships primacy, homosexual ones had to become abominations.

Contemporary medicine is rife with examples of discipline, and the medical system is fraught with biopower.[23] Many fat patients experience being treated as people who take up too much space rather than as patients in need of care and diagnosis.[24] The American Medical Association's decision to label "obesity" as a disease was decried, then, as a nonmedical judgment of size rather than a description of a particular condition.[25] This approach to weight, further, can exacerbate other issues, such as delayed care, because health problems are often dismissed as matters of body size.[26] Similarly, the use of genome-wide association studies and polygenic risk scores, which is meant to assign a predictive value to genetic traits, has been criticized for using data—particularly race as a biological category—in identifying genetic predispositions toward violence or homosexuality.[27]

The moralization of abnormal bodies also appears in the treatment of transgender patients. In gender-affirming care, damaging myths about hormone therapy and claims of "mutilation" of minors stem from transphobia and gender-based discrimination, thus blocking what is often life-giving care.[28] Meanwhile, many cis-gendered women seeking hysterectomies (removal of the uterus) because of severe menstrual pain are refused by doctors who believe that one day they will want to have a child.[29] This imposes a moralized vision of what a "normal" woman should desire, rather than listening to or respecting her interests. Biopower is not wielded, however, by any one political party or perspective. Discussions of Foucault and biopower also arose during the COVID-19 pandemic to criticize mandates on mask-wearing, vaccination, and school and business closings.[30]

Regulation, as discussed in chapter 3, is intended to safeguard the public against industry, so it becomes implicated in the project of biopower when it is used to support market-based policies. Neoliberalism generally refers to the rise of governmental infrastructure that protects a certain group of actors operating within the "free market." In contrast to the stated goals of liberalism, which is freedom-protective, neoliberalism "rejects the idea of positive freedom, seeing freedom only in relation to negative liberty," or when the government's abstention from banning an action permits that freedom.[31] As Nadesan argues, the regime of neoliberalism did not concern itself with the goals of distributive justice or identifying the just distribution of resources and systems of inequality, such as those articulated by John Rawls.[32] Instead, the market is seen as an impartial arbiter of distribution, maximizing individual liberty and utility.

Building on the work of Friedrich Hayek and Milton Friedman, the "neoliberal regime" in the United States in the latter half of the twentieth century identified individual choice as requiring this negative freedom or noninterference of the government.[33] As Nadesan writes further, "All neoliberal proponents share the belief that capitalist market freedom is the ultimate medium for the expansion and realization of individual liberty."[34] Economic interests at the time were thought to be served through free market mechanisms, and the state's role would be to create the conditions wherein that would flourish through "expert knowledge."[35] Therein neoliberalism holds that the market was never meant to truly be self-sufficient, as Adam Smith idealized, but instead would require state intervention.

These activities include market techniques that relocate authority to the regulators and market experts.[36] Achieving this market freedom required political rationality, knowledge, and power, as well as the rule of law and maintaining the mantle of freedom, liberty, equality, and universality.[37] As Wendy Brown writes, neoliberalism is not merely an economic policy but "activates the state on behalf of the economy, *not* to undertake economic functions or to intervene in economic *effects*, but rather to facilitate economic competition and growth to economize the social, or, as Foucault puts it, to 'regulate society by the market.'"[38] "Marketization," Nikolas Rose further writes, became the new impartial arbiter of knowledge, and the market became a source of freedom "where the relations between citizens and experts are not organized and regulated through compulsion but through acts of choice."[39] As Paul Patton also describes, the organization and maintenance of markets operates as one of the most influential ways in which "governments have sought to regulate the behavior of those whom they govern" over the twentieth and into the twenty-first centuries.[40]

This authorization, importantly, does not require democratic accountability. As Benjamin McKean writes, "Neoliberalism does not require states to be authorized by a *demos* in order to be legitimate and the internal logic of markets calls for them to grow beyond state borders in order to be more efficient and beneficial."[41] In this way, market knowledge and regulatory power are also forms of disciplinary power. One of the clearest examples today is the proliferation of what Shoshana Zuboff refers to as "surveillance capitalism."[42] The experience of 9/11 and the subsequent "War on Terror," Zuboff argues, gave impetus to the rise of personalized devices and the mass collection of data in the "digital age."[43]

This account has important implications for regulating biotechnologies such as gene editing: While market competition is important for medical advances, it can also lead to problematic priority setting by following market incentives rather than community-based interests.[44] This is reflected in the tragic failures of health care and medical insurance systems to care for and treat many patients due to limited coverage and inordinately high prices.[45] It also contributes to the shortage of antibiotics, oral contraception for men, and treatments for numerous gynecological conditions like endometriosis, particularly when these disproportionately affect different groups based on gender, race, and socioeconomic status.[46] As Dorothy Roberts warns, increasing the amount of medical testing sometimes only serves to generate profit for health care providers. Thus, "letting health trump social justice does not really improve the welfare of most people; it supports the interests of big business and the most privileged members of society . . . [by creating] a false dichotomy between health and justice that hides the social factors that determine health not only for individuals, but of the entire nation."[47]

In the case of CRISPR, many worry about "neo-eugenics," a regime (similar to liberal eugenics mentioned in chapter 2) in which "undesirable" or "abnormal" members of society might be prevented from reproducing or existing going forward.[48] This worry is compounded by the rise of "genetic surveillance," wherein companies are given greater liberty to design and market their health care technologies and governments have greater access to genetic information about their citizens.[49]

Biopolitical power and biotech markets, however, are neither totally beneficent nor totally evil. Indeed, Foucault's analysis is helpful in framing the problem of state and market power in influencing medical decisions. However, it can also be too totalizing and does not distinguish between *justified* and *less justified* forms of power, *good* and *bad* forms of medicalization, or *productive* and *harmful* manifestations of pharmaceutical freedoms. While Foucault depicted biopower as intrinsically problematic, one can also see how biopower can be productive, leading societies to successfully implement public health standards, including the immunization campaigns that eliminated smallpox and, in many areas, polio.[50] The discovery of germ theory and improvements in sanitation, after all, led to the reduction of cholera.[51] Pharmaceutical advances can also be transformative, such as the creation of preventive medicine like HIV PrEP, which prevents a condition that used to be a death sentence.[52] And finally, Casgevy, the CRISPR-based therapy for sickle cell disease (SCD), has been welcomed by many as a needed treatment for a vulnerable population.[53]

Domination

In chapter 2, I argued that technologies like PGD and IVF contribute to a "liberal eugenics" program by harboring "hidden costs" and barriers to entry for those seeking reproductive health care. This results in systemic inequalities, where individuals are *excluded* from medical advances rather than *disciplined* by them (as biopower would suggest). So, whereas being disciplined refers to conformity to "normal bodies," being dominated refers to living in states of systemic inequalities built on arbitrary rules. These inequalities lead, for example, to white and wealthy populations reproducing and thriving far more easily than others because they have more time, resources, privileges, and know-how to access and navigate health care systems (such as in assisted reproductive technologies and genetic testing).[54] In the case of CRISPR, inequality perpetuates domination whereby some people can pay for Casgevy and others cannot. So, too, climate refugees are dominated when they have no recourse against interventions intended to offset environmental catastrophes using CRISPR.

Over the years, domination has also come to be understood in different forms: Foucault even used the terms "power" and "domination" interchangeably, and nineteenth-century social and political theorist Karl Marx used "domination" to refer to economic conditions that create alienation and exploitation.[55] In the contemporary era, domination most often refers to a totalizing experience of power as interference in one's life through exclusion from accessing resources or fulfilling rights for "arbitrary" and nefarious reasons. Domination, then, as arbitrary interference, manifests in moments when individuals do not have recourse or representation in the system. This is often due to forces that are nontransparent and work to others' favor, such as through capture, systemic racial or gender inequalities, and oppression. Domination is similar to, but still distinct from, other kinds of unjust power relations, such as marginalization and oppression. Iris Marion Young posited that oppression is more diffuse, an unconscious shaping of institutions that denies some social groups the capacity to understand and express themselves.[56] Domination, for Young, occurs in contexts of action, reflecting asymmetric power relations.[57] Marginalization is even more clear-cut: It is the inability of a group to advocate for or gain needed resources for themselves, thus treating them as insignificant or less than the dominant or privileged group.[58] One is dominated, for example, when that person presents at the hospital as a Black patient with COVID-19 and receives worse treatment than a white patient.[59] One is dominated when facial recognition software algorithms cannot distinguish between different people of the same race because the developers failed to provide enough data input to ensure representation.[60]

Like biopower, domination is a kind of unjust or morally illicit social power that illegitimately curtails freedom and prevents as Frank Lovett writes, "human flourishing."[61] As such, the state of non-domination is itself a good, and the avoidance of it is the ultimate obligation of government. Domination is not necessarily linked to mistreatment or harm, but it does reduce one's ability to engage in meaningful areas of life.[62] Unlike discussions by theorists of distributive justice, such as Rawls, which are predominantly concerned with equality and the just distribution of resources, domination, as Ian Shapiro writes, can be "a by-product of the distribution of resources and can be embedded in structural relationships."[63] Elizabeth Anderson's theory of "private government" similarly holds that employers dominate when their rules and policies make it impossible for workers to exercise rights and freedoms that they would otherwise have and when workers are unable to leave or quit.[64] The control and power of the private government of employers, such as dictating recreational activities, diet, speech, dress, restroom breaks, and so on, are far less limited than our own government, and that constitutes domination.[65]

Captured institutions often dominate because private interests blind "good reason" and forestall public accountability and public interests. K. Sabeel Rahman directly applies his theory of domination to regulatory agencies as "the concentration of arbitrary, unchecked influence on another's opportunities and actions."[66] Interested primarily in financial market regulation, Rahman writes that "the fundamental problem of the modern economy is best understood not as a matter of income inequality or distributive justice, but rather as a broader problem of power and domination—manifesting in the concentration of economic power in large corporations, or the power relationships baked into the very structure of the diffuse market economy."[67] For Rahman, the primary problem of politics is not disagreement; rather, it is living outside politics and being excluded from important moments in politics that shape one's life and ability to operate in the system. Rahman's work effectively critiques the construct of the technocratic, managerial state as one that favors markets as sources of knowledge, as Foucault also lamented. His argument is that "corporations, economic elites, and even market forces themselves all exercise a kind of unchecked power over others in the economy," and in this way, they "arbitrarily interfere with the life chances of others."[68] Offering his own critique of the neoliberal managerial state, he argues, not unlike Foucault, that government intervention in the market through regulation in the name of "efficiency" came at the cost of protection of vulnerable people.

Vulnerability, then, is baked into systems of domination. After all, domination can only exist in systems that create vulnerability, such as poverty and poor health, or its potentiality, such as falling into a "poverty trap."[69] In environmental applications, consider that poorer and more dominated communities already experience the dangers of climate change more acutely and thus may also be subject to the harmful impacts of a gene drive experiment gone wrong.[70] Dominating laws need not *intend* to oppress vulnerable populations, but by barring them from holding power, systems of domination are perpetuated. As Lovett writes, "One example would be a law that effectively blocked legal actions in defense of the members of some defined group: this would allow other people to arbitrarily harm the members of that group with impunity. Under Jim Crow laws, for example, the result was more or less effectively obtained by excluding blacks from the jury pool."[71] For Lovett, even if Black citizens were not explicitly excluded from the jury pool or the polling booth, the system of knowledge exams and oppression generally ensured that in practice they were.[72]

To be dominated also means having little recourse to change the laws and system. Iris Marion Young argued that individuals who are dominated are unable to reciprocate the conditions of their actions, "either directly or by virtue of the structural consequences of their actions."[73] She offered a detailed account of the rise of bureaucracy and meritocratic systems of governance meant to control the market while also asserting a certain "hierarchy of authority, where each position is bound by rules" that transfer down to other groups and individuals.[74] As Philip Pettit also writes, "In the absence of some forums and procedures for dealing with minority grievances, a mainstream government may well dominate those various marginalized groups."[75]

Domination is helpful, then, in understanding bioethical concerns because it describes the pervasive problem of exclusion due to arbitrary forces (those lacking rationality and purpose) embedded in unequal power relations, rather than the purposeful creation of policies meant to control the behavior of individuals.[76] Patients are dominated, in this sense, when they distrust their own experiences in deference to the perspective of the physician, who tells them it is not possible for them to be experiencing pain when, in fact, they are. And we can start to see how biopolitical power, neoliberal forces, and dominating bureaucracies and laws thrive in democratically deficient landscapes. The solution, then, is to nourish democratic accountability through greater representation and testimony.

Democratic Accountability and Testimony Against Illicit Power

Eradicating problems of disciplinary power, biopower, and domination requires the testimony and participation of those with firsthand knowledge of these forms of illicit power. Democratic accountability requires this to better understand, for example, the ways in which social systems either preclude those with disabilities from participating or seek to eliminate their groups' reproduction. Within the non-domination scholarship in particular, there is near consensus that democracy, as opposed to any other regime, is preferable for mitigating the problem of domination because it at least strives for political equality.[77] Not as much has been written about democratic solutions to the problems of disciplinary power and biopower, in part because Foucault did not think there *was* an institutional solution to the problem of disciplinary power in particular, as it was an endemic feature of all institutions. However, I suggest that non-dominating testimony can similarly redirect the problems of disciplinary power and biopower, for example, by harnessing the tools of medicine and public health for more productive purposes, such as disease interventions.

First, concerning non-domination, Rahman argues that democratic participation acts both against market acquisition of power and within institutions that are supported by neoliberal systems. The financial collapse of 2007 to 2008, for example, signaled vast economic domination and the imperative of democratic control over those with concentrated power in the market. As Rahman writes, "The task of democratic theory and institutional reform, in this view, is not to reconcile problems of pluralism or rationality—challenges that inform much of modern democratic theory—but rather to help catalyze more (and more productive) forms of collective political action, expand our capacities as individuals and communities to remake our world through politics, addressing the issues affecting all of us in common."[78] Rahman elaborates that he supports a vision of deliberative democracy in which "social movements, civil society actors, and interest groups . . . interface with regulators in a dense ecosystem of networks and coalitions, resulting in political clashes and conflicts that over time generate innovations in public policy and moral commitments."[79] The form of participation he endorses is not "necessarily contestatory" but rather facilitates "considered judgments and common understandings through debate and experiment."[80]

Henry Richardson similarly offers a public reason framework specifically to address the "problem of potential bureaucratic domination."[81] Through "reasoning together," he argues, we can attain "democratic autonomy." In particular, "administrative agencies in institutionalizing democratic reasoning is what will finally make it possible to describe what would prevent bureaucratic domination."[82] For Richardson, "democratic reasoning can be distributed across the legislative and administrative institutions of representative government in such a way as to satisfy the populist requirement that the people rule."[83] In so doing, Richardson also critiques, as I have in chapter 3, the notice-and-comment rulemaking for failing to sufficiently achieve "public participation."[84] His worry is rooted in whether the people's "voices will then influence the revision of proposed rules."[85] Richardson's solution is to create opportunities for "negotiated rulemaking" through "diverse delegation of political judgment."[86] His ultimate aim is that "democratic deliberation will combine both of those aspects in a process of mutual accommodation."[87]

Others have proposed different collective action solutions to achieve non-domination.[88] Zuboff, for example, argues that historically, collective bargaining and unionization have been the primary route for non-domination.[89] Others in social movement theory similarly indicate the importance of progressive voices applying pressure to stagnant institutions to change the status quo as mechanisms of democratic power.[90] These were particularly important messages, for example, during #MeToo, Black Lives Matter, and climate crisis activism.[91]

In applying lessons learned from collective action and unionization of the twentieth century, some scholars argue for workplace democracy to prevent employer-based domination. Anderson, for example, offers that workplace democracy is an effective mechanism against the problem that "orders may be arbitrary and can change at any time, without prior notice or opportunity to appeal."[92] Though the case for workplace democracy has been criticized for failing to change broader systems, its goal to more equally distribute power within an institution that controls significant aspects of workers' lives reflects the importance of diversity of interests, expertise, and social goods.[93] It is thus an instructive model for innovating cloistered technocratic institutions.

Democratic accountability also requires leadership representation from diverse and marginalized groups to combat domination. As Julie Dolan and David H. Rosenbloom argue, democratizing the bureaucracy to make it "socially representative" is morally important because of the "linkages between social origins, life experiences, attitudes, and administrators' decision making."[94] Who is in the room, or who holds power, has important implications for resulting policies and social justice outcomes. When diverse or historically underrepresented people hold leadership positions, more resources and rights flow to those communities, and policies are created to reduce exclusion.

One area of research in which this is most clearly seen is financial regulation. Echoing Rahman's work, scholars have shown that there has been a "near total exclusion of African Americans from roles as senior policy staffers in current regulatory agencies, regardless of the political affiliation of the political appointees making hiring and staffing decisions."[95] Kyle Farmbry shows that the dearth of minority leadership affected resource distribution after Hurricane Katrina, noting that "issues of difference have played a significant role in the formal and informal discourse in policy design and implementation process."[96] Farmbry presents a succinct and thorough account of administrative law as largely creating and cementing a construct of "the Other" as those undeserving of administrative attention and resources.

Representation matters increasingly because when minorities and diverse leadership are allowed to make bureaucratic decisions on how to create and implement policies, "the Other" becomes less "othered." As Brian Feinstein, Peter Conti-Brown, and Kaleb Nygaard show, "racial diversity on Reserve Bank boards is associated with Fed-regulated banks' increased lending to underserved groups."[97] Similarly, a number of studies have found that when there are more women involved, financial institutions experience "higher financial performance" by "reducing male over-confidence" that accompanies higher levels of "risk tolerance."[98] This finding mirrors work done in the judicial space that shows that judges' backgrounds influence their rulings, including for gender and sexual identities.[99] Research on representation in mentorship is instructive here as well.[100] Academic mentorship contributes to students' success by allowing them to relate to and feel supported by people less likely to discriminate against them.[101] The diversity of editorial boards as "key decision makers" also influences what literature receives more attention.[102]

Correspondingly, we see important trends in medicine when physicians and medical care providers are representative of their patients. For example, in the United States, Black pregnant patients are three times more likely to die during pregnancy or childbirth than non-Hispanic white pregnant patients. Further, the maternal mortality rate for Black patients has doubled since 1987 and now surpasses many other developed nations. The main explanation for this is structural racism and implicit racial bias, which impact the stress on their bodies and the care they receive from medical practitioners.[103] Black doulas play an important advocacy and support role for Black pregnant patients, leading to a noticeable decrease in the mortality rate.[104] Black patients are also more likely to ask for "preventatives, particularly those which were invasive" when seen by a Black doctor.[105]

Part of the problem of implicit bias and divergent treatment stems from the lack of representation in the US health care workforce.[106] These findings are similarly reflected in the pharmaceutical and biotechnology space. In a 2022 survey, researchers found that only "1% of executives, 1% of board members, and 5% of CEOs identified as Black or African American. Similarly, 3% of executives, 3% of board members, and 3% of CEOs identified as Hispanic/Latinx."[107] Finally, we also see this pattern in science, technology, engineering, and mathematics (STEM) fields generally, where racial domination results in fewer minority participants in these fields, leading to a smaller pool of scientists to draw from.[108]

In the case of CRISPR, the background and diversity of decision-making leadership, which sets the research agenda for developing certain therapies, can be shaped by encounters with medicine and family treatment history. Someone's experience of disability, of navigating the complex and dominating health care system with a genetic disorder, or of managing a family farm that has suffered from the impacts of climate change and monopolization of seed GMOs (see chapter 2) can all contribute important testimonies of injustice that are required for creating the conditions of democratic equality.

From Power to Knowledge

Some worry, however, that democratic systems that take seriously the testimony of "the people" may compromise the epistemic quality of our policies.[109]

In other words, there is a trade-off between justice and knowledge. Perhaps heightened during the "post-truth" moment and amid the pervasive problem of misinformation, such as around vaccines, the concern is that because the people are not experts, identifying them as testifiers of injustice may come at a cost to the quality of our information.[110] I argue that it does not. Rather, to mitigate illicit uses of power through testimony and participation, we must take seriously those who are dominated as sources of knowledge as well. This is for two main reasons: (1) the deeply embedded relationship between power and knowledge, where those with power determine knowledge and those with knowledge hold power, and (2) knowing conditions of injustice requires certain kinds of wisdom—the dismissal of which constitutes a form of injustice itself, known as epistemic injustice.

First, returning to Foucault, power not only harnesses knowledge (such as to create the instruments of population-level control through biopower), but power becomes tantamount to knowledge. The idea that *knowledge itself is power* (*ipsa scientia potestas est*)—originally coined in the modern era by English philosopher Francis Bacon and endorsed by Thomas Hobbes—has played important yet divergent roles in justifying the authority of the state, its institutions, and its administration.[111] In his utopian fantasy *New Atlantis*, Bacon depicted knowledge as power in his justification for the scientist's authority to rule and create institutions based on the experimenter's discoveries.[112]

But whereas for Bacon "knowledge is power" creates freedom and flourishing, for Foucault, it curtails it.[113] As José María Rodríguez García writes, in contrast to Bacon's desire to create a "well-organized community of technocrats who have tested their knowledge empirically and acted on the utilitarian principle of procuring the greatest good for the greatest number of people," Foucault's poststructuralism detailed that "for the modern state to survive as a central, self-reproducing power, it has to develop a series of mechanisms of control that prevent individual subjects from exercising their freedom of choice, or even their intellectual freedom."[114]

Toward this end, Foucault differentiates between two forms of "subjugated knowledge": the first, "erudite archaeologist of knowledge," and the second, "popular knowledge."[115] "Erudite knowledge" refers to the professionals and the psychiatrists—that is, those with trained or demonstrated knowledge. A more recent variation is having "epistemic privilege" and "epistemic authority."[116] "Popular knowledge" refers to the psychiatric patients, the ill, and the "delinquents," or those with subjective or lived experiential knowledge. It is that form of knowledge that Foucault laments is largely ignored or discredited, and what I argue one of the goals of deliberation should be to reveal. When we reimagine the power-knowledge relationship as one of inclusion rather than one of hierarchy, we end up harnessing and learning from lived experiences.

Second, justice and epistemic goals are mutually reinforcing. Consider the US history of all-white juries convicting Black defendants: The ultimate judgment of the jury itself is flawed, tainted by homogeneous viewpoints informed by one's racial identity.[117] In the contemporary era, juries are more diverse, but problems of bias through homogeneity continue to proliferate in the criminal justice system. Court reporters, for example, who are predominantly white, "frequently fail to accurately transcribe and comprehend testimonies . . . resulting in crucial inaccuracies."[118] The result is not only an injustice, given the unequal treatment of Black versus white defendants, but also an *epistemic* injustice. In other words, the defendants are wronged procedurally both in that they did not receive fair treatment and that their testimony was overlooked.

Miranda Fricker coined the phrase "epistemic injustice" in 2007 to describe the wrong done in denying a person's status as a "knower."[119] Building on her earlier work on "epistemic oppression" and "epistemic privilege," she distinguishes two kinds of epistemic injustice.[120] Testimonial injustice is when a hearer's prejudice causes them to devalue the speaker's credibility. Hermeneutical injustice occurs beforehand but also separately, when a gap in "collective interpretive resources" prevents someone from understanding or recognizing their own social experiences as forms of oppression or domination because those experiences are under-theorized or "ill-understood, perhaps even by the subjects themselves."[121] "Marginalised groups [that] encounter barriers to being accepted as knowers" include Indigenous, racial, and gender-based groups.[122] I offer that perhaps the most pertinent version of epistemic injustice, in keeping with José Medina, is "hermeneutical injustice."[123] In this formulation, the system continues to dominate *despite* marginalized groups drawing attention to patterns of injustice.[124] As James Bohman writes, "There can be no systematic forms of epistemic injustice, testimonial or hermeneutical, that do not also involve domination."[125]

For example, #MeToo shed light on widespread experiences of sexual harassment and assault where survivors often did not have the language or knowledge to express that what they were experiencing was harassment until there was a broader societal reckoning.[126] Further, the rallying cry of #MeToo, "believe women," reflected a long history of discounting women's testimony.[127] Consider that it was not until 1993 that "marital rape" was declared a crime in all fifty states, meaning that a law had to replace the accepted notion that rape could not occur within marriage.[128] Identifying and listening to testimony of sexual assault not only recognizes the survivor as a knower of the assault but is required for any justice to occur.

Epistemic injustice is embedded throughout medicine as well. As Ian Kidd and Havi Carel write, "Stereotypes and structural features render ill persons especially vulnerable to [Fricker's] two types of epistemic injustice."[129] This can be seen when Black SCD patients present at hospital emergency departments in sickle cell crisis (extreme pain) and are not taken seriously.[130] Or when women's gynecological pain is treated as "normal" rather than as requiring serious medical attention.[131] These actions dismiss these patients as knowers of their life's desires, experiences, and aims. When contemplating whether to edit an embryo using CRISPR, excluding certain groups from the regulatory decision would not only perpetuate a system of domination, but it would also mean losing valuable insights into how people encounter genetics in their lives and make choices around procreation. Much of the justification then relies on understanding theories of knowledge and those who have epistemic authority beyond the rational technocrat, as detailed in chapter 3.

Feminist Epistemology

The objectivist, rationalist version of epistemic authority—which has largely guided regulatory agencies as well as science, philosophy, and concepts of expertise—is incomplete. I contend that we can create a richer account of knowledge when we have greater awareness of the role that identities such as gender and race play in providing and interpreting information about the world. This is because non-dominant groups can account for knowledge unfamiliar to dominant and well-represented groups.

While the tools of the Enlightenment that created scientific methods and rationality are important modes of knowledge acquisition, such as in creating and evaluating clinical trials, knowledge writ large must reflect other realms of experience, thinking, and being in the world. Feminist epistemology, for example, maintains that gender plays an important role in shaping our conceptions of reality as informed by our situatedness and experience. Dominant theories of knowledge (such as objectivism) often disadvantage women by devaluing this form of inquiry and cognitive style, describing it as irrational, emotional, and therefore necessarily inferior. As Anderson writes, "Feminist epistemologists explore how gender norms distort the dissemination of testimony and relations of cognitive authority among inquirers and how the social relations of inquirers could be reformed, especially with regard to the allocation of epistemic authority, so as to enable more successful practices of inquiry."[132] Feminist epistemology shows, then, that due to the position of women as subordinated members of society, their condition of oppression makes them knowers of that oppression in ways that scientific and "objectivist" metrics cannot capture. Jane Addams, a Progressive Era thinker noted in the introduction for her contributions to democratic theory, was also an early adopter of feminist epistemology. She emphasized the importance of recognizing women's domestic labor as testimony in her social theory of democracy, writing in 1902: "We know at last, that we can only discover truth by a rational and democratic interest in life, and to give truth complete social expression is the endeavor upon which we are entering. Thus, the identification with the common lot that is the essential idea of Democracy becomes the source and expression of social ethics."[133]

The feminist epistemology literature has expanded to other realms of philosophy of science, critiquing modes of inquiry that discount the role of personal experience in trial design, for example, as well as ethics and philosophical expertise, arguing that the role of ethics adviser also requires significant attention to background and experience as knowers of moral theory.[134] It has also taken a broader and more inclusive role in incorporating critical race theory, stating that the experience of race also situates individuals as the best knowers of racial oppression and the needs and values of that community, and it has become less universalizing around, for example, white middle-class women's experiences. As Lorraine Code writes, "Since the early 1980s, feminist epistemology has developed into a vibrant area of inquiry which challenges many of the taken-for-granted assumptions of traditional, mainstream theories of knowledge to work towards developing theories and practices that close a persistent gap between theories of knowledge and knowledge that matters to people in real situations."[135] As Code further explains, the field has fully embraced that "knowing is a political activity" for it dictates who has power over the truth of the world, and who does not.[136] Relatedly, epistemic oppression, Kristie Dotson writes, is the "persistent epistemic exclusion that hinders one's contribution to knowledge production."[137]

The power-knowledge relationship yet again emerges here. Some refer to the approach of moving away from more "hierarchical" or "elite" forms of knowledge as "decolonizing knowledge," because it moves us past *assumptions* about who knowers are: how they may look, speak, dress, and so on.[138] Knowledge, after all, is political: We often judge knowledge as "legitimate" or "true" based on who is providing that knowledge or whether we trust them (more on this in chapter 6). "Epistemic authority" reflects this notion that what we think of as knowledge is rooted in preexisting power hierarchies that are often contingent on in-group association and identification.[139] As Harry Collins writes, "Far from being the apolitical study of truth, epistemology points to the ways in which power relations shape who is believed and why."[140]

One way to show this is to highlight the problems of science policy when we *exclude* certain groups. For example, the domains of science and medical research have been historically dominated by men, and the male body has been seen as the standard bearer of the human body, leading to an "information gap" in what we know about diseases as they present in nonmale bodies.[141] As detailed in her 2019 book *Invisible Women*, Caroline Criado Perez shows that the exclusion of women in positions of science and engineering has created a persistent and pernicious lack of information gathered about women in medical and drug trials, social science surveys, and general social data collection.[142] For decades, car companies crash-tested their vehicles using only "male" dummies, which led to women dying at higher rates than men in car crashes of equal severity.[143] Even after 2003, when car companies began using "female" dummies, the dummies were extremely unrepresentative of the average woman's body because the companies opted to use "5th percentile female" bodies for height and weight, equating to "a 12–13-year-old girl."[144] In another example, there are a plethora of medications to enhance the "male sex drive" and only one (noneffective) option available for women.[145]

Policies and drugs reflecting a male normalized construct of health result in gaps in medical treatments and care for women because they are excluded during their development. Consider that it was not until 1993, as part of the National Institutes of Health (NIH) Revitalization Act, that clinical trials were required to include women. Before this, women were rarely included in clinical trials, at first because of an aversion to women's hormonal fluctuations, which might "distort the data."[146] This stigma against women's menstruation even played a role in the exclusion of female mice from animal trials until 2016, when the NIH began requiring that both male and female animals be used in vertebrate studies.[147] The Trump administration "archived" this policy in 2025.[148]

Feminist epistemology shows that women, and those in the non-dominant power-knowledge situatedness, have unique knowledge of their lived experiences *and* that their perspective on questions of science and inquiry are also important for a more comprehensive understanding about the world. Michael Polanyi refers to these "counter-Enlightenment" forms of knowledge as "personal knowledge," and Thomasina Borkman uses the term "experiential knowledge."[149] As Borkman writes, this form of epistemology is a kind of "truth learned from personal experience with a phenomenon rather than truth acquired by discursive reasoning, observation, or reflection on information provided by others. . . . Insights learned from direct participation in a situation are truth, because the individual has faith in the validity and authority of the knowledge obtained by being a part of a phenomenon."[150] Lived experience offers a window into a different state of truth about the world, one that cannot be accessed by rationalization.

In the case of CRISPR, the lived experiences of individuals from diverse backgrounds are valuable in ethics, policy, and rulemaking. This includes those who are members of a social group that has experienced the trauma of nonconsensual genetic research, such as the Havasupai, or have been part of communities identified as having "undesirable traits," such as many disability groups (see chapter 2). Other experiences may include being part of a family with inherited illnesses, experiencing an invisible disability, serving as a caretaker, working as a health care provider, or living in an ecological system that is vulnerable to the climate crisis. Lawmakers, regulators, scientists, and experts alike also bring their own experiences to bear. Consider that the one person who voted against the House ban on human genome modification in 2019 was Debbie Wasserman-Schultz because she carries the BRCA2 gene that raises her risk of breast cancer.[151] Thus, I offer that there are additional sources of knowledge that are importantly informed by lived experience and that the goal of an epistemic *system* is to cultivate diverse groups to "co-produce knowledge."[152]

Social Epistemology and Diversity

It may seem now that we face a new problem: How does one accept both the authority of the expert and of the "nonexpert"? Rather than replace or even downgrade the important role of experts, I argue that nonexperts can complement their important work in the proper deliberative and democratic space. Both modes of situatedness provide important perspectives on science policy by aiding in value judgments, risk analysis, and the interpretation of uncertainty. The goal of our institutions, then, should be to cultivate various sources and varieties of knowledge from different life experiences and perspectives by creating multiple avenues for participation, entry, and group-based deliberation. This is because diversity aids, rather than hinders, the co-production of knowledge.

The goal of policy creation is not just to identify one specific mode of truth-bearing but to identify broader systems in which information can better "approximate" the truth.[153] Social epistemologists refer to this endeavor to understand modes of knowledge as group-based rather than individual. This epistemology reflects an "anti-reductionist" move, a critique that shows that knowledge cannot be reduced to a particular individual or experience.[154] As Rachel McKinnon writes, "No knower is an island: we depend on social structures and other people for the acquisition, creation, and dissemination of knowledge."[155] Philip Kitcher, famed twentieth-century philosopher of science, similarly rejected the "primary source of knowledge" as "individual human beings," arguing instead that we "should be concerned with the organization of communities of knowers and with the processes that occur among knowers within such communities that promote both the collective and the individual acquisition of true belief."[156]

Social knowledge, then, uses diverse sources of truth production. In addition to historical cases of lessons learned from exclusion, there has been an "explosion" of contemporary research on the power of diverse groups that upholds the importance of their wisdom.[157] Earl Lewis and Nancy Cantor, for example, situate diversity as a "compelling interest" for instrumental reasons of prosperity, such as "economic prosperity," and as a "strategic advantage" through the "national project of integration."[158]

This requires a kind of Socratic willingness to acknowledge uncertainty and ignorance inherent in science policy and technology regulation.[159] A crucial part of regulating gene editing requires not just admitting uncertainty but also being able to articulate when and where predictions are expressed as percentages and how they are calculated, considering that great anxiety and complex politics are involved. Ian Graham and Peter Llewelyn Jones refer to this as the "management of uncertainty."[160] Cynthia Townley further writes that "ignorance is often described as a lack of knowledge," but we should instead consider it an asset for it allows us to identify where others should fill in the gaps.[161]

Experts, importantly, suffer from their own "illusion of predictability" and are often swayed by social circumstances in their interpretation of urgency, risk, and time.[162] David Lindley reminds us that science requires some amount of uncertainty because "uncertainty, discrepancy, and inconsistency are the stock-in-trade of any likely scientific discipline."[163] Recall from chapter 1 that uncertainty plays a prominent role in CRISPR applications, such as the undetermined impact of gene drives on the environment and the threat posed by off-target genetic changes.

Steven Sloman and Philip Fernbach identify the "knowledge illusion" as a kind of ignorance that emerges when scientists offer justifications that bolster their own success over the truth. In particular, when asked to justify their work publicly, experts often downplay the uncertainty.[164] As Marjolein B. A. van Asselt writes, "The decisions of modern life are complex, giving rise to often intractable uncertainty in decision making. This uncertainty can take several forms. Different perspectives lead people to perceive and seek to manage the world in different ways."[165] Her approach is to enhance "awareness of this pluralism and . . . avoid being over deterministic and positivistic in their practice."[166]

Insofar as regulations entail uncertainties, anxieties, risk analyses, and an evaluation of evidence, diversity of standpoints and expertise on a subject better "tracks the truth" than homogeneous groups. As Hélène Landemore argues, "In groups of problem solvers it is often more important to maximize cognitive diversity—i.e., a diversity of ways of seeing and interpreting the world—rather than individual competence."[167] For example, climate activist Greta Thunberg, speaking at the R20 Austrian World Summit in Vienna on May 28, 2019, emphasized the need for "neurodiversity" of thinkers in order to prevent the occurrence of expert group's tendency for "double think."[168] Part of the argument for neurodiversity inclusion is that "larger systems by which we organize the training of inquirers and the circulation, uptake, and incorporation of individuals' epistemic contributions to the construction of knowledge may need to be reformed to ensure that justice is done to each knower, and to groups of inquirers."[169]

This echoes Elizabeth Anderson's 2010 book *The Imperative of Integration* in upholding both the intrinsic value of equal association of peoples in civic society and educational settings as well as its instrumental value, such as societal welfare and flourishing by protecting rights.[170] Anderson builds on Fricker's concept of diversity as a "virtue of social systems" that leads us to demand the "integration of diverse institutions and persons engaged in inquiry."[171] Rather than take up Fricker's individualism and virtues of recognizing individuals as knowers, Anderson's vision is more social, not unlike Alvin Goldman's "systems-oriented social epistemology," and thus concerns entire systems, such as the legal tradition and academia.[172]

Mathematical modeling has also shown that mixed rather than homogeneous groups are preferable for epistemic outcomes in problem-solving areas.[173] In particular, Lu Hong and Scott E. Page found certain conditions under which, "in the limit, a random group of intelligent problem solvers will outperform a group of the best problem solvers."[174] As the group of randomly selected individuals becomes larger, and as the problem solvers become more diverse, the "group performs relatively better."[175] One of the implications they draw from this is that "even if we were to accept the claim that IQ tests, Scholastic Aptitude Test scores, and college grades predict individual problem-solving ability, they may not be as important in determining a person's potential contribution as a problem solver as would be measures of how differently that person thinks."[176]

Scott Page in particular has gone to great length to articulate the value of "instrumental diversity" in improving the quality and strength of various kinds of groups in society, arguing that diversity trumps both "*homogeneity*" and "*ability*."[177] Page argues that "often, diversity merits equal standing with ability and that sometimes, although not every time, it even trumps ability," such as when "constructing a welfare policy, designing a physics experiment, cracking a secret code, or evaluating post–heart attack treatment."[178] In so doing, he warns that it matters *how* institutions are designed to best use the value of diversity in both "*problem solving* and *prediction*" tasks because "diverse perspectives increase the number of solutions that a collection of people can find by creating different connections among the possible solutions."[179] And while Hong and Page speak of diversity as pertaining to differences in "how people represent problems and how they go about solving them," Page supplements this view by arguing that "identity diversity can also indirectly influence how people tap into their cognitive repertoires. *Evidence from controlled experiments shows that the presence of a person from a different identity group causes people not from that identity group to generate more ideas and construct more complex arguments.*"[180]

Hong and Page readily acknowledge that their model does not take into consideration the role of "communication and learning," such as deliberation, and that the problem-solving tasks diverse groups are better equipped to handle are those that have definite answers rather than those involving judgments of risk and uncertainty.[181] However, I offer that their findings can still be applied to regulatory questions related to gene editing, insofar as diversity of thinkers improves risk analyses due in part to the important social perspective that risk analysis requires (see chapter 3). Rahman, for example, postulates that if women had been more involved in financial decision-making at the major banks prior to 2007, then there would have been fewer risky loans issued which led to the collapse of the housing market.[182]

One of the concerns of this line of thinking is that "diversity talk" can create a reductive role for individuals, because it "essentializes supposedly intrinsic female qualities in order to 'activate' those qualities and transform them into particular forms of economic agency."[183] The worry here is that selecting for a diverse group reiterates a kind of "tokenism" and the assumption that identity is constitutive of ideas.[184] I address this more thoroughly in chapter 5 by offering that deliberative groups can have different kinds of representation, including attitudinal rather than identity-driven ones.

Further, institutions need more than just diverse representation: They need to be held accountable for the efforts and participation of diverse groups. Jonathan Haidt confirms the significance of institutional structure in maximizing the epistemic benefits of diversity: "If you put individuals together in the right way, such that some individuals can use their reasoning powers to disconfirm the claims of others, and all individuals feel some common bond or shared fate that allows them to interact civilly, you can create a group that ends up producing good reasoning as an emergent property of the social system."[185]

I argue, similarly, that diverse groups, or ones that embody varying sources of knowledge and abuses of power or injustices, are better at judging and understanding the role of uncertainty, risk analysis, and evidence in regulatory decision-making. Experiential wisdom as communicated through the testimonies of affected groups combined with scientific expertise, or those whose knowledge is based on a particular kind of training in rational inquiry and experimentation, hold better political judgments than homogeneous groups. As W. E. B. Du Bois wrote in 1920 in describing the importance of including racial minorities in decision-making, "the real argument for democracy is . . . that in the people we have the source of that endless life and unbounded wisdom which the rulers of men must have."[186]

Epistemic Democracy

"Epistemic democracy" refers to the idea that democracy is the preferred system of governance because it allows the rulers to "tap" into the available knowledge, intelligence, or wisdom of the "*demos*."[187] Though they differ on the precise mechanism and rationale, epistemic democrats hold that democratic systems are procedurally preferable to other forms of government because, as James Bohman writes, "the results are better."[188]

For proceduralists, epistemic democracy is the idea that certain procedures of democracy better "track the truth" than other systems.[189] For example, Alfred Moore argues that democracy is ideal for epistemic inquiry because when experts are required to testify and present evidence, the people can operate as a kind of filter, creating greater accountability to the truth.[190] Cheryl Misak refers to this modality as creating a "culture of justification."[191] Michael I. Räber offers a "pragmatist approach," similar to those discussed in the introduction, which helps to fuse the power-knowledge gap: Knowledge is produced by those in charge, and the task "of the people" is to engage in political judgment.[192]

The aim of cultivating a more "knowing democracy" often features two mechanisms: (1) aggregation and (2) deliberation. First, aggregation methods are those wherein knowledge is understood or collected through methods that do not "require actual communication or an exchange of views among the participants," such as crowdsourcing, market observations, polling, voting, and commenting.[193] As articulated by James Surowiecki, the "wisdom of the crowds" argument emerged from the observation that averaging a crowd's guess is more likely to be accurate than the guesses of its individual members, even if those individual members are experts.[194]

However, these formulations also get critiqued by those who hold that the "crowds' data" reflects trending preferences (consider, for example, restaurant review sites) or biases and polarization within social networks or online platforms.[195] Crowdsourcing also suffers from similar preselection problems as those seen in consensus conferences and draft rule comments. In addition, these data collection methods require centralized power, usually held by experts or leaders of industry, and as such, the content, interpretation, and communication of the results is still controlled by those lacking adequate diversity. Critics also point to examples where the wisdom of crowds has merely turned out to be populist mob rule, often aggressive or oppressive in nature and involving harassment and bullying online.[196] Further, given how fast (mis)information spreads today, skeptics doubt whether aggregating information from these sites can truly track the truth as opposed to the perspectives of the most passionate or active participants.[197] It is these concerns, I suggest, that public health officials are recreating when they express fear of a backlash against emerging technologies like GMOs and vaccines (see chapter 2).

For other epistemic democrats, *deliberation*, or the talking part of democracy, serves as the crucial mechanism that enhances a democracy's substantive policy outcomes. As Thomas Christiano argues, the "legitimacy of democratic institutions are justified to the extent that they resemble or approximate the ideal deliberative procedure."[198] Through "talking politics," we are able to truly explore the varied sources of experience, identity, and cognition that create considered judgment and co-production of knowledge.[199] For Simone Chambers, part of what is crucial to this is that citizens are presented with expert information "with a stated attempt at neutrality or fairness where frames are made explicit."[200] As a result, participants will experience their engagement as respecting their autonomy, which is the "background normative assumption underlying the idea of a well-working public sphere."[201]

William A. Gamson's theory of deliberation reflects this process wherein citizens can engage in conversations to combine "experiential knowledge" and "popular wisdom."[202] Archon Fung has similarly described this as "privileged access to [citizens'] own preferences and values."[203] Because power structures are complicit in creating knowledge, the more diffuse, diverse, and mixed the structure is, the better the knowledge creation becomes, including by giving the citizens "agenda setting power." Drawing on Jürgen Habermas's defense of deliberation as instrumental to setting the agenda, Chambers argues that "epistemic quality and equal participation" are vital to a deliberative process precisely because this "division of labor" allows for greater "agenda setting."[204] This, then, has the added benefit of reducing stakeholder influence and capture.[205] The presence of, and accountability to, a body of the people also reduces the chances of those in power manipulating data for their own purposes and lessens the corrupting influences of industry and conflicts of interest. And because regulators need to act even in the face of great uncertainty, diversifying the voices around the table better aids in this task of evaluating risk and safety, particularly through better alignment with risk propositions.[206]

Epistemic democrats often encounter two main points of criticism. The first concerns general skepticism regarding the capabilities of laypeople to participate. The second concerns domination within deliberative groups and co-option by the loudest voices.

First, the "anti-capabilities" critique casts doubt on the capacity of lay participants to deliberate against, among, or instead of experts. The critique relies on the idea that "the people" are unqualified to weigh in on certain matters of governance. As Jason Brennan infamously argues, individuals do not have the adequate "political knowledge" to participate in deliberations. He maintains this is not because they are "stupid . . . they just don't care."[207] Voters are irrational and do not know "candidates' policy platforms, how candidates are likely to vote in congress, what policies the candidates are likely to support," and so on.[208] Brennan then cites psychologists Amos Tversky and Daniel Kahneman to explain "availability bias" or the "availability heuristic" wherein individuals suffer from "framing effects" that deliver different results depending on "how questions are posed," which has been shown to have a "major effect on what opinions people form."[209]

As we have seen, there are prejudiced and false narratives embedded in this theory of who counts as a knower. Further, the problem with these kinds of arguments is that their evidence of "bad decision-making" often relies on voting or election results, rather than *deliberation*. Lastly, the critique largely concerns individuals in isolated tests and studies rather than group-based deliberative experiments. Rather, as Hélène Landemore writes, "Under the right conditions, including everyone in the decision-making process simply makes the group more likely to get the right (or at least better) answers."[210] On technical issues that deliberators lack knowledge of, Jane Mansbridge makes the case for how thorough deliberation can still be useful for self-clarification and self-understanding even when there is no objectively correct answer.[211] Landemore's "Numbers Trump Ability" theorem shows that while "delegation to experts might be necessary on a range of other issues," it is not enough to point to one or two examples of a particular task and ask whether a nonqualified person can accomplish that task.[212] The tasks defined are not whether they can fly a plane, but rather whether a community is willing to participate in an experimental environmental intervention, like a gene drive. These are questions that require some technical knowledge, such as understanding what a gene drive is, as well as experiential knowledge.

As we will see in chapter 5, the empirical literature on epistemic democracy also refutes the anti-capabilities claim by showing that "learning" occurs among deliberation participants.[213] This has been studied perhaps most thoroughly in deliberative polling and more recently by Huihui Ding and Marcus Pivato, who found that people *learn* over time in smaller, more resource-enriched, and moderated environments where deliberators are asked to persuade one another, creating the conditions for better epistemic outcomes.[214] John Dryzek has described this as "social learning" because of the ways in which individuals not only attain further information but do so in moments of great "identity conflict."[215]

In addition, Tversky and Kahneman's heuristics should be understood as a universal human phenomenon, rather than particular to nonexperts. Kahneman even wrote that he and Tversky examined whether "other researchers suffered from the same affliction," noting further that biases of intuition "do not denigrate human intelligence, any more than the attention to disease in medical texts denies good health."[216] In other words, they warn against using the heuristics they identify to prove or disprove anything about how society should be arranged or authority granted. Other research shows that groups of experts and industry representatives are perhaps even more constrained in their heuristics and decision-making methods than mixed or hybrid groups are.[217] It is unreasonable to expect that scientists are somehow exempt from problems of insufficient knowledge in certain areas.[218]

Second, Ian Shapiro (and others) worry that deliberative groups can be co-opted or manipulated by charismatic or overly passionate participants.[219] While that is entirely possible, there is no less risk of it happening in deliberative forums than in legislatures, elections, or backroom, closed-door, expert-driven decision-making environments, and the risk can be mitigated by using properly trained moderators. The design considerations for creating "idealized" deliberative environments are explored in much greater detail in chapter 5.

Furthermore, the diversity of the deliberative groups is more likely to reduce the problem of domination because it facilitates, rather than exacerbates, engaging with "the Other" in ways that truly explore each other's similarities and differences in lived experience and understanding of expert testimony. This creates more productive conditions for innovative and thoughtful political judgment formation than previously possible. Deliberation does not work well when there is *insufficient* diversity because the more dominant group has more power over the small minority.[220] As Archon Fung argued early in the COVID-19 pandemic, creative policy solutions required "more democracy, not less."[221]

Consider the case of how we measure kidney function. The estimated glomerular filtration rate (eGFR) is a numerical value based on input data that indicates how well the kidneys are filtering blood by removing waste and excess water to make urine (i.e., whether they are doing their job). The number a patient receives determines whether they get dialysis and, eventually, a kidney donation. Until 2021, there were two separate eGFR calculations: one for Black patients and one for non-Black patients. Rooted in a racialized trope, Black patients were assumed to have more muscle mass, which contributed to higher levels of creatinine—a waste product generated from muscle tissue breakdown that is filtered by the kidneys—leading to the eGFR adjustment for Black patients. This overestimation of kidney function contributed to significant disparities in Black patients receiving follow-up care and being placed on the organ transplantation list.[222] Inspired by Black Lives Matter protests and the COVID-19 pandemic, which pulled the curtain back on significant racialized health disparities, the American Society of Nephrology and the US National Kidney Foundation formed a joint task force in July 2020 to review the eGFR calculations. The task force recommended the adoption of a new race-free equation in 2021, and in 2022, the Organ Procurement and Transplantation Network approved a policy, effective in 2023, that requires transplant hospitals to use the race-neutral eGFR.[223]

There are several notable takeaways from this case. First, much of the resistance to changing the system came from medical experts.[224] Some argued that instead of solving problems of racial injustice, this would exacerbate them.[225] Others voiced concern that health care systems would need to do more than just change the equation to get Black nephrology patients the follow-up care they needed. Disagreements among experts do not necessarily indicate that expertise is an illegitimate source of authority, but rather signal the need for institutional arrangements that highlight the virtues of different sources of knowledge. Second, only 6 percent of nephrologists are Black, emphasizing the importance of racial diversity and the epistemology of diverse groups in decision-making.[226] And finally, using racialized biomarkers (to return to the moral considerations of gene editing), as we saw in chapter 2, is not accurate because race is not a biological construct.

Thus, expert-based groups can suffer from epistemic deficiencies when operating in isolation due to silo effects, problems with forecasting, and incomplete interpretations of uncertainty. To be clear: Experts play an invaluable role in democratic society, such as when reviewing the safety of vaccines and recommending vaccine schedules. My argument is not to replace them; rather, it is that nonexperts play an important role as well, and that the faults of nonexperts are often shared by experts, emphasizing the importance of diversity of knowledge and experiences in creating more robust epistemic and non-dominating outcomes. The next chapter concerns how and in what form "*more*" democracy should take shape in a regulatory agency in order to enhance the non-dominating and epistemic outcomes of CRISPR rulemaking.

5 Deliberative Minipublics

This chapter builds on the moral considerations of chapters 1 and 2, the institutional critique of chapter 3, and the justice and knowledge considerations of chapter 4 by arguing for the potential of deliberative forums to increase the US administrative state's democraticity and substantive regulatory outcomes. Deliberative minipublics (DMPs) are a promising mechanism for mitigating regulatory agencies' democratic deficit by infusing the system with greater diversity and representativeness of the people. DMPs are meant to act as a kind of "proxy" for the public. Random selection creates a more "fair cross-section of the people" to help foster what the people *could think* given adequate time, resources, and space.[1] DMPs, as Michael MacKenzie and Kieran O'Doherty write, "can be used to inform policy development processes and expand the range of policy options by giving policymakers insights into public discussions which have not yet happened."[2] Including DMPs early in the rulemaking process—before and for the purposes of drafting rules—offers a promising institutional mechanism for rulemaking. They can help to create rules that reflect important evidence reviews and risk conversations due to the more direct and deliberative input from individuals with diverse experiential wisdom and without allegiances to special interest groups or other overrepresented entities. Further, experimenting with incorporating DMPs within a regulatory agency will hopefully lead to broader DMP inclusion in governance.

In upholding DMPs as a promising intervention in the regulatory process, I examine the current and potential uses of DMPs in cases related to gene editing to advocate for their use in both the rulemaking process generally and in regulating gene editing specifically. As such, I begin with an overview of the different types and examples of DMPs, particularly those that address similar problems to gene editing, such as genetically modified organisms (GMOs), gene therapy, and allocation of scarce resources. In the second section, I explore DMP procedural details such as the selection of participants, the role of experts and moderators, the decision-making mechanism, and whether they should be authoritative or solely provide

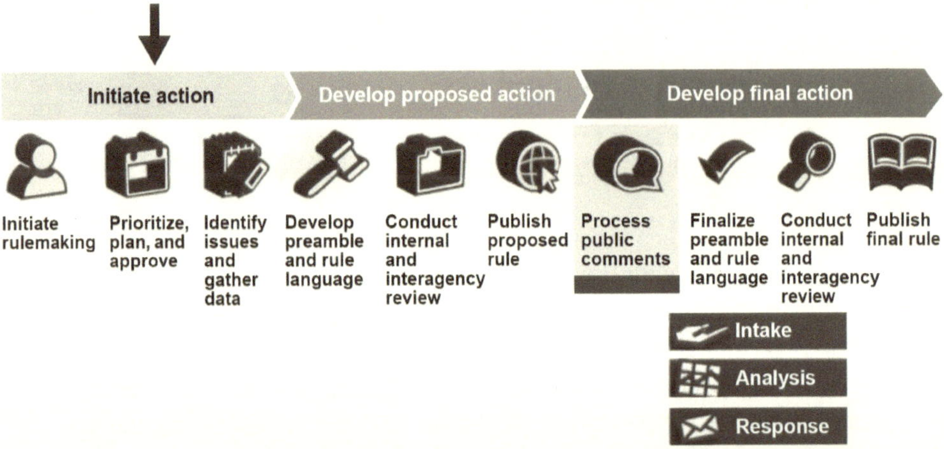

DMPs should be incorporated during the "prioritize, plan, and approve" phase of formal rulemaking. Source: US Government Accountability Office.

guidance. I then make recommendations for how a regulatory DMP should function and identify important trade-offs in design, noting that each choice will likely require modification. Finally, I offer an example of how a DMP could be used in the context of gene editing.

Deliberative Minipublic Models and Examples

A DMP is a forum in which participants with different viewpoints and backgrounds are randomly selected to participate in facilitated deliberation and reach considered judgment.[3] In Nicole Curato and colleagues' book *Deliberative Mini-Publics: Core Design Features*, DMPs are defined "as carefully designed forums where a representative subset of the wider population comes together to engage in open, inclusive, informed and consequential discussions on one or more issues."[4] Ian O'Flynn similarly defines them as institutions "designed to enable small groups of people, chosen more or less at random from the public at large, to reason together about some important matter of law or public policy."[5] The main throughline of DMPs is that the members are randomly selected (also referred to as sortition), they have opportunities to hear expert testimony, their deliberation is moderated (or facilitated), and they reach some sort of considered judgment (such as in a written report or voting on a policy question). O'Flynn adds that "what they share is a commitment to creating a safe space within which people from diverse backgrounds can deliberate together under good conditions—conditions that typically include access to balanced information, expert witnesses, facilitated small-group discussion, a promise of personal anonymity, financial assistance (e.g., for carers or those with disabilities who might otherwise not be able to attend) and the time to reflect."[6] This section provides an overview of four main types of DMPs that have been developed over the years: consensus conferences, citizens' juries, Deliberative Polls, and citizens' assemblies. Other minipublic adjacent models—planning cells, citizens' initiative reviews, and participatory budgeting—are also addressed in Table 5.1.

Consensus Conferences

The Danish Board of Technology, an independent advisory institution associated with Denmark's Ministry of Science, Innovation, and Higher Education, began convening consensus conferences in the 1980s to learn about public perspectives on policy related to science, the environment, biotechnology, health, and energy. Danish consensus conferences were founded on the philosophy that if technology should protect the public interest, then those tasked with directing the use of technology must consider the views of those affected. Conferences were convened also in part due to the public's widespread concern that experts and scientists were too isolated and nontransparent in confronting "cases characterized by radical uncertainty, fact-value overlap and intractable controversies."[7] Denmark first used a conference to assess "gene technology in industry [and] agriculture" in 1987; since then, it has engaged consensus conferences on such topics as human genome mapping (1989), transgenic animals (1992), gene therapy (1995), genetically modified foods (1999), and gene testing (2002).[8]

Consensus conferences are composed of ten to sixteen laypeople, all selected from an applicant pool. Organizers aim for a "cross-section of the population in terms of age, social class, and political preferences."[9] The structure of the conference allows participants to question a panel of experts for two consecutive four-day weekends. For four days following, the participants deliberate, create a "consensus statement," and hold a public conference to disclose their "findings."[10] Their reports are shared with members of Parliament, scientists, and the public.[11]

Other countries, including Brazil, Canada, France, Germany, India, Israel, Japan, the Netherlands, New Zealand, Norway, South Korea, Switzerland, the United Kingdom, and Zimbabwe, have adopted the consensus conference

TABLE 5.1 Deliberative Minipublics Typology

DMP Type	*Definition and Description*	*Examples*
(Danish) Consensus conference	• 10–16 participants • 3–4 days of deliberation • Created by the Danish Board of Technology in 1987 to address a particular technological question and create a "consensus statement" delivered to the government and public alike. They originally did not use random selection but rather open invitation and selection for diversity among their participants.	• Human genome mapping (Denmark, 1989) • Gene therapy (Denmark, 1995) • Genetically modified foods (Denmark, 1999) • Gene testing (Denmark, 2002) • Genetically modified organisms (Australia, 2000)
Citizens' jury	• 12–25 people • 4–5 days of deliberation • Citizens' juries have been used worldwide to address a particular question, ranging from providing candidate information to delineating the moral guiding principles of a particular technological question.	• Citizens' Jury on GMOs (France, 1998) • Citizens' Jury on GM Crops (Brazil, 2000) • Citizens' Jury on Genetic Test Results and Life Insurance (Scotland, 2004) • Citizens' Jury on Genetically Modified Plants (Denmark, 2006) • French Mini-Public on Healthcare and Big Data (2016) • Climate Change, Air Quality, and Environmental Policies (numerous examples, including Leeds, 2022; Lancaster, 2022; and Kendal, 2020) • Australian Citizens' Jury on Genome Editing (2021)
Deliberative Poll	• 100–600 people, with deliberations in smaller groups (10–12 people) • 1–4 days of deliberation • Proposed by James Fishkin in 1988, Deliberative Polls' trademarked feature is that they are polled before and after deliberation to measure opinion change and assess "learning."	• America in One Room (Texas, 2020) • What's Next California (California, 2011) • Unification of Korea (South Korea, 2011) • London Power 2010: Countdown to New Politics (UK, 2010) • What to Do with the Euro Cup Stadium in Poznan? (Poland, 2009) • Public Servants Careers Reform (Brazil, 2009) • Policies Toward the Roma (Bulgaria, 2007)

DMP Type	*Definition and Description*	*Examples*
Planning cell	• 25 people per "cell," several cells that deliberate in parallel • 4–5 days of deliberation • Conceived by Peter Dienel in the 1970s in Germany, a planning cell (which operates like a citizens' jury) works separately on a particular question and then, together with other cells, creates a "Citizens' Report" to be delivered to the commissioning body, often a governmental office or municipality.	• German Ministry of Research and Technology: planning cells to study energy policy options (1982) • New Jersey Department of Environmental Protection: planning cell to evaluate sewage sludge at a Rutgers University research farm (1988) • Bavarian consumer planning cells project (2001–2)
Citizens' assembly	• 50–250 participants • 20 or more days of deliberation, at times spanning several months • Beginning in 2004 in British Columbia, these DMPs are larger in scale and at times nationwide projects, asking participants to weigh in on electoral and constitutional reforms; they incorporate consultation, such as through public hearings, open submissions, and online deliberations.	• Icelandic Constitutional Assembly (2010–11) • Irish Citizens' Assembly (2016–18) • Camden Citizens' Assembly on the Climate Crisis (2019) • UK Climate Assembly (2020) • French Citizens' Convention for Climate (2019–20) • Scotland's Citizens' Assembly (2022) • Netherland's Climate Assembly (2025) • Global Citizens Assembly on Genome Editing (proposal only)
Citizens' initiative review (CIR)	• 20–30 participants • 3–5 days of deliberation • Modeled on the citizens' jury, CIRs began in 2008 in Oregon; they are tasked specifically with providing information and recommendations of ordinary citizens on ballot measures or candidates.	• Oregon's Citizens' Review (2008) • Arizona CIR (2016) • Colorado CIR (2014) • Massachusetts Citizens' Review (2016, 2018) • Finland (2019) • Switzerland (2019)

TABLE 5.1 *(continued)*

DMP Type	*Definition and Description*	*Examples*
Participatory budgeting	• Groups of 20–100 participants broken into small groups of 6–10 people • 2 hours to a full day of deliberation • Participatory budgeting focuses specifically on how to allocate money. Some are well-established programs, such as in Chicago's 49th Ward, while others are one-off events for a specific fund or purpose.	• Municipalities throughout the US, including Chicago, Boston, New York City, and San Francisco • Porto Alegre, Brazil (1989) • Ontario, Canada (since 2001) • Berlin Lichtenberg (since 2005) • Seoul, South Korea (2013) • Paris, France (2020)
Citizens' summit	• 1,000 participants • Nationwide projects to revitalize citizen participation in politics and discuss a wide range of policy questions, including unemployment benefits, pensions, child support, health care, urban design, and transportation.	• G1000 Citizens' Summit (Belgium, 2011–12) • G1000 Citizens' Summit (Netherlands, 2018)
Other	• Differently named deliberative forums that share a DMP's essential features: random selection, expert guidance, facilitated deliberation, and considered judgment formation.	• BC Biobank Deliberation (Canada, 2007) • Public Deliberation on Genetically Modified Salmon (Canada, 2008) • Australian Citizens' Parliament (2009) • Implications of Whole Genome Sequencing for Newborn Screening Public Dialogue (UK, 2020) • Citizens' Panel on Vaccine Skepticism (France, 2021) • Public Dialogue on Genome Editing and Farmed Animal Breeding: Social and Ethical Issues (UK, 2022)

model to address their own technology issues. In the United States, the National Institutes of Health (NIH) began using a similar structure, the "Consensus Development Conferences," in 1977 to assess medical evidence in clinical practice and to develop a standard for diagnosing gestational diabetes.[12] However, a notable difference between the NIH's model and the Danish one is that the NIH conferences were composed of a diverse array of experts rather than citizens. The NIH retired the model in 2013, citing other organizations that accomplish the same goals.[13] Private institutes and universities conducting biological and medical research have also implemented the model in the United States, including in 2008 to discuss "human enhancement, identity and biology" at Boston University's School of Public Health.[14] In a final example, the Organ Procurement and Transplantation Network (OPTN), the Scientific Registry of Transplant Recipients (SRTR), and the United Network for Organ Sharing (UNOS) have over the last two decades used consensus conferences to better understand organ transplantation needs of various hospitals, systems, and communities.[15]

Overall, consensus conferences have had positive policy impacts. Ida-Elisabeth Andersen and Birgit Jæger reported that after the "urban ecology" conference, the Danish Minister of Environment established a national committee on urban ecology. Later, the Urban Ecology Committee established a Danish Centre of Urban Ecology, the Green Foundation to finance its activities, and the Ecological Council and Association of Green Families. Andersen and Jæger also noted overall increased public awareness of the issues presented at the conference and a sense of civic responsibility toward urban ecology.[16] In addition, Maria Powell and Daniel Lee Kleinman concluded from examining nanotechnology consensus conferences, "even if consensus conferences have little or no influence on policy or policymakers, they may empower citizens by improving their perceived abilities to participate meaningfully in technoscientific issues."[17] In 2011, in a "series of conferences on acute care in Africa to engage stakeholders in the development of consensus statements for the region," researchers found that the African Federation of Emergency Medicine recognized significant understanding of local needs and promised to increase the level of care for participating countries.[18]

In some cases, the evidence regarding the efficacy of the deliberation format and structure is less positive. Researchers studied eight of the NIH's consensus conferences from the 1980s and found that the format limited panelists' ability to learn due to "selection bias, particularly with respect to the choice of questions and panelists."[19] However, the authors also concluded that this was part of an ongoing quest to perfect the process of the conference, rather than a dismissal of the conferences' potential.

Consensus conferences successfully embodied a theoretical framework for participation and had some politically useful outcomes. However, this model ultimately suffered from flaws of selection bias because participants had to apply to participate. It also suffered from forced consensus because considered judgments should include minority views. The primary weakness of the consensus conference, such as in the NIH model, is that these conferences are too "self-selective," reproducing similar problems of overrepresentation of dominant groups and underrepresentation of subordinate ones.[20] Even though the selection strategy was intended to be representative, application processes create self-selection problems, leading to barriers to entry. As Archon Fung argues, "The difficulty with voluntarism is those who show up are typically more well-off—wealthy, educated, and professional—than the population from which they come."[21]

Citizens' Juries

Trademarked in the 1970s by the Jefferson Center, citizens' juries (CJs) were inspired by consensus conferences. Adapting the idea of a "group of one's peers" for criminal justice, Ned Crosby conceived of juries for answering "value questions."[22] CJs are composed of roughly twelve to twenty-four people and deliberate for a set time period (four to five days).[23] They are tasked with a particular policy question, ranging from providing candidate information (such as through citizens' initiative reviews, see Table 5.1) to delineating the ethical principles of a technology (drawing on their origins as consensus conferences). They are meant to provide a space for equal participation and operate using "rules of conduct" that encourage "mutual recognition and respect and [are] orientated towards the public negotiation of the common good."[24]

The Australian Citizens' Jury on Genome Editing (2021) is perhaps the most important CJ for our conversation on CRISPR. The twenty-three participants were asked to consider nonheritable and heritable applications, research on human genome editing, and mitochondrial donation. Among their main recommendations, they emphasized that (a) if genome editing is to be used, it should be used to promote welfare and alleviate suffering; (b) "meaningful consent and equity of access should be the overarching ethical principles"; and (c) oversight is crucial.[25] For heritable human genome editing (HHGE), they emphasized that its use should be limited to "clinical applications" in which "both . . . the severity of the disorder or disability (such as life-threatening conditions), and the number of people affected should be considered."[26] They also noted that while they "hold differing views on the use of HHGE for non-clinical applications, including enhancement," the majority felt "that these should be prohibited."[27] In addition, any decisions should include "community consultation," particularly with those "affected" and that those with "inherited disease or disability should have input into the approval process."[28] Lastly, patients should be informed of alternatives when having children.

Australia has convened other CJs in the domain of biomedical research and clinical applications, including the Citizens' Panel on Mitochondrial Donation in 2019 that facilitated consultation alongside an expert working committee to draft pending legislation. The final report of the citizens' panel conveyed the importance of mitochondrial donation to different stakeholders and outlined the risks and participants' reservations.[29] The panel stressed that the majority of participants thought that mitochondrial donation should be permitted, while a small minority felt that there were too many unresolved concerns, including the uncertain impact on future generations.

CJs have been found to be overall a helpful mode of gauging public input on health policy, including on health care priority setting and resource allocation (generally and during the COVID-19 pandemic), consent protocols for data sharing in precision medicine, climate change policy, genetically modified (GM) food safety and regulation, genetic testing, and the moral limits of genome editing.[30] Jackie Street and colleagues analyzed sixty-six health policy–oriented juries and found that "citizens' juries are a useful tool for engaging citizens in health policy decision-making: they are small enough to permit effective deliberation, relatively inexpensive compared to the larger deliberative exercises of planning cells and consensus conferences, yet sufficiently diverse that the citizens engaged are exposed to a broad range of public experience and perspectives."[31]

CJs have been used by numerous governing institutions and researchers worldwide to gauge input on risks and trade-offs for GMO foods and agriculture products. They were predominantly framed as attempts to better understand what participants saw as the risks and benefits of this new technology in settings where they could learn, discuss, and weigh arguments with those who may be directly affected by GMOs or those who were representative of the wider public, such as small farmers. For example, in

France in 1998, the government, which was largely GM-friendly, convened a CJ when it encountered public opposition. The jury recognized the economic benefits of GM products alongside concerns of diminished "farmer independence," "overproduction of certain products in Europe," and "ecological risks," including those that may become "uncontrolled."[32] According to John Dryzek and colleagues, the content of their recommendations confirmed a precautionary approach and a desire for more nonexpert participation in regulatory oversight.[33]

In Denmark in 2006, a CJ of sixteen people was convened by the Danish Board of Technology for five days. They met with twenty-four experts (including scientists and representatives from industry, nongovernmental organizations, and public administration) and explored arguments for and against growing GM plants in Danish fields, using GM plants for medical, industrial, and aesthetic purposes, and any further "handling of new GM plants."[34] The majority of the sixteen participants ultimately recommended that it would be beneficial to grow GM plants in Denmark, largely for the financial opportunity. However, they warned, this should be done only under strict regulations, such as monitoring pollution and groundwater. They also endorsed GM labeling and stressed the importance of public debate and transparency.

Citizens' juries have also extended beyond European and English-speaking spaces. In 2001, CJs were held in agricultural settings in Africa, Asia, and South America. In India, a CJ composed of "marginal-livelihood citizens from all three regions of the state" who had never heard of GM crops was convened in Andhra Pradesh for six days. Participants articulated a fear of Western foreign influence in controlling crop access and selection and entreated policymakers to preserve "indigenous" agriculture and biodiversity.[35] In Zimbabwe, the government gave participants several scenarios to consider on the role of GM crops after which the CJ called for greater government accountability and transparency.[36] In Brazil, three CJs were convened to address whether GMOs posed a threat to the environment and whether the testing of commercial uses was "'democratic, transparent and careful enough.'"[37] Finally, in Mali (West Africa) in 2006, forty-five Malian farmers were gathered to deliberate on GM cotton. This deliberation, referred to as Espace Citoyen d'Interpellation Démocratique (Citizen's Space for Democratic Deliberation), is notable both for using the CJ format while also incorporating "indigenous methods for debate and dialogue, including the traditional African *palaver*," and serving as "one of the very few mini-publics that has directly influenced policy making to date" in regulating the introduction of GMOs.[38]

Though not technically a CJ, a public dialogue event on "genome editing in farmed animals" was held by the UK Nuffield Council on Bioethics in May and July 2022 to discuss the potential of editing farmed animals. During four online workshops, eighty participants expressed animal welfare and environmental concerns and took a precautionary approach to the technology and its oversight. They further articulated a desire for "positive policy interventions in the food and farming system, not simply minimal protections beyond which it would be left to be shaped by market forces."[39]

Deliberative Polls

In 1988, James Fishkin also innovated on the consensus conference model with the concept of the Deliberative Poll (DP). This model uses random sampling to invite a large number of participants (up to five or six hundred) to deliberate in small groups of ten to twelve people and vote on agenda items that are often relevant policy concerns or even constitutional matters. Referred to as the "gold standard" of deliberation by Jane Mansbridge, DPs have been developed as tools for governments and society to stimulate discussion and educate the public.[40] As described on Stanford University's Deliberative Democracy Lab website: "Deliberative Polling® is especially suitable for issues where the public may have little knowledge or information, or where the public may have failed to confront the trade-offs applying to public policy. It is a social science experiment and a form of public education in the broadest sense."[41]

By polling participants before and after deliberation to measure changes in perspective, researchers have been able to show that participants are *capable* of learning, engaging with experts, and deliberating.[42] Before a meeting, participants are sent "carefully balanced briefing materials,"[43] which are also made available to the public. At the beginning of the DP, participants are polled about the topic. They then engage with an array of experts and political leaders; these meetings are followed by "small group discussions where people share their own reasons for support or opposition to the various proposals, and [have] access to competing experts who can answer questions arising from the small group discussions."[44] The small groups are guided by "trained moderators," often locally recruited teachers. For Fishkin, the essential feature of deliberation is this "weighing of reasons under good conditions in shared discussion about what should be done."[45] DPs are also often broadcast to the public, and their reports are made publicly available.

After the deliberations, participants are polled again on the original questions to test for learning. Fishkin and others have found shifts in opinion, signaling learning and listening. As Fishkin explains it, the quality of a deliberative process is conditioned on information ("the extent to which participants are given access to reasonably accurate information that they believe to be relevant to the issue"); substantive balance ("the extent to which arguments offered by one side or from one perspective are answered by considerations offered by those who hold other perspectives"); diversity ("the extent to which the major positions in the public are represented by participants in the discussion"); conscientiousness ("the extent to which participants sincerely weigh the merits of the argument"); and equal consideration ("the extent to which arguments offered by all participants are considered on the merits regardless of which participants offer them").[46] For Fishkin, the last point about equality is key: Random sampling ensures that anyone *could* participate, which fulfills the condition of political equality at the heart of democracy. Another part of Fishkin's vision is that the practiced deliberative methods of communication transcend the bounds of the institution by scaling up to civil society.

Deliberative Poll topics tend to be upstream questions of governance. Between 1996 and 1998, the Public Utility Commission in Texas convened to deliberate "about energy choices for the state."[47] In 2007, Protestant and Catholic members in Northern Ireland debated the future of a small town's schools. In 2009, Polish citizens met to discuss what to do with the Bulgarska St. Stadium after the completion of Euro Cup 2012. In 2011, the European Union held deliberations about the upcoming elections for the European Parliament. Also in 2011, California participants engaged in the "What's Next California" project to weigh in on thirty-nine proposals "from a broad coalition of reform groups, vetted by an extensive and politically diverse advisory group."[48] In South Korea in 2011, participants discussed Korean unification. In 2012, Japan used DPs to discuss Japan's future energy policy in the wake of the Fukushima disaster. In 2014, participants in Bududa and Butaleja, two towns in Uganda, discussed "resettlement, land management, and the population pressure" in light of "recurrent environmental disasters."[49] Mongolia employed DPs in 2015 for participatory budgeting projects. DPs were convened in 2020 in Grapevine, Texas, to discuss the election in an event called "America in One Room," and in 2024 in Taiwan to deliberate on the use of artificial intelligence (AI) "to enhance information integrity."[50]

Taiwan, in particular, has extensive experience crowdsourcing laws and harnessing deliberative democracy, including on AI. Examples are "Taiwan's Pathway to 2050 Net-Zero Emissions," a DP from 2022, and in 2023, "Responsible AI Innovation."[51] The 2024 Taiwan Deliberative Poll used AI for various parts of the deliberation while also deliberating on the ethics and governance of AI to directly inform policymaking. Taiwan's Ministry of Digital Affairs partnered with the Collective Intelligence Project (an international nongovernmental organization) and the Institute of Science, Technology and Society at Yang Ming Chiao Tung University to convene a deliberation on AI regulation and ethics, with a focus on democratic values.

Using Stanford's Deliberative Polling platform, the government sent out 200,000 random invitations and eventually selected 447 to participate in a six-hour online deliberation using stratified sampling on March 23, 2024. The large group was divided into forty-four smaller groups; twelve of them were "digital practitioners," and thirty-two were lay citizens.[52] AI bots moderated the online discussion, "automatically managing speaking time to ensure everyone had a fair chance to speak . . . while interruptions or verbal attacks were muted, preventing discussions from being dominated by the loudest voices."[53] Deliberation content featured questions pertaining to AI self-regulation and government oversight, as well as large platforms that used AI to help citizens understand threats and to "analyze," "maintain," and "enhance" integrity standards.[54]

The post-participation poll revealed increased AI system learning, and participants held "positive" attitudes toward their experience. In addition, the main recommendations included developing government infrastructure to directly communicate with the public on threats to information integrity, harnessing AI to evaluate threats to information, including through "AI-generated content detection technologies," and recognizing fraudulent content.[55] Their findings ultimately informed Articles 30 to 32 of the "Fraud Prevention and Control Act" enacted on July 31, 2024.[56]

Citizens' Assemblies

Citizens' assemblies (CAs) have become an increasingly popular site of democratic deliberation convened by governmental bodies, particularly in Europe and Australia. They have been organized for a wide range of mandates and agendas, including housing, welfare benefits, childcare access, and constitutional revision, such as in the Irish (2012 and 2016), Scottish (2021), and Icelandic (2016) assemblies. The poster child of CAs is the Irish Citizens' Assembly (2016–18), which paved the way for legalizing marriage equality and abortion rights in a traditionally conservative country. CAs have, accordingly, become popular responses for recognizing the demands of the people, as they are often situated at a national governing level and receive significant media attention and even support from social movements, such as Extinction Rebellion. In addition, there are increasing calls for the creation of CAs as a mode of mass mobilization of democratic power, as in the G1000 in Belgium.[57]

Early uses of CAs focused on electoral reform, such as in British Columbia (2004), Ontario (2006), and the Netherlands (2006). During the British Columbia Citizens' Assembly (BCCA), 158 voters from British Columbia met regularly for over a year to "learn about and deliberate over changes to the electoral system."[58] The CA deliberated, consulted the public, issued a report, and presented its recommendation to the legislature "that the electoral system of BC be changed to a form of proportional representation (PR) called the single transferable vote" from the mixed member proportional system.[59] As Michael Rabinder James reports, the participants were acutely aware of the trade-offs, capable of grasping the technical matters of voting systems, weighing competing values regarding descriptive representation in their deliberations, and controlling the agenda.[60] The decision was ultimately put to a public referendum, but fell short of the 60 percent requirement to pass, so the BCCA recommendations were never adopted, even after they tried again in 2009.[61]

Two years later, the Ontario Citizens' Assembly, modeled on the BCCA, also examined matters of electoral reform. It recommended, in contrast to British Columbia's CA, a mixed member proportional representation system. However, the selection process for the citizens' assembly and the end result of a referendum vote were similar to the BCCA in that it required a supermajority, a 60 percent margin, to succeed, which it did not receive.[62] Also as in the BCCA, the "Ontario version was created largely because the leader of an opposition party believed that one of the most fundamental institutions of democracy ought to be examined by an independent body and he had campaigned for office on the issue."[63] This structure was also later used in the Netherlands in 2006, in which the citizens' assembly recommended changing the system, but it was met with even less success when the ruling government rejected it. A similar case of political intervention transpired in Iceland's Constitutional CA: While the Parliament did not take up the suggestions for a variety of political reasons, the draft that the assembly came up with "enjoyed reasonably strong support."[64] The ultimate impact on policy has been mixed; more often, policy is the result of political maneuvering rather than the CA itself.

One of the most pressing and popular CA topics, and with great relevance for CRISPR and risk governance, has been climate change.[65] Examples are the French Citizens' Assembly on Climate (2019), the German Citizens' Conference on Climate Policy (2015–16), the Irish Citizens' Assembly (2016–18), the UK Assembly on Combating Climate Change (2020), the Danish Citizens' Parliament on Climate (2021), the Austrian Climate Citizens' Assembly (2022), the Belgian Citizens' Assembly on Climate (2023), and most recently, the Netherlands Citizens' Assembly on Climate (2025). In June 2019, the UK Parliament announced a citizens' assembly on climate change in its efforts to "adopt a target of reaching net zero emissions by 2050," in part responding to the growing public pressure from Extinction Rebellion (a global environmental movement) and other efforts in the United Kingdom, European Union, and Australia to address the climate crisis.[66]

In 2018, France's Yellow Vest Movement arose to protest "perceived unfairness of government policies—environmental in particular," including an increased fuel tax.[67] Partially in response to this, in 2019, President Emmanuel Macron convened the French Citizens' Convention for Climate, a CA of 150 randomly selected French citizens. In June 2020, it released 149 recommendations for how the government should combat climate change, including "banning the construction of new airports, taxing corporate dividends to pay for green programs, and revising the constitution to make damaging the environment a crime."[68] The CA also called for reducing greenhouse gas emissions by 40 percent by 2030 compared to 1990. In 2021, France's legislature passed a large climate change law that aimed to curb gas emissions and cut waste by setting quotas on short-distance flights and requiring vegetarian school meals. However, "the High Council on Climate, an independent body, warned . . . that these efforts were 'insufficiently' in line with its pledge to curb greenhouse gas emissions by 40 percent by 2030," as was called for by the CCC. Rather, it was reported, "a whopping 53% of the recommendations made by the recent French Citizens' Convention for Climate were rejected by the French government while another 37% were watered down or otherwise modified."[69]

In another French case, the French Citizens' Assembly on Assisted Dying was convened over the course of several weekends in late 2022 and early 2023 to seek public input on medical aid in dying (MAID) legislation. The assembly members voted on certain issues. To the question, "Should access to active assistance in dying be opened?" the majority answered "yes." Assembly deliberation time focused on detailed considerations of the process, timing, and overall details of the requests for MAID, as well as medical and psychological support and assessment of capacity.[70] The final report was submitted to President Macron on April 3, 2023, who subsequently announced a bill in April 2024 legalizing MAID in France to be debated by the National Assembly on May 27 of that year.[71] However, the debate was interrupted when Macron called for snap elections in June 2024.[72] The bill was reintroduced in May 2025, which the French National Assembly (the lower house of the legislature) passed, and the legislation is expected to pass in the Senate.[73] This case illustrates a rare but welcome alignment between popular opinion (90 percent of the French public support legalizing MAID), the recommendations of a CA, and government responsiveness in lawmaking.

The precedent for CAs has led to calls for convening them to address gene editing. In 2020, John Dryzek and colleagues endorsed the creation of a "Global Citizens Assembly" composed of at least one hundred people selected from throughout the world to meet for a week or more, hear from experts and advocates, deliberate in smaller groups, and produce a report summarizing their recommendations. The Australian Citizens' Jury on Genome Editing served as a platform for this global assembly, which would explore the question of what regulatory principles should govern genome editing technologies. For Dryzek and colleagues, the assembly design is the preferred institutional platform for engagement for three "essential reasons." It (1) creates legitimacy through increased "public confidence"; (2) helps to narrow the gap between "expertise and public values" by solving information problems; and (3) solicits the feedback and judgments of lay individuals who are better situated than ethicists at conveying "public interest."[74]

Design Choices and Trade-Offs

Having explored models of past and current DMPs, as well as precedents for future ones, I next present a more detailed analysis of DMP design choices and philosophical trade-offs. The list that follows in Table 5.2 highlights the most pertinent procedural details but is not at all meant to be exhaustive. Further, each decision entails trade-offs: Opting to prioritize one element of the DMP usually comes at the expense of a different value or priority. As a result, I offer some recommendations while emphasizing that different DMPs may call for different directions and experimentation to prevent stagnation or adapt to changing populations and topics, for example. Thus, rather than arguing that I have discovered the perfect DMP formula, I present empirically grounded discussions that illuminate institutional modifications intended to incorporate the wisdom of diverse groups into regulatory spaces for CRISPR rulemaking. Importantly, DMPs thrive best in a superstructure of DMPs; as such, this is only the blueprint for where to start building.

The design choices and trade-offs are organized into the following categories: deliberation design, facilitation considerations, and decision-making details. For deliberation design, the important decisions are the participant selection mechanism, the number of participants (the size of the DMP), the recruitment strategy for selecting participants, the role of participant consent, participant payment, the frequency of DMP meetings, whether to hold meetings in person or online, whether DMPs should be temporary or permanent parts of the convening institutions, who the convening institution should be, when the DMP should be convened in the course of rulemaking, and the degree to which the DMP should have control over its agenda. Next, there are four main facilitation considerations: the role of experts, the role of special interest groups, the role of moderators, and the role of wider public engagement. The third category is the final decision-making details: What, ultimately, should be the outcome of the deliberation? Should the DMP's outcome have authority (i.e., should it have an impact on resulting regulation)? Finally, to what extent should the DMP report be available to the public or be transparent?

Deliberation Design

Selection Mechanism: How should the DMP participants be selected? The participant selection mechanism has deep and abiding implications for the DMP's procedural legitimacy and substantive output.[75] Who is selected directly translates to who has authority, thus the procedure and methodology for that matters a great deal. Thus far, I have referred to the selection as "random," but what does this actually mean, and how does it work in practice? There are generally two options: The first is "simple random selection," in which participants are selected by lottery from a large group. The second is "stratified sampling," in which a predetermined number of participants are randomly selected within designated categories or selection

TABLE 5.2 Deliberative Minipublics Design Options

Category	*Design Question*	*Option A*	*Option B*
Deliberation design	Selection mechanism	Fully randomized	Stratified sampling
	DMP size	Large	Small
	Recruitment strategy	One-stage	Two-stage
	Participant consent	Required	Optional
	Compensation and benefits	Minimal payment	Payment, childcare, travel costs, food, lodging, etc.
	Frequency of meetings	Short	Repeat
	Online or in-person	Online	In-person
	Institutional relationship	Temporary	Permanent
	Convening institution	Agency	Outside organization/ civil society
	Timing in rulemaking process	Beginning/During	End
	Agenda-setting power	None	Some/complete
Facilitation considerations	Expert role	Advisors/ consultants	Deliberators (i.e., a hybrid forum)
	Special interest groups	Expert role	Deliberators
	Moderator role	Passive	Active
	Public role	None	Consultation
Decision-making details	Deliberation outcome	Report with majority views and disagreement detailed	Consensus statement
	Ruling impact	Guidance	Authority
	Transparency	Transparent	Confidential

criteria. Simple random selection often uses descriptive (social/demographic) representation, whereas stratified sampling can better utilize discursive (viewpoint/attitudes) representation. The trade-offs here primarily implicate modes of representativeness or diversity. I argue that a DMP on gene editing within a regulatory agency should use stratified sampling to attain discursive representation of people's attitudes toward and experiences of genetic illnesses and technologies.

First, descriptive representation, also referred to as "mirror" representation, is the idea that the representative group of people resembles the public's demographic. As Hanna Pitkin shows, this form of representation as "standing for" the public often underlies the idea of "proportional representation," wherein the legislature or the representative institution has the same, or at least similar, percentages of the population on the outside according to descriptive characteristics such as race, gender, and socioeconomic status.[76] If this is the desirable form of representation, as Archon Fung argues, "randomly selecting participants from among the general population is the best guarantee of descriptive representativeness."[77] Descriptive representation can also be attained through stratified sampling, wherein one selects certain percentages of people from identity groups. The original Danish consensus conferences were built on a similar model—random selection was used to invite a broad swath of the public, and the percentages were then narrowed to reflect the broader Danish public by demographic groups.

Descriptive representation is justified because those with similar identities often hold similar experiences and thus what Carolyn Hendriks refers to as "social perspectives."[78] As such, what it means to be a representative in a DMP—say, of a racial minority—would be someone who is authorized to speak on their behalf. This presumes that their views are at least prevalent in that group. For example, the forty-five participants in the Mali CJ on GMOs were selected from a pool of farmers (269 total) in all the districts of the Sikasso region.[79] The Irish Citizens' Assembly (2016–18) was also stratified based on gender, age, socioeconomic status, and place of residency. Further, this method may be a fairer democratic procedure insofar as everyone has an equal chance to hold power without worries that the selection criteria are being manipulated by those who aim to design a DMP to support a certain outcome.[80] This is reflected in the idea of sortition systems as civic "lotteries."[81]

The main concern with using descriptive representation is that it has the potential to become tokenistic and unproductive. There are validity problems in assuming that there is a single Indigenous view of genetics research or a single disability stance on gene editing, as every group has a diversity of views and experiences. Assuming uniformity of group perspective denies the important role of individual variation, including life experiences and intersectional identities. This can then silence voices because of overreliance on what the person looks like as opposed to what they may think. Descriptive representation, for this reason, is sometimes referred to as "symbolic" representation because the representation is understood to stand in for others. As Volkan Gül writes, "Relying on descriptive representation too much carries the risk of essentialism," which Jane Mansbridge defines as "the assumption that members of certain groups have an essential identity that all members of that group share and of which no others can partake."[82] Importantly, descriptive representation often serves as a proxy for other things, such as lived experiences of racism or personal experiences of medical mismanagement and exploitation.[83] Thus, if we are concerned specifically with those experiences or attitudes, then we should select directly for them, rather than through imperfect and potentially damaging alternative means.[84]

The second formulation of representation, then, discursive representation, refers to the distribution of participants across a position, or in a "discourse map." As John Dryzek and Simon Niemeyer write, discursive representation refers to assigning representative standing based on "a set of categories and concepts embodying specific assumptions, judgments, contentions, dispositions, and capabilities."[85] And as Carole-Anne Sénit explains, "It is not the types of individuals that need to be represented, but instead the discourses that frame perspectives and positions on complex issues."[86] This mode of representation is also referred to, or at least is closely related to, the idea of a "substantive" and "principal agent goal" of representation.[87]

By selecting stratified sampling for descriptive representation, the DMP will have a greater diversity of views. If the goal of DMPs generally is to encapsulate a certain view of the world that is missing in majoritarian or more homogeneous systems, like a technocratic one, then stratified sampling can expose researchers and regulators to views absent in groups with power.[88] In British Columbia's Biobank Deliberation in 2007, Holly Longstaff and Michael Burgess argued that it was important to not merely replicate majoritarian rule through proportional representation but also to select for diversity using stratified sampling. Longstaff and Burgess used "demographic stratification and recruitment from health regions as proxies to enhance diversity of experience and background on the explicit assumption that this would lead to diversity of views in the final sample."[89] They argued that selecting for particular groups proved necessary for inclusivity, for a "diverse sample that includes such citizens is more likely to avoid capture by like-minded majorities. . . . Such participants can also be encouraged to introduce a diversity of narratives, materials, and discursive styles in addition to a wide range of views."[90] Similarly, participants in the 2020 UK Climate Assembly were selected for a diversity of views on climate change.

The Australian CJ on Genome Editing in 2021 also used stratified sampling and discursive representation: "In the case of descriptive criteria, the quotas were set to reflect the distribution of each category within the Australian population, with a different approach taken for Discursive Representation. Selection also sought to achieve distribution across several secondary categories (religion, political spectrum), as well as excluding limiting the number of participants who were either trained or worked in areas related to genomic research."[91] Participants were randomly selected from a range of views that conveyed greater and lesser degrees of skepticism and optimism toward the potential of science to better our existence. The position descriptions ranged from "scientific progress," "principled concern," "agnostic," and "profound concern," and the report presents a scatter plot to ensure equal representation in all but "agnostic" views, which were in the smallest minority.[92]

One of the main critiques of DMPs is that they are either not representative enough or representative in the wrong way, which has implications for democratic legitimacy. As David Ryfe writes, "Random selection may ensure inclusivity, but it does not grant representativeness."[93] Another critique of DMPs in this vein is that even if they start out as representative, they often diverge from the views of the public because the deliberation itself leads to viewpoint change. In other words, because the goal of deliberation is to create a semblance of the outside world, engaging in a process wherein considered judgment is formed separately from the outside world necessarily decreases its representativeness.

However, the goal of the DMP is to develop judgments, not necessarily to parrot ones from the wider public. The DMP should serve as a proxy for what the people *would* think, given the proper resources and conditions. The aim, then, is *not* necessarily to serve as an ideal representative body, but rather to attain greater diversity on a range of views and ideas on an issue. Further, as we saw in chapter 3, including a DMP selected for diversity within a regulatory agency increases the *representativeness* of that agency writ large when it suffers a significant deficit of representativeness and is not built on that philosophical foundation in the first place. Lastly, diversity itself is an important goal in democratic accountability and representation of views, leading to greater representativeness even if not perfect representation.

Thus, some of these concerns are also resolved by pulling back on the goal of representativeness and aiming instead to maximize diversity. For example, "diversity—and not representativeness—was the central principle governing the G1000" wherein "the diversity of social characteristics and plurality of initial points of view in the larger society are substantially present."[94] The Brussels Regional Parliament organized a citizens panel in 2017 called "Make Your Brussels—Mobility" that was tasked with formulating recommendations related to mobility (e.g., pedestrian, travel using cars, public transportation, or bicycles). The conveners decided to diversify participants according to "mobility behavior."[95] And in 2016, "participants in the French mini-public on health care and big data were recruited, in part, based on their usage of online public service platforms."[96]

Another concern in stratified sampling is the potential for manipulation by those in charge to choreograph a particular outcome. While this is a ubiquitous concern through DMP literature, the threat of it can be mitigated, as noted in the World Health Organization's 2024 "Guide to Mini-Publics," by choosing reputable companies, of which there are increasingly many, that perform the selection work for the convening institution.[97] Further, the promise of DMPs being able to mitigate the already semi-captured institutions of the regulatory space in the first place outweighs this concern.

For a gene editing DMP within a regulatory agency, the aim then should be to select for viewpoint diversity modeled on the Australian Citizens' Jury on Genome Editing, wherein attitudes toward gene editing (such as whether it is an important pathway to pursue, for whom, and under what conditions) are used as categories for selection. In gene editing, it may not be entirely clear if a particular demographic group has a specific perception of gene editing, like Black patients do with sickle cell disease interventions or disabled individuals do when it comes to altering future generations to eliminate their disability. However, identity *does* play an important role in judgment formation and experiential wisdom. Thus, the second step is to select for demographic diversity or to weight traditionally underrepresented groups, particularly around disability, race, and Indigenous status. There may even be good reasons at times to *over*-sample from these groups to better represent their views and not have them dominated by the majority groups, as was the case in a public deliberation on salmon genetic modifications that took place in Canada, where First Nations had greater representation.[98] Selecting for diversity of viewpoint, and within each difference group weighting underrepresented groups' participation, balances competing claims of representation and diversity while also preventing tokenism and presumed identity-based attitudes.

Weighting the representation of traditionally underrepresented groups is also important in combating problems that arise in majoritarian systems, particularly the tyranny of the majority, wherein minority rights and views are routinely and systematically subordinated to the majority.[99] One way this is achieved is by weighting underrepresented groups, which improves on the deeply nonrepresentative status quo of the outside world, and particularly, as seen in chapter 3, within regulatory agencies. DMPs, in turn, are less likely to face concerns about the tyranny of the majority than outside institutional dynamics, because it is not necessarily clear what the majority viewpoint or stance will be prior to deliberation. Moreover, the deliberation itself affords minority groups the opportunity to voice opposition throughout the process and to write their disagreements into the final report (more on this later in the chapter). In addition to combating tyranny of majority worries, DMPs have been shown to be effective not only at avoiding a takeover by extremists but also, in some ways, at combating the problem of extremism generally.

Discourse mapping to attain diversity of views toward the promise and peril of gene editing also reduces the risk of extremist views taking over the DMP. Within the CRISPR space, such extremist views may include neo-Nazis who believe that we should edit all embryos to proliferate the Aryan race, transhumanists who hold that every person is entitled to alter their humanity to transcend the human species boundaries, or techno-pessimists who warn that all technology is existentially threatening to humanity ways.[100] Selecting for discursive views within a "reasonable" range of disagreement reduces the chance of extremist views being represented or, at the very least, constituting a majority. The selection mechanism allows for a diversity of viewpoints in ways that, for example, town hall or online chatrooms do not attain because of the proliferation of the most passionate and extreme voices that can overtake the conversation and silo their views even further.[101]

Further, even if extremist voices had substantial presence within a DMP, DMPs are able to bring together people from different extremes in a productive and depolarizing setting. In their 2024 book *Deliberative Democracy for Diabolical Times: Confronting Populism, Extremism, Denial, and Authoritarianism*, Andrée Bächtiger and John Dryzek write, "Bringing people from different sides of deep conflicts in small-scale forums can lead to an increase in mutual understanding and recognition (if not necessarily approval) of the identity of the other side."[102] They argue that deliberation can bring together participants previously positioned "across deep divides," such as "former leftist guerillas and right-wing paramilitaries in Colombia, Bosniaks and Serbs in Bosnia, poor people and police in favelas in Brazil."[103] Deliberation can do this by affording participants the space and opportunity, "often in the form of personal stories" to be able to recognize and learn from their adversaries' narratives.[104] As Bächtiger and Dryzek write, "The lesson from micro forums in deeply divisive and polarized contexts is that positions can be moderated and recognition of different others advanced if extremists participate in deliberation, and especially if that deliberation involves people from the other side of a deep political conflict."[105] In polarized societies, "if extremes can be induced to participate, they can moderate their positions."[106]

One of the reasons for this is the role of moderation (as will be discussed more later), which allows for equal opportunity for all to speak and creates dynamics that emphasize civil discourse and compassionate listening. Further, the process of learning from experts and writing reports allows for a more robust examination of issues rather than further entrenchment in extremist silos. Bächtiger and Dryzek discuss this in the context of engaging "deniers," such as on climate change and vaccine safety. They argue there are plenty of "vaccine skeptics" or "vaccine hesitant" individuals who can "be drawn into more productive engagements via the other discourses they inhabit," such as concern for children's welfare or fear of adverse reactions. Even if denial is part of their identity, such as by tying it to government corruption and capture, Bächtiger and Dryzek offer that "identity is not primordial and fixed."[107] Nevertheless, there may indeed be times when "lay citizen forums generally do well in promoting reflection on the merits of different arguments and considerations, though whether that reflection can encompass hardcore deniers in a forum is a different question, for which we are aware of no evidence."[108]

Further, as was seen in chapter 4 on epistemic democracy, democratic mechanisms may still be the best modes by which to achieve greater understanding in the long run, both due to the selection mechanisms and moderation, and thus the risk of extremists participating in DMPs may be outweighed by the risks of barring them from engaging with experts and those who may disagree with them and have alternative narratives to offer. Ultimately, Bächtiger and Dryzek conclude, there will be limits to the deliberative approach in its ability to reach deniers and other kinds of extremists, but deliberative forums may be the most productive path forward in some polarized environments because they can be designed to prevent the forum from collapsing into divisive extremism: "We do not claim that these measures will be sufficient in overcoming the pernicious effects of denial—only that they can help."[109]

DMP Size: Stratified sampling can be successful with either a larger group (several hundred people) or a smaller (twelve to twenty-four people) DMP. Smaller DMPs can facilitate more in-depth deliberation through narrative telling and greater listening time. In addition, they may be more adaptive to a convening institution's limited resources and serve as a good starting point for experimental regulatory DMPs. Small groups also have drawbacks, such as reinforcing the status quo by moving "collective opinion in the direction of the preexisting views of the majority," perhaps because they may lack adequate diversity to encourage new forms of thinking.[110] However, in a gene editing DMP, the status quo is not necessarily apparent. Most likely it is to limit the use of genome editing technologies, but significant questions remain that lack clear consensus about the moral limits of these technologies outside of embryo modifications.

Large DMPs are often more representative but may lack cohesiveness. As mentioned earlier, part of the reason that the Deliberative Polls are so large is that they aim to reflect the wider population. As Ian O'Flynn writes, "Deliberative Polling seeks to be statistically representative—to represent the population in microcosm."[111] And, as Hélène Landemore argues, this is advantageous because it "is conceptually the most parsimonious way of ensuring the presence of a key ingredient of successful deliberation-as-problem-solving, namely, a diversity of ways of interpreting the world and addressing problems in it."[112] Thus, a large DMP would potentially yield a more discursively representative body.[113] Further, a larger DMP may seem more legitimate, particularly in a system that has never used a DMP, such as the United States's regulatory process, because trusting a smaller group to make complex decisions seems novel and risky.

Larger sizes, however, have downsides. Consider that the largest of the DMPs, such as the Belgium G1000, which had over seven hundred participants, usually break into small groups to deliberate. This allows individuals to share grievances and conduct narrative or storytelling.[114] However, this practice also dilutes the benefits of greater diversity by limiting the participants' exposure to whomever else happens to be within their group.[115]

Recruitment Strategy: While the selection mechanism and DMP size are crucial to determining the kind of group one wants, the recruitment strategy is just as important for ensuring the DMP convenes. Given the problems of preselection (many individuals do not have the time, resources, or energy to participate or lack the organizational power of a corporation, such as in chapter 3's discussion of the comment process), how an institution recruits participants can contribute to, or eliminate, certain barriers to entry. According to Nicol Curato and colleagues, "the most common approach is a two-stage process."[116] In the first stage, invitations are sent by "letter, email, telephone or some mix of these approaches" to a large number of people, or "at least 100 times larger than the desired number of actual participants in the DMP."[117] In the case of the 150-member French Citizens' Climate Assembly, 255,000 people were initially contacted.[118]

The second stage entails the stratification selection process, followed by making sure those invited "actually show up."[119] This often involves a "multi-platform approach, using several channels such as mail, email and even phone calls or messaging."[120] The media may become involved, which can raise enthusiasm. The process can be lengthy. In the case of the British Columbia Citizens' Assembly in 2004, recruitment through this method took months.[121]

Single-phase approaches are also possible, primarily for nonstratified selection, as in the case of Deliberative Polls as well as many CJs and the Irish Citizens' Assembly. The latter assembly "cold-called door to door until" they had "filled the requisite quotas."[122] This has the advantage of simplicity, while also running the risk of excluding certain populations or underrecruiting "difficult to reach" groups.[123]

One concern raised here is that the people reached will misrepresent their views in the interest of participating in the DMP. While this is indeed a worry generally in selecting groups of people, I do not anticipate this being a significant problem in a regulatory DMP because it is on a novel topic, and recruited participants will have limited information about the selection criteria. Thus, they should be less likely to be able to game the system or figure out how to maneuver their way onto the DMP.

Participant Consent: Should participants have the option to participate or be compelled, such as through law? This question is not unlike the long debate around whether voting should be optional or mandatory, with some nation-states levying fines for nonvoters. As we begin to integrate

DMPs into an American system, voluntary participation and consent are crucial, particularly given that many people may not be able to take off work in ways that jury duty permits, and we should minimize barriers to entry. However, one of the original conceptions of deliberation as a civic duty comes from Bruce Ackerman and James Fishkin, who called for a "Deliberation Day," when everyone would be given time off from work and school to participate in deliberation events throughout the country.[124] Similarly, many model deliberation on the jury system, where participation is required by law and seen as an important civic duty.[125] This would contribute further to the goal of integrating DMPs into a wider system of deliberation.

Compensation and Benefits: As with recruitment, participation compensation reduces barriers to entry. Other structural support systems that reduce barriers to entry include reimbursement for travel, covering childcare and eldercare costs, and assistance with other logistical hurdles. Ian O'Flynn similarly argues that low or no payment skews the DMP to be composed of participants with higher-than-average education. Participant remuneration allows those who were previously prevented from participating in governance due to problems of preselection (e.g., inability to attend town hall meetings, lack of internet connection, or working three jobs and being unable to commit the mental energy to the topic) to participate. It also fairly compensates individuals for their time and energy. Regulators and government officials get paid, and so too should citizens selected to deliberate on rulemaking. One may worry that payment can be tantamount to undue influence or even coercion, as some worry happens in other contexts, such as clinical research trials, gestational surrogacy, and plasma "donation," for example.[126] However, in this case, the advantages of fair compensation and incentivization for disadvantaged populations to weigh in on regulation outweigh the concern of individual instances of overt pressure.

However, compensation is not yet the norm. In a 2020 report on 172 DMP cases, researchers found that participants were compensated "in one way or another 57% of the time."[127] In 44 percent of the cases, remuneration took the form of payment, and in a smaller number of cases, transportation and other expenses were reimbursed. In DPs, participants are paid $75 to $200 USD per day but are not given extra reimbursements.[128] The cases without remuneration tended to be local, so costs may not have been as great a barrier to participation. In contrast, during the French Citizens' Convention for Climate (2019–20), participants were given a daily allowance of 84 euros; lower-income earners were given an extra benefit on top of that using a progressive model. Travel and accommodations were covered up to 110 euros, and meals and childcare up to 18 euros per hour were covered.[129] However, even with high levels of compensation and payment, some preselection inequalities will remain because DMP participation is not mandatory, nor is it commonplace or well understood by others (such as employers).

Frequency of Meetings: The number of days and repeated meetings of the DMP can range quite a bit. For example, while DPs are a one-time opportunity, some CJs and CAs meet multiple times over the course of many months. For a new and emerging technology, where there is great uncertainty and obscurity, taking the time to carefully consider risk and evidence and to engage with relevant stakeholders and participants is preferable for decision-making and judgment formation, particularly when it occurs within a novel institution.[130]

One concern, however, with repeated meetings is the risk of capture of the members of the DMP. Recall that one of the goals of DMPs is to offset regulatory capture by checking for conflicts of interest and requiring greater transparency. The concern around participant capture, then, is that repeated meetings could expose participants to potential influence by moneyed corporations. Temporality could mitigate this risk.

Another consideration here is the finite resources available to facilitate small group sessions, expert-interviews, and report writing. For example, many DMPs hold a plenary session and small group breakouts, requiring not only sufficient time but also space unless the DMP is held online (see next section). Further, the amount of time required to explain, for example, genetic technologies and concepts can vary. The most important point, Curato and colleagues argue here, is to ensure that participants do not feel rushed and that there is flexibility embedded within the design, including opportunities to "take a break, get fresh air, engage in informal chats with fellow participants and deliberate from within. . . . At a bare minimum, participants must not feel alienated from the process because of its speed or format."[131]

Online or In-Person Meetings: As in other contexts, there are trade-offs to whether the deliberation takes place online or in person. Before the COVID-19 pandemic, most DMPs were held in person,[132] but due to DP experiments as well as the necessity of online spaces during times of disease outbreak, we are seeing more and more evidence to suggest that online is as good as, if not preferable to, in-person, such as in the 2024 Taiwan AI

DP.[133] Online meetings are also significantly less expensive, so in a time of unprecedented regulatory budgetary cuts, they would be the more feasible option.

Institutional Relationship: While most DMPs are a one-time event (they are convened for a particular question or purpose and then disperse), some may serve a more permanent institutional mechanism. For example, the Ostbelgien Model (referred to in German as "Permanenter Bürgerdialog," meaning Permanent Citizens' Dialogue) is a citizens' council that serves alongside the legislative and executive branches in its namesake region of Ostbelgien ("Eastbelgium"). Twenty-four members are selected at random to serve for eighteen months, with a third of the members being replaced every six months. Their job is to oversee the questions posed to, the organization and implementation of, and political follow-up to the CA in the region.[134] Second, the "Observatorio de la Ciudad" (the City Observatory), which was implemented in Madrid in 2019, uses stratified sampling to convene about fifty people and whose job is to analyze and "pass citizen proposals and suggest them for referenda."[135]

The advantages of permanence include greater learning and experiential wisdom. Consider that regulators are invaluable themselves precisely because of their expertise and longevity in regulating certain industries. Further, through practice and engagement, the deliberators grow to know each other better, trust their deliberative process and each other, and are better able to learn from others.

However, as with regulators, members of a permanent DMP would also be exposed more to the risk of capture, or "elite manipulation."[136] Given that one of the purposes of DMPs is to reduce the problem of regulatory capture, the risk that they themselves may become captured is particularly concerning.[137] Further, this could lead to institutional bias for the status quo, something that is potentially good but also potentially deleterious. Gene editing comprises something new, but as we've seen, it is built on the foundations of decades-long moral considerations surrounding genetics and medicine writ large. A permanent or longer-standing group may be able to connect these threads more deftly, but a temporary group may be able to do so as well.

Convening Institution: While the convening agencies in the case of gene editing would be the Food and Drug Administration (FDA), the US Department of Agriculture (USDA), and the Environmental Protection Agency (EPA), as noted, they should outsource the job of organizing and selecting DMP participants to organizations experienced in this regard.[138] One reason for this is that whoever selects participants plays a crucial role in ensuring that the DMP is as free from bias and capture as much as possible. Allowing agencies to select whom to hear from could potentially corrupt the results—for example, by curating outcomes that are politically advantageous—and thus exacerbate problems of capture.[139] Partnerships with private organizations such as Healthy Democracy, which specialize in selection mechanisms, organization, and the facilitation of DMPs, are thus prudent paths forward. For example, the British Columbia Biobank Deliberation was a collaborative research effort between scientists and social scientists at the University of British Columbia who hired a professional firm to help organize and recruit participants.[140]

Timing in the Rulemaking Process: In a nonregulatory setting, a DMP may not necessarily be tied to a particular legal procedure because the DMP does not have authority and instead provides guidance. Thus, convening the DMP is often voluntary and relies more on public support. A regulatory agency, on the other hand, is bound by the Administrative Procedure Act (APA), as discussed in chapter 3, so DMP inclusion would need to be in compliance with legal procedures. Namely, the APA would need to be modified to convene DMPs as part of rulemaking, ideally prior to the draft rulemaking. As of now, the APA's main opportunity for public inclusion, commenting, emerges too late in the regulatory process to garner adequate public participation. Part of the reason is the problem of preselection and the lack of deliberative opportunities for substantive judgment formation. In addition, the comments occur after the rule is already drafted, and the agencies most often still promulgate it unless there are *major* objections.[141]

Prior to the notice-and-comment period, the "rule development stage" is when agencies compile "preliminary policy initiatives." Researchers have found that earlier comments at this stage are more likely to influence the final rule in part because they are more often from the stakeholders and interest groups. As Keith Naughton and colleagues write:

> Public commenters who participate early in the rulemaking process are able to help frame the scope of the facts considered, the policy problem, and the preliminary policy solutions. . . . Issue definition occurs relatively early in the regulatory process, and therefore early commenters, who are, by definition, participating in the rule development stage, are well situated to assist in the creation and formalization of the political debate surrounding any particular regulation. . . . The end result . . . is that the whole frame-work of

the rule may be preset toward policy outcomes that are beneficial to these "early birds."[142]

Thus, to have greater impact on the outcome, a regulatory DMP should take place earlier in the rulemaking process. Part of the rationale is that it gives the participants greater control over the "agenda," a crucial role in influencing policy.[143] Indeed, providing space for citizens to identify and define problems and have an institution that is directly accountable to their prioritization is crucial for democratic accountability.[144]

Agenda Control: To what extent should a DMP have control over the agenda? For example, should DMPs be prescribed a specific list of questions or deliberation goals, or should they be able to create their own priorities? While in the regulatory setting, it is necessary to set the DMP's topic, allowing flexibility for the DMP to introduce new problems that arise in deliberation is also important for the democratic process and its substantive outcomes. Agenda-setting of the DMP can be done by building flexible DMP schedules, such as by using "open-format measures."[145]

In democratic theory, "agenda setting power" refers to determining the priorities of a law or policy.[146] For Edana Beauvais, this is intimately tied to equal distribution of power: "Democratic processes always *begin* with inclusion, since collective agendas and decisions are only 'democratic' to the degree that those affected by collective outcomes are empowered to influence them."[147] William Smith similarly argues that agenda-setting becomes particularly important when the questions pertain to "agendas or regulatory regimes, such as climate change, biodiversity protection, trade and finance, refugees and migration, or security and defense."[148] For Smith, inserting citizens in moments where they can have more conversations on these downstream matters enhances their ability to contribute.[149] This is similar to my argument in support of using a DMP in the International Commission on the Clinical Use of Human Genome Editing to aid the work of the commission. In the space of gene editing regulation, helping to shape the value-based ideas of risk governance should occur early and often throughout the process.[150]

Sometimes, however, control over the agenda may detract from the work of the DMP. In the cases of CIRs and the British Columbia Citizens' Assembly, for example, absolute control over the agenda such that they deviated from discussing a new electoral system would not have been productive. In other contexts, however, the point is very much to create an agenda, such as in the Australian Citizens' Parliament, where the aim of the DMP was to collect "the people's" legislative priorities.[151] Similarly, the Ostbelgien model was created to give the people control over the agenda of their CAs. There is also a hybrid approach: In 2012, the Canadian National Research Council hosted an "advanced biofuels" event wherein it gave participants a "combined master list of 71 concerns and 51 benefits as part of the agenda setting exercise at the end of the first weekend."[152]

In the case of a regulatory DMP on gene editing, a combined approach would work well. While participants should focus on discussing gene editing and the use of CRISPR in certain cases, the prioritized questions within that discussion should be determined by them. The basis of the Australian Citizens' Jury on Genome Editing rested on a similar notion—that "agenda-setting should not be left only to scientists, medical practitioners, specialist government agencies or international committees."[153] However, it is important to note that a DMP on gene editing would necessarily involve scientists and medical practitioners, as well as others, as expert witnesses.

Facilitation Considerations

Expert Role: All DMP models require expert guidance and involvement, but most do not endorse experts deliberating with, or alongside, participants (except in hybrid models). For example, DPs circulate briefing books and open with panels of expert testimonies that aim to be balanced and thorough in providing evidence and points of disagreements.[154] The involvement of experts fulfills important goals of educating lay participants so that they may have sufficient technical knowledge to deliberate effectively. This must be done effectively, otherwise, as Mei-Fang Fan writes in the context of Indigenous deliberation, "inappropriate formatting of information, impregnable jargon, and thousand-page-long impact studies often leave discussions in the hands of a few technical and legal experts on both sides of the table."[155] Further, the aim of expert inclusion is to provide the participants with the opportunity to engage with their ideas, not to replicate them necessarily.

Thus, curating a diverse and accessible expert session is essential. In studying the 2016 to 2018 Irish Citizens' Assembly, researchers found that the quality and kind of communication mattered most: "It is not . . . information per se, but how it is communicated that matters for participants' uptake of climate measures."[156] For example, storytelling and good presentation skills, such as repetition and uniqueness, were able to gain more traction and engender more robust deliberation. This has become particularly important in contexts such as Indigenous deliberation about genetics research, where the goal of experts is not just to impart information but to do so in a way that can generate discussion and experience-sharing rather than epistemic domination.[157]

Finding the sweet spot between expert guidance that does not dominate or oppress minority views is an ongoing challenge for many researchers. Indeed, one of the main risks to the success of a deliberation is "elite framing," wherein the participants are influenced by experts such that they reiterate or reinforce the status quo.[158] Aubin Calvert and Mark Warren warn that these "framing effects" organize cognition by bundling claims into a framework so that any particular claim brings with it unreflective judgments about other claims.[159] As Mikko Leino and colleagues write, "Expertise may foreclose the prospects of democratic deliberation and violate key virtues, such as openness to different viewpoints and critical reflection."[160] On the other hand, a DMP without expert influence would diminish the potential for substantive outcomes and would certainly suffer the anti-capabilities critique (see chapter 4).

DMPs can mitigate expert framings and support participant competency through the recruitment of diverse experts and the careful presentation of balanced information.[161] During a DMP in Finland in 2021 on measures taken to contain the COVID-19 pandemic, researchers found that ensuring a diversity of experts on the same matter, such as different fields of expertise, was important for participants to fill "gaps in their knowledge."[162] And in an examination of three CJs on onshore wind farming in Scotland (2013–14), researchers found that the selection of "witnesses" (experts) mattered just as much as the selection of participants in ensuring diversity of representation.[163] Allowing expert witnesses to engage with deliberators in a "teaching-style" format was preferable to a model that felt like they were imparting knowledge and giving presentations of the issues.[164] Further, as Simone Chambers writes, "Many of the epistemic problems associated with framing can be mitigated in small-scale deliberation . . . [and] frames are themselves the object of deliberation and require justification."[165] Thus, the facilitated deliberation part can at times recover even with elite framings.

Special Interest Groups: As in the case of experts, the role of special interest groups in a DMP invites considerations of how to weigh potentially valuable perspectives with overt influence on the deliberations. Special interest groups refer to organizations that have a particular policy stance or goal they want to achieve. These may include patient advocacy groups that participate in FDA advisory meetings to discuss certain drug approvals and environmental organizations that petition the EPA to bolster, or eliminate, restrictions on pollution. In the deliberative democracy space, questions around special interest groups arise when it comes to whether they should have a "seat at the deliberation table." In other words, should we include patient advocacy groups as representatives of patient interest alongside other lay participants?

I argue that we should not because (a) special interest groups today often have ties to industry, which can corrupt their patient interest protections; (b) they are often imperfect representatives of patients, with different organizations representing different needs and demands of their patient communities; and (c) the purpose of the DMP is to include minority perspectives that are not necessarily well represented by these groups in the first place. However, special interest groups do hold specialized knowledge and expertise and thus can be important experts in educating participants. Thus, to eliminate both capture and the problem of loss of diverse views within these groups (since not all patients reflect the views of the advocacy group) while preserving their valuable role in regulatory decision-making, advocacy and other special interest groups should be considered expert witnesses who help to train the deliberators, but they should not actually deliberate among them.

Patient advocacy organizations have a long history of pushing for and accomplishing important successes for the patients they represent and the causes for which they care deeply. Even from its inception in the Progressive Era, the FDA was formed in large part in response to popular demands for public health (as seen in chapter 3), due to the dangers of drugs and food on the market. The 1938 Federal Food, Drug, and Cosmetic Act was also in large part a response to the outcry from the 1937 Elixir Sulfanilamide disaster, which was responsible for over a hundred deaths, mostly children, "who had received the antibiotic sulfanilamide dissolved in diethylene glycol, a sweet but highly poisonous solvent."[166] The FDA had identified that bottles were mislabeled (elixir sulfanilamide did not contain alcohol, as elixir would suggest), but it was unable to act because it lacked the authority over drug safety. The 1938 law then gave the FDA authority to regulate safety, but not yet to require certain standards of testing. Then, Dr. Francis Kelsey's blocking of thalidomide led the FDA to add the 1962 Kefauver-Harris Amendments to the Food, Drug, and Cosmetic Act, despite "vigorous industry opposition."[167]

Lewis Grossman notes that the 1960s and onward witnessed the "rise of the patient rights movement" that both overlapped with civil rights, gay rights, and women's rights movements in many ways and also pushed against the establishment that was seen as "gate-keep[ing]" their medicine.[168] For example, the women's health rights movement advocated for such rights as safe and effective birth control and empowered patient knowledge with the publication of *Our Bodies, Ourselves* as a pamphlet in 1970 and later a best-selling book in 1973. As Sharon Batt and coauthors write, "What bound the women's health movement together was an emphasis on not just protecting the public health, but empowering women to reject a paternalistic medical system while deploying a wide-ranging critical assessment of medical interventions, government regulation, and pharmaceutical company claims."[169]

In many ways, the AIDS advocates of the 1980s who followed the women's rights movement continued this tradition, but in this case, "their public advocacy focused on their needs as patients with that specific [fatal] disease."[170] AIDS organizations formed to "urge action for creating and approving drug treatments, and they demanded to be included in the process, particularly in the design of clinical trials."[171] Instead of viewing the scientific industry as the "enemy," they saw them as needed collaborators, recognizing the need for safe as well as quickly developed drugs. Sharon Batt and coauthors further note: "For both the women's health and AIDS movements, marginalized groups spoke authentically about personal experiences on behalf of a community to address bad practices and injustices in healthcare. They embraced science as a means of advancing knowledge but sought to change the rules of science to correct its male, heteronormative biases and professional exclusivity. Through effective advocacy, these groups significantly changed how their communities participated in and received treatment from the healthcare establishment."[172] Part of this, however, required not just an FDA that was responsive to the patients' movement and advocacy influence, but also an industry willing to develop the drugs and pharmaceuticals deemed important.

This has led to a now complicated story of industry-patient group collaborations and funding that continues today. Batt and colleagues show that beginning as early as the 1980s, many of the women's rights groups were approached by industry with funding opportunities. Some, particularly the "grassroots" ones, turned them down, but many accepted the money, launching decades of intertwined relationships between patient advocacy groups and industry that continue today, where "thousands of patient advocacy groups operate within the United States and studies estimate that at least two-thirds accept funds from the pharmaceutical industry."[173]

In a 2017 study published in the *New England Journal of Medicine* of 104 large patient advocacy organizations, researchers found that 83 percent received financial support from drug, device, and biotechnology companies.[174] Concerns of the "revolving door" are also relevant here. In a study published in 2023 in the *Journal of the American Medical Association*, researchers identified that 75 percent of the fifty highest-revenue patient advocacy organizations in the United States had "board members, senior paid staff, or executives with prior or current ties to the pharmaceutical and medical device industries."[175] Further, there are no standardized reporting practices for advocacy organizations to disclose their funding to the public or their patient population, creating transparency concerns.[176]

Industry alliances are morally and politically contentious. Some defend them by arguing for their necessity and utility in developing needed pharmaceuticals, or they contend that their influence on advocacy organizations' actions is "overstated."[177] Others fear that they erode the very "ethics and epistemic foundations of advocacy," leading to problems such as loss of independence, distortion of agendas, silencing of harms, high prices, and the erosion of public trust.[178] Patient organizations with industry funding run the risk of carrying out activities to lobby for certain treatments or medications that may not be in the best interests of the patients but for the company developing them.[179] Thus, just as in chapter 3's discussion of capture in regulatory institutions, here too concerns around corrupted influence of the advocacy's main interests (the patients) apply, as they weaken the very foundations of public health advocacy from which they originated. As Ray Moynihan and Lisa Bero write, regulators "should seek out and engage with genuinely independent patient and citizen voices, entirely free of the distorting impacts of industry funding, to help shape the future of health care."[180] To do so, I offer, a DMP should not include patient advocacy organizations as deliberators at the table representing their own interests (and potentially the industry's), but should consult them as experts.

In addition to the capture worry, individual patient advocacy organizations are also often imperfect representatives of patient populations given the vast disagreements and tensions that often emerge within groups. In the case of autism advocacy groups, for example, some of the conversation echoes that of the Deaf community (see chapter 2)—many, particularly parents and physicians of autistic children, advocate for such research on the causes of autism and early intervention strategies, while others, such as autistic adults, push back against this medical model of disability by asserting that research priorities should not necessarily focus on "curative" measures for the autism but other medical and supportive measures.[181] In a 2021 systematic review of studies reporting on priorities in the autism community, researchers found that among autism-interested parties (parents, doctors, therapists, etc.), priorities for research vary widely, reflecting disagreement about whether research should focus on "treatments and interventions for autism characteristics" or "interventions for physical and mental health."[182] The researchers also note that among the "stakeholders" interviewed in the studies examined, only 9 percent had autism, and the largest representative group was composed of family members (61 percent). Thus, the authors of the study conclude that there is a great "need to expand the diversity of stakeholder groups contributing to the research priority studies."[183] In particular, Autism Speaks, a parent-focused advocacy group, has received heavy criticism that it is not fully representative of the autism community and that its approach to "treat" autistic people has alienated many, widening "the division between traditional autism organizations founded by parents and caregivers and the growing autism rights/neurodiversity movement."[184]

One place where this tension plays out is in genetic research. Autism Speaks sponsors an autism spectrum disorder genomic database, MSSNG, that enrolls minors with autism and aims to facilitate research into the potential causal genetic factors of autism.[185] By 2019, it had sequenced ten thousand whole genomes of people with autism.[186] Such enrollment raises ethical considerations around the consent of minors with disabilities who "may develop a perspective on their condition that is different from or even at odds with the perspective that motivated their parents' enrollment decisions" and whether re-consent at adulthood is required.[187] It also sparks debates over research priorities and who gets to make decisions about autism research. In a 2020 examination of the pushback this project got on Twitter, researchers identified themes that "autistic people are missing from the conversation," and "attempts to develop a cure for autism are equivalent to attempts to eliminate people with autism."[188] Thus, autism should be understood as a diverse and multifaceted group with numerous interests and priorities, leading to more participatory involvement in decision-making.[189]

Further, because one of the important goals of DMPs is to uncover perspectives not readily found in groups that have other means of advocating for their views due to resources and organizational capacity, special interests should not occupy specific roles in the deliberation.[190] The Irish Citizens' Assembly reflected a similar logic: "Individuals who were members of advocacy groups on either side of the abortion debate were excluded on the basis that such groups would have the opportunity to make submissions and/or presentations to the Assembly."[191] This allowed for a more measured and facilitated conversation wherein individuals could form their own ideas rather than be unduly influenced by organizations. Further, if not at the deliberation table, special interest groups, like experts, can play important roles by supporting the convening of a DMP, as in the case of Extinction Rebellion, which has called for climate change CAs, and by representing the views of various rights-based groups like disability and rare disease populations.[192]

Moderator Role: Once participants are selected, present, and learn from experts, how then does deliberation proceed? As Curato and colleagues argue, "trained facilitators are crucial to the success of a DMP."[193] They play a strong role in setting the conditions for all voices to participate fully and can help reduce polarization between participants by increasing the encounters that people have with each other, which are often cited as the source of the more marginal and radicalized views in the first place.[194]

Given that one of the aims of deliberation in this case is to enhance its non-dominating outcomes, it is important to ensure that the DMP itself does not dominate more marginalized voices during deliberation. The role of the moderator, then, is to "spot inequalities in speech and implicit hierarchies that distort the process of free and equal communication."[195] For Nicole Doerr, the "truly democratic deliberation involving diverse groups depends on . . . [recognizing] the need to reconcile inequality and misunderstandings based on differences."[196] Equalizing political power requires more than just situating participants in a room; it also necessitates an environment conducive to people truly contributing to and producing collective judgment. Lynn Sanders warns that we should not automatically filter "those who are less likely to present their arguments in ways that we recognize as characteristically deliberative," such as women, racial minorities, Indigenous groups, and the socioeconomically disadvantaged.[197] Similarly, Curato and colleagues write that moderators should protect "speech cultures [that] depart from 'rationalist' forms of discourse that privilege dispassionate argumentation, logical coherence, and evidence-based claims as practiced in the most exclusive kinds of scholarly debates, parliamentary procedures, and judicial argumentation."[198] For Doerr, this role of the "translator" is similar to that of mediators of speech for the entire group, for they ensure that what John Rawls would have understood as "passionate" or "emotional" language takes its place within the hierarchy of speech power.[199] Because "deliberative democracy involves multiple sources of communication," it is essential to have moderators who are trained in acknowledging different sources as valid.[200] Given the technocratic and Enlightenment-based constructs of knowledge present within regulatory agencies already (see chapter 3), incorporating the previous chapter's goals of experiential wisdom and diverse life experiences requires this mediation.

One may push back against this role for the moderator by arguing that the moderator should instead remain impartial, and that the moderator's job is only to enforce the rules of deliberation.[201] Others critique the role of moderators when they act as a "referee and information gatekeeper," fueling the concern in the case of expert and stakeholders around framing effects and overt influence.[202] However, the danger of *not* managing dominating speech is far greater than the danger of gatekeeping. Indeed, a more active moderator is important in "bridge-building across political divides."[203]

Public Role: Should the DMP be required to engage the public, such as through crowdsourcing information or even through final approval of their recommendations? Should public consultation sessions take place? To what extent should the DMP participants choose a wider public engagement strategy? The benefits to broader public inclusion are that the DMP participants consider views not otherwise present, increasing their democraticity and accountability. On the other hand, these modes of knowledge and preference solicitation also suffer from the problems of preselection in who participates in aggregative measures (such as commenting).

There are various options for engagement that the wider public could have in participating in the decisions of the DMP, such as online voting, commenting, and conducting surveys.[204] The Icelandic Constitutional Assembly case used crowdsourcing online.[205] Similarly, in the Irish Citizens' Assembly, they received over "13,000 submissions online when it discussed abortion."[206] In the case of the regulatory process, the comment period at the end of the rule-drafting process could remain in place, meaning that there would be an opportunity for continued public involvement in conjunction with the DMP influence through an already existing participation mechanism. Thus, I do not suggest that the regulatory DMP use crowdsourcing methods. However, if the DMP determines, using its agenda-setting power, that polling or soliciting feedback from "the people" is important on certain questions, then that would be sufficient reason to do so.

Decision-Making Details

Deliberation Outcome: Most DMPs produce reports that include summary statements, minority viewpoints and main disagreements, outcomes of votes on key issues, and suggestions for next steps. Similarly, I argue that the creation of a proposal for rulemaking would be the most helpful to regulatory agencies when drafting their own rule to post on the registrar's website for commenting, as it provides succinct information on the findings of the DMP and shows the important role of disagreement and nuance within their deliberations. Further, in combating the problem of majoritarianism, which was one of the aims of the DMP in the first place (to prevent outside dominant narratives from taking over within a DMP and instead unearth important underrepresented and diverse perspectives), disagreements on *each* recommendation or matter of voting should be documented in the final report.

This is preferable to "consensus statements," such as used in consensus conferences. As discussed earlier, the goal of reaching consensus may be problematic because it can convey a forced decision rather than robust group agreement. In other words, consensus may not reveal any deeper considered judgment than plurality or majority would.

However, the deliberative democracy literature generally is split on the goal of consensus. Amy Gutmann and Dennis Thompson argue that it is a primary goal of deliberation.[207] Michael Fuerstein argues that though it may not be the central aim, it is an "essential byproduct."[208] Still others support a Rawlsian "overlapping consensus," or the idea that people may agree but use different frameworks, such as those from religion or secular ethics.[209] My worry, shared by others, including scholars of deliberation and social movements, is that consensus can also be a feature of a dominating system that aims more for social control.[210] This actively works against the benefits of diversity and pluralism, which are meant to challenge the status quo.[211] Consensus, then, may instead constitute ambiguously disguised forced agreement rather than present a convergence on the truth, leaving conflict unresolved or perhaps even exacerbated.[212]

"Consensus building," however, is an admirable goal insofar as it brings together divergent communities that may not otherwise speak to each other. As Martin Hébert articulates, "The aim is always to build consensus. . . . The challenge of reconciling all points of view and all interests increases as the constituency becomes larger. In this sense, voting is a practical expedient that creates its own framework for deliberation in regional and national politics."[213] In theorizing a model of "indigenous deliberation," Hébert maintains that consensus building may seem desirable, but it is also dangerous to assume that the members of a particular group, because they are united by one factor, will all hold the same viewpoint (as discussed earlier in this chapter in the section on representation and selection mechanisms). In reality, Hébert writes, there is great diversity and complexity within particular groups, such as "ecological groups, feminist groups, and human rights organizations."[214] Thus, consensus does not convey much information regarding "truth" or "knowledge" of a group.[215] As a result, governments and institutions would be wise to deploy various strategies over time that may build toward consensus but do not require it for conflict resolution.[216] As David Estlund and Hélène Landemore argue, although majority rule can serve as an artificial stopping point for deliberation—meaning that it does not necessarily reflect the full capacity of the deliberation's output—it can also be an important pragmatic tool given limited time for deliberation.[217]

Ruling Impact: To what extent should the DMP hold authority? Most DMPs convened by governments have served a consultative role, leading to minimal impacts on policy.[218] For example, as noted previously, a year after the 2020 French Citizens' Convention for Climate released its recommendations for how the government should tackle climate change, the legislature passed a large climate law that addressed only a small percentage of the proposals.[219] Even when the governing body declares a commitment to its findings, political pressures and tides can prevent it from enacting many of the recommendations. In the Icelandic Constitutional Assembly case, the citizens were charged with making recommendations to the new constitution, and the legislature sidestepped their legal obligations to consider them.[220]

However, there are also many examples of when a DMP did have a direct impact on the resulting policy.[221] In cases of participatory budgeting, elected institutions usually remain faithful to the resource allocation decisions of the body. In the British Columbia (2004) and Ontario (2007) Citizens' Assemblies, their output resulted in a "binding referendum."[222] In the Ostbelgien model, the elected assembly is supposed to respond to the reports and recommendations of the Citizens' Council. And in some cases, such as the Irish Constitutional Convention (2012–14) and the Irish Citizens' Assembly (2016–18), the governing officials felt public pressure to give more "weight to the DMP recommendations than the formal rules imply."[223] Thus, there are many reasons why a DMP may or may not have a political impact, including public pressure and political tides that can sway one way or another.

In the case of gene editing, I argue that DMPs should *not* serve only consultative roles for three main reasons. First, DMPs are substantively beneficial; they can offer valuable insights into the risk assessments and community histories implicated in a regulatory question. As this book has argued, the considered judgments of a diverse deliberative body have justice, epistemic, and trust-enhancing benefits.

Second, authorizing the DMP to rule will enhance the quality of the deliberators' engagement. Evidence shows that deliberators will more seriously engage in mutually respectful deliberations or good-faith negotiations when they believe their work has binding measures.[224] As Denis Thompson writes, "If participants believe that they have a stake in the outcome and will have to live with the decision and with their fellow decision makers, they may take the discussion more seriously and try harder to reach a decision that is mutually acceptable."[225] And, if there is little "possibility of making a real practical difference," then deliberation is more likely to attract what Shawn W. Rosenberg refers to as "hostility," which "abstract or hypothetical" deliberative bodies evoke.[226] This was confirmed by Kim Strandberg and Kimmo Grönlund in their study of online deliberations: "In order to engage citizens in time-consuming deliberation in which different policy options are reflected upon, governments must show that the outcomes of public deliberation really have an impact, not only in the deliberation but also on actual policies."[227]

Third, having authority may reduce the problem and potential of being used by politicians as popularity mechanisms—that is, for politicians using the findings of DMPs to serve their own political gains, such as by cherry-picking the results or designing them with a particular structure to achieve certain outcomes.[228]

Establishing a "deliberative system" will aid not only in streamlining the emergence of DMPs but also in creating a culture of accepting their findings. Through continued experimentation with DMPs, the hope is that they will gain both formalized legal authority and informal popular legitimacy.[229] In the case of the regulatory rulemaking process, the agencies should be committed to responding to the findings of the DMP, just as they are with the comments. And just as with the comment period, this procedural mechanism would be enforced through the courts.[230]

Transparency: Transparency has value in democratic theory writ large. In representative politics, transparency is necessary for holding our elected officials accountable, as it allows us to judge whether they are acting on our behalf as opposed to their own or private interests.[231] In regulatory agencies, transparency similarly aids in reducing the corrosive politics of capture and generally promotes having well-informed citizens.[232] The same requirements should apply then for DMPs, including declaring funding sources and selection mechanisms, as well as publishing briefing materials, agendas, and the final report.[233] For example, as noted previously in the discussion of Deliberative Polls, the media is often invited to witness and report on the deliberations, which may be televised or recorded.[234] In addition, "DMPs have websites that consolidate resources and data on the deliberative forum, which allows scrutiny from researchers, journalists and other interested parties."[235]

However, as John Parkinson points out, relying on the media "to bear the weight of responsibility for generating links between insiders and outsiders in public deliberation" carries its own risks, such as biased reporting and failure to properly portray the deliberation.[236] Further, there is reason to worry that greater public scrutiny may diminish the quality of the deliberation. As Daniel Naurin and Christine Reh note, privacy may reduce the likelihood of "posturing and demagoguery in defense of narrow (national) interests . . . and actors enjoy more leeway to communicate freely and to make the offers and generous concessions necessary for integrative agreements."[237] Thus, to counter the corrosive influence of private interests outweighing public ones, DMPs need to understand and report on relationships between experts and industry and be insulated from them in their own deliberations to avoid capture.

Regulatory DMP on CRISPR

Based on this discussion, my design recommendation is that a DMP on CRISPR should have twenty-four to fifty participants, selected randomly for a diversity of views (with oversampling for disability and other affected groups), and they should deliberate for one or two sessions, either online or in person, using briefing materials, expert testimony, and moderated deliberation. They should produce a report that highlights shared values and disagreements to provide nuanced judgments that help to inform the regulatory agency's actions.

What, then specifically, would a deliberative minipublic in a regulatory agency ultimately deliberate about when it comes to CRISPR-based regulations? Up to this point, we have seen in chapter 4 that diversity is instrumental in mitigating problems of power and enhancing knowledge-based outcomes, and in this chapter, we have painted a picture of the procedural mechanisms and compositions of this diverse group. I now want to explore the question: What regulatory content would they deliberate about? First, as we have seen, DMPs are not necessarily the best fit for every policy question (i.e., Swiss cheese holes). Another example would be drug approval itself: It is not clear that a DMP would be helpful to the FDA's work of interpreting and asking for further clinical trial evidence to show a drug's safety and efficacy, such as for Casgevy (the sickle cell disease treatment).[238] However, as discussed in chapter 1, in the space of CRISPR regulation, agencies are still creating their guidance toward industry and grappling with the moral extent of the use of gene editing, particularly for heritable conditions (editing the germline) and disability, and the introduction of edited crops or insects to create gene drives.

The most effective and appropriate time to convene a DMP on CRISPR within a regulatory agency would be when drafting guidance documents for industry, outlining what they expect to see from industry and how they will be evaluating risk. A DMP could supplement that process by providing the social and value-based perspectives I have argued may be lacking. And, to reiterate from chapter 3, even though the agency is not legally required to evaluate the morality of a particular technology itself, its decisions on risk and safety are necessarily value-laden, so public interest plays an important role.

For example, the FDA's 2024 draft guidance "Heritable Intentional Genomic Alterations in Animals: Risk-Based Approach" delineates categories of animal products based on risk. In category 1, developers are not expected to consult with the FDA prior to marketing an animal containing Intentional Genomic Alterations (IGA). In category 2, the FDA would require developers to submit an application for approval if they had previously determined that there are risk factors and special review is needed; if not, then no application is expected. In category 3, the FDA will require developers to submit data showing that the IGA is "proportionate to the risk associated with the particular product."[239] This format of providing guidance to the industry developing CRISPR-based therapeutic interventions could be illustrative of how a DMP could provide insight into a broader realm of thinking and considerations on human genome editing, while simultaneously not negating the work of the regulatory agency's experts.

Thus, the DMP could be tasked with establishing social risk tiers based on the history, moral implications, community experiences, and individual aspirations of different disease, race, and disability groups (for example).

Therein, they would have an opportunity to hear from patient advocacy groups, geneticists, and other experts about the role that gene editing plays for different groups and its implications, including the necessity and desire for treatment, the fear of neglect and being left behind, and the feelings of over-targeting for elimination that different communities may experience. And though this would be draft guidance, which is already unenforceable, it would carry important weight within the institution and for industry going forward. Ideally, then, the guidance would also impact the ways in which DMPs are generally integrated into governance and democratic institutions as models for soliciting public input and translating it for both policy and industry practices.

Finally, in addition to mitigating problems of power (such as domination and disciplinary power) and enhancing the substantive outcomes of the guidance or rule, another instrumental social outcome is increasing the trustworthiness of the institution that engages with a DMP and takes its recommendations seriously. In chapter 6, I turn to analyzing the necessary role of trust in discussing ideas related to DMP systemic approaches, authority, legitimacy, democratic governance, and regulation.

6 Democratic Trustworthiness

Cultivating Healthy Skepticism

Over the past several decades, we have witnessed a steep decline in trust in our government, its institutions, experts, physicians, and in one another, creating a "crisis of trust."[1] This is notable, and lamentable, because trust underlies the plethora of our political and moral commitments—in acting as a guiding social force, it helps to explain our engagement with certain groups as opposed to others, the information we hold as true, and the values that we cherish. Dynamics defined by trust feature prominently in medicine and technology, mediating relationships between patient and physician, researcher and participant, hospitals and marginalized communities, the chronically ill and caregivers, the public and their insurance companies, and so on. Whom one trusts as sources of medical wisdom and authority can determine whether one vaccinates their child against measles or consents to participate in clinical drug trials for a CRISPR-based therapy. In situating the profundity of trust's role in our moral and political decision-making, Sissela Bok wrote in her 1978 book *Lying*, "Whatever matters to human beings, trust is the atmosphere in which it thrives."[2]

This chapter explores the meaning and significance of trust in democratic governance, how and why deliberative minipublics (DMPs) are a promising means for cultivating trustworthiness in institutions, and why this is important for both democratic legitimacy and the instrumental benefits of DMPs in enhancing knowledge and mitigating problems of power and injustice in gene editing. I begin by analyzing the DMP evidence on trust, showing that "in an era of declining trust in government," DMPs can help to cultivate trust among DMP participants and nonparticipants alike.[3] Second, I show why these findings matter: Trust is conceptually linked to "perceived legitimacy," which is significant for democratic validity and social cohesion. Third, cultivating trust is also important for realizing the DMP's instrumental benefits of mitigating abuses and enabling co-production of knowledge by identifying injustices and different sites of epistemic authority. Finally, I apply this thinking to CRISPR to show how trust, and trustworthy institutions specifically, are essential in the cases introduced in chapter 1—sickle cell disease (SCD) therapy, deafness, and gene drives.

Democratic Trust

DMPs, as institutional mechanisms for including diverse voices, can help to cultivate *healthy trust* in our democratic institutions. Healthy trust is marked by individuals having good reasons to trust, as opposed to performing compliance. Trustworthy institutions, in turn, are those that are accountable to the interests of the people. Cultivating healthy trust and trustworthy institutions requires a political process in which governing institutions are held accountable to the results of a diverse deliberative process. As Christoph Niessen argues, "The aim of mini-publics is to create more inclusive, diverse, rational and egalitarian spaces of public reasoning, that lead to a fairer policy process, better outcomes and, ultimately, to greater public trust in an era of increasingly critical citizens."[4] Trust, in acting as a kind of social capital, facilitates greater democratic stability because it allows for the efficient distribution of resources and public confidence in government.[5]

It is perhaps paradoxical, then, that democracies are built on foundations of distrust in one's rulers, particularly the monarchy or aristocracy. Many of our contemporary democracies are similarly built on a pervasive skepticism that our leaders need institutional restraints on their power such as through term limits on their time in office. Competitive democratic theory, for example, featuring elected representatives and fair voting systems can also be reframed around trust: We do not trust our legislators' "ambition," thus we must "counteract it" through checks and balances and separation of powers.[6] In parallel, representative systems are founded on a lack of trust in the people. From Plato's ancient warning of mob rule in *The Republic* to the drafters of the US Constitution, who instilled a "filtered" representative system in which the people do not have direct power to make laws but rather to select the people who make the laws on their behalf, most architects of governing systems did not and do not fully trust the people to make wise decisions for themselves or for the greater good.[7] As a result, even as we champion the values of the people's self-determination, we develop governance systems that give power to the people, but in truncated ways, such as allowing them only to comment on regulations rather than participate in drafting them. This is also why many even dispute calling our system "democratic" and instead prefer "constitutional republic" or even just "republic."[8] Thus, even as we lament the loss of trust by the populace in figures of authority, it is helpful to realize both that *some* degree of lack of absolute trust is necessary for our system *and* that figures of authority have similarly lost significant trust in the people.

It is also helpful to recall that democracies are created precisely because we think that every other system is even *less* trustworthy. As Pierre Rosanvallon writes, the purpose of "democratic distrust . . . is to make sure that elected officials keep their promises and to find ways of maintaining pressure on the government to serve the common good."[9] Demands for transparency or "openness," for example, reflect deeply rooted public skepticism toward rulers and the importance of reason-giving by officials.[10] Transparency, a strong investigative media, fair and frequent elections, and accountability are all features of democratic politics that we often view as healthy and that also exist in environments of alert weariness.[11]

Thus, in some ways, a lack of trust can be *good* for democracy insofar as it keeps governments obligated to the interests of the people, even if only for fear of losing power. As Rosanvallon further writes, this can ultimately benefit society because it creates a kind of "counter-democracy" in which the strategy of "institutionalizing distrust" serves as "a protective barrier, a guarantee of the interests of society."[12] Rosanvallon further argues for upholding the authority of scientists, noting that the precautionary principle of regulatory governance embodies this skepticism toward trusting new scientific developments. Just as scientists engage in a scrupulous method (i.e., the scientific method) created to ensure the trustworthiness of their findings, so too should our governance systems, such as the FDA, follow rigorous procedures in checking their work.

Mistrust, then, is also baked into regulation, for it fuels oversight of industry, allowing agencies to catch errors or cases of fraud or industry deception, such as in the thalidomide case. More recent examples include Purdue Pharma's stifling of evidence of the addictiveness of opioids, Theranos's faked blood test machines, Johnson & Johnson's tainted baby powder, and Monsanto's poisonous chemicals.[13] As Dietram A. Scheufele and coauthors argue, "simply to foster trust or consumer acceptance has serious pitfalls. Not only is this approach disingenuous from the spirit of equitable deliberation, but it is also problematic with respect to societies being able to meaningfully assess all potential benefits and risks of emerging technologies."[14]

Philosophies of science and epistemology are replete with examples of the importance of skepticism. In John Stuart Mill's theory of fallibilism, progress necessitates free and open discourse to correct or fine-tune outdated theories with newer competing ones. This requires that we do not fully trust the old/current theories of science and that we remain open to realizing we are wrong and need to endeavor to find a better answer.[15] Mill uses this theory to justify freedom of speech and expression, for it is the silencing of the free flow of ideas that prevents societies from progressing. As Mill wrote in *On Liberty*, "Only through diversity of opinion is there, in the existing state of human intellect, a chance of fair play to all sides of the truth."[16] Hence, according to Nadia Urbinati, Mill supported the creation of a deliberative assembly where members could engage in "self-revision and learning."[17]

The danger of fallibilism, however, is that it creates what Willard Van Orman Quine feared would be a perpetual state of skepticism that, though necessary for deeper inquiry, could also fuel ambivalence or outright distrust. Distrust of experts can hinder institutions' epistemic robustness as progress can be both born and killed in moments of skepticism.[18] Regulatory agencies can respond to the challenge of managing the right balance of skepticism and expert authority through deliberation. In theorizing the role of citizen participation alongside administrators, Cheryl King and coauthors observed, "Citizens and administrators in our study defined the key elements of authentic participation as focus, commitment, trust, and open and honest discussion. . . . Another administrator talked about listening and trust: 'The first step is to make it clear that you're going to be receptive to their comments. But also I think a critical second step to maintaining their trust is to demonstrate to them that they're being heard . . . and that their ideas are shaping whatever you're developing.'"[19]

Deliberative democratic theory similarly describes trust as an essential goal of deliberation or as a crucial benefit because it can cultivate institutional accountability and positively transform participants' relationships with one another. DMP trust-based outcomes are significant for two main reasons: The first is that trust is tightly related to concepts of legitimacy, which in turn is important for justifying or maintaining the validity of democratic institutions (discussed in the next section). The second is that trust is essential in biotechnology regulation for facilitating the instrumental ends of mitigating abuses of power and enhancing knowledge (covered in the last section).

In this section, I offer the philosophical and empirical evidence for how DMPs can cultivate trust and trustworthy institutions. As Mark Warren writes, trust can be cultivated through a deliberative process because "political deliberation is a distinctive way of making public judgments, the characteristics of which can combat subjective cognitive biases toward the past. Deliberation enlarges the perspectives of individuals by introducing them to new perspectives, voices, experiences, and knowledge."[20] He explains that "*warranted* trust"—or trustworthiness—would ideally then emerge from deliberation's ability to "transform and align interests."[21] Trust is generated, or "facilitated," through deliberation because the process requires transparency and communication with the wider public. This tends "to mitigate perceptions of betrayal by providing reasons that justify compromised interests and identities."[22] In conjunction, Warren shows that deliberation can chip away at the foundations of distrust by providing "opportunities to explain oneself, one's group, one's problems. Because mutual respect and tact are more likely in face-to-face relations, narratives are more likely to focus upon self-characterization and to do so in ways that function as assurances, focusing not simply on differences, but also on shared commonalities and predicaments. These in turn may check the imagined vulnerabilities . . . that political situations inevitably magnify."[23]

Deliberation, Warren argues further, is possible *due* to trust because "trust would *directly* enable deliberation," for it indicates a willingness of participants to be vulnerable and accept the duties of "political engagement."[24] This is not a simple task. Indeed, Warren is acutely aware of the difficulty of bringing strangers together from different walks of life and with no preestablished relationships. As such, he stipulates, there may not be enough time to root out systemic distrust during a short DMP meeting or even one that may be more prolonged. In addition, he qualifies that the outcome of increased trust does not mean that the deliberation itself was successful because it may not have achieved its other purposes, like substantive policymaking.[25]

In response to these worries, Warren makes a compelling case for the continued practice and civic culture of deliberation as worth striving for. As John Gastil and Robert Richards further argue, the social goal of building "strong trust in neighbors and fellow citizens" can generally be achieved through deliberation, but not necessarily through one-off DMP events.[26] Warren writes that over time, tensions "can be eased if we pay special attention to the structure of protections and assurances within political relations necessary for the fragile powers of discourse to do their work. If institutions can be designed in ways that mitigate these tensions, then we shall have taken an important step toward meeting the challenges of the late-modern/postmodern era."[27] This cultural shift therefore requires a systemic change in how institutions writ large incorporate deliberative opportunities.

The systemic approach to deliberation that I have endorsed in this book holds that building a robust culture of DMPs, rather than relying solely on singular DMPs, can help solve divisions and institutional failings. As Maija Setälä writes, "If routinely used, deliberative mini-publics could help to increase citizens' trust in policy-making processes."[28] Simon Niemeyer and Julia Jennstål similarly explore whether the trust formed within the DMP between participants can be scaled up to the macro political environment.[29] They "anticipate that improved trust in public institutions, which are better discursively regulated, will have a positive scaling-up effect in terms of public deliberation."[30] These findings were confirmed in a study of Switzerland where "citizen participation is already substantive and frequent."[31] Researchers also found that citizens supported "replacing existing institutions by deliberative mini-publics" due to the higher cultivated rates of trust citizens held in "fellow citizens' ability to make good political decisions."[32] Thus, the practice of engaging in deliberative environments can have outside effects on the civic life of a demos, creating bonds that are required for cooperation elsewhere even if not directly within the deliberative experience.[33]

"Minipublic trust," as described by Michael Kenneth MacKenzie and Warren, can help to "guide policy-makers in rapidly developing policy areas that may become contentious in the future but which do not (yet) have public opinion attached to them" and can also bolster the stability of regulatory institutions.[34] Trust can enhance agencies' ability to safeguard the public interest both by reducing systemic distrust (through mitigating captured interests' corruptive influence) and by simply making better regulation. As noted in chapter 3, capture is a significant impediment to trust because it stands in the way of the agency acting in the interests of the public. Deliberation can cut away at captured influence by cultivating citizens' considered judgments through a process that strengthens public adherence to mandates, guidance, and rules as opposed to private interests. To do so, DMPs help to create, as Jane Suiter and colleagues argue, a kind of "facilitative trust" to "make judgments on [others'] behalf."[35] This aim is ultimately an important public goal for democratic institutional legitimacy and stability and has deep implications for the trajectory of public acceptance and development of CRISPR and other biotechnologies.[36]

The experimentation literature on trust shows that there is increased trust among both DMP participants and nonparticipants, or the outside public. First, interpersonal trust among DMP participants has been found to increase during deliberation.[37] While in some ways this may not be remarkable, given that the participants enter the deliberation as strangers and depart with some familiarity, researchers found that trusting bonds are formed not just because of natural progression but due to the mutual exchange of reasoned ideas and learning, signaling growth and appreciation for others' views.[38] For example, in a study of a DMP held in Finland in November 2006 to address the question of whether Finland should build another nuclear power plant, researchers found an "increase of political trust and readiness for collective action" after deliberation.[39] The authors also found that "political trust may increase through deliberation as people learn to understand the processes of democratic decision-making, which can make them more trusting toward institutions and actors of representative democracy."[40] In another case, an Estonian Citizens' Assembly in 2012, "participants' social trust increased following their participation in the ECA."[41] Christian List and John Dryzek further found that in a series of laboratory games, "discussion-induced cooperation" occurs when "there is a positive payoff in being seen by others as co-operative and trustworthy."[42] This trend, they note, will probably only increase in real-world settings where the stakes are greater, such as when the DMP has real authority or an impact on policy.

Indeed, giving the DMP authority, rather than treating it merely as a source of guidance, aids in cultivating interpersonal trust. As James Fishkin and Robert Luskin argue, it is "reasonable to conclude that some part of the increase in political participation and public trust among participants arises from their belief that the results of their deliberation have significant 'recommending force' for policy makers."[43] Recall from chapter 5 that a main worry about the widespread use of DMPs today is that their popularity can backfire if the DMP becomes merely a theater for staging political wins rather than sincerely earning the public's trust. Furthermore, accomplishing these outcomes is contingent on the institutional environment and setting. Discussion rules and facilitation play a crucial role in navigating the trajectory of deliberation toward an atmosphere of authenticity and mutual respect. A DMP with experienced moderators, balanced expert guidance, and engagement opportunities is a more promising environment for creating trusting relationships than during deliberations or exchanges that occur outside the DMP.

Findings here support, then, not only that treating the DMP as a source of authority improves the DMP quality of deliberation, but also that it boosts public perception of the DMP as a productive mode of facilitated viewpoints. Thus, another important factor for achieving trustworthy institutions is the wider acceptance of deliberation by the public, for if there is "low deliberative capacity in the system—low levels of trust, engagement, and deliberative stance—then it is difficult for mini-publics to gain purchase and effect change in the system."[44] Indeed, evidence suggests that the DMP "trust mechanism potentially works, but only to the extent that citizens take heed of it, which in practice has been limited."[45]

As such, the second main set of findings concerns external effects: Nonparticipants view the members of the DMP as "trusted information proxies."[46] In connection, this cultivates the DMP's convening institution's trustworthiness.[47] Increased trust in DMPs is well documented in citizens' initiative reviews (CIRs) in particular.[48] The Citizens' Jury on Referendum Options in Korsholm, Finland, for example, modeled after CIRs, was asked to present referendum information to voters. Maija Setälä and colleagues found that "the facts and the arguments summarized by a mini-public can help voters rise above the personalization of campaigns as well as manipulations and accusations put forward in campaign rhetoric."[49] For the researchers, these results are "rather optimistic in terms of the capacity of deliberative mini-publics, and CIR-type procedures in particular, to enhance public trust. When properly connected to mass publics, mini-publics can help enhance trust in public institutions, including those on the 'other side,' in polarized contexts."[50] Researchers also found that simply learning about policymaking processes as part of a deliberative exercise can increase overall trust in institutions.[51]

The British Columbia Citizens' Assemblies (see chapter 5), which addressed voter reform rather than candidate information, had similar results. In the BCCA, researchers found that outsiders viewed the process as trustworthy either because they were impressed by the ability of the participants to deliberate or because they were a group of nonexpert citizens who had found agreement.[52] Here again, researchers argued this happens because nonparticipant citizens perceive the DMP as a trusted and legitimate source of information and seek to subsequently update their own knowledge about the issues.[53] Similarly, in analyzing the Oregon Citizens' Initiative Review, researchers describe this kind of trust as a "trusteeship model": "Minipublics can act as facilitative trust agents when such bodies are high in competence but low in motivated reasoning. The wider citizenry might use these facilitative trustees to make judgments on their behalf when they offer a consensus recommendation."[54]

The process of communicating the work of the DMP to the wider public creates the conditions that allow more people to engage with the system using their own judgments.[55] In a 2022 study of a CIR in Switzerland, Alexander Matthias Geisler found that the CIR replicates the "facilitative trust," described by Jane Suiter and colleagues, by easing "citizens' cognitive expenses in forming opinions."[56] Geisler explains that the ideal of facilitative trust is that it gives people good reasons to trust others as sources of information and as "decision proxies." Geisler writes that "facilitative trust [is] a type of political trust that helps voters reduce their cognitive effort needed to reach a reasoned conclusion to a politically relevant decision."[57] Drawing on MacKenzie and Warren's work as well, Geisler argues that DMPs can "maximize the democratic impact of their political resources by enabling warranted forms of trust."[58]

Other studies suggest that trust is present even when the participants of the DMP are different from them and highly polarized, such as during political crises. In a study of fifteen citizens' assemblies, scholars investigated the conditions for the public to "support" these assemblies, finding that "supporters of citizens' assemblies selected through sortition seem to be rather disengaged or politically dissatisfied: they have low education levels, a negative opinion about political elites, and do not feel politically competent."[59] They point out that this is interesting given that the profile of supporters is the opposite of the profiles of those who "generally accept to participate in deliberative assemblies."[60] Further, "respondents are more supportive of deliberative citizens' assemblies selected through sortition when they know that their position is shared by their co-nationals. In other words, their support for the institution is at least partially by an expected favourable policy outcome."[61] Even in high political conflict zones, such as Northern Ireland in 2018, researchers found "general public support" for "decision-making by a citizens' assembly, even when the decision reached is one they personally disagree with."[62]

Thus, trustworthiness as the goal of institutions can in many ways have a tremendous impact on public policy and can be facilitated by DMP inclusion.[63] Here as well, the degree of trust is determined by how the convening organization takes up the DMP's institutional recommendations, and not just the fact that the assembly represents a swath of themselves and their peers. Even if the public trusts the process and outcome of the DMP, if the host institutions, such as regulatory agencies, disregard much of the advice, then the wider public may discount the use of the DMPs. Institutional change, achieved not just by creating DMPs but also by creating well-designed ones that have fair selection processes and transparent outcomes, is instrumental to cultivating trust in the wider public. This trust by the wider public is closely related to the idea of "perceived legitimacy," or whether the public views the DMP as proceeding through a fair or reasonable process that can protect public interests.

Perceived Legitimacy

Another way that scholars of DMPs have studied trusting effects is by measuring perceptions of legitimacy. Trust is tightly connected to perceived legitimacy insofar as individuals are more likely to believe in the veracity and soundness of an institutional process they recognize as, for example, fair or representative.[64] Democratic *trustworthiness* is a form of perceived legitimacy. As Daniel Gauys and coauthors argue, perceived legitimacy is bound up in procedural fairness: "Citizens can accept democratic decisions if and only if they can trust the democratic political system to function according to egalitarian *and* epistemic procedural logics. Their trust in the functioning of these logics, however, may not be equated with and is not dependent on trust in the deliberative capacities of fellow citizens."[65] In examining several citizens' juries in Spain, Joan Font and Ismael Blanco argue that the goals of legitimacy and trust are "different but complementary. . . . If these methods are to be able to create a certain degree of political trust, then there is probably a necessary previous step—that of gaining enough procedural legitimacy by themselves. They must become clearly accepted as fair, convenient and adequate methods for listening to citizens and for involving them in public decision making."[66] Font and Blanco ultimately argue that for political trust to be formed, the DMP first needs to gain "public acceptance and eliminate some of the problems that still generate reluctance," such as regarding their "capacity to represent citizen interest, the role that information and deliberation play, and the perceptions that they can be biased in favour of the interests of local governments."[67]

Perceived legitimacy effects are present even in polarized conditions or in situations where there are lower levels of trust in one another and institutions. Lisa van Dijk, Emma Turkenburg, and James Pow, for example, studied perceived legitimacy among polarized citizens in Northern Ireland after a DMP was held in March 2019 on the topic of Northern Ireland's constitutional future and the various options of a united Ireland compared with staying in the United Kingdom. They found that "higher levels of ideological polarization and, to an extent, higher levels of affective polarization are associated with lower levels of perceived minipublic legitimacy among the general public; those who did not participate in the minipublic themselves. At the same time, levels of perceived minipublic legitimacy were lower but not necessarily low among polarized citizens."[68] The researchers describe intergroup trust as a mechanism of perceived legitimacy for polarized citizens who are particularly "less comfortable interacting with those who hold different political viewpoints or affiliations."[69] Polarized individuals, they argue, are more likely to not trust others with divergent views. The DMP that they surveyed received no media attention before or after the event and was part of an academic study rather than one where the government was the convening institution. The authors postulated that this is related to trust: "Results seem to indicate that respondents' willingness to participate and trust in other citizens may be associated with perceived minipublic legitimacy, but not with polarized attitudes."[70] In other words, given the proper conditions and procedural mechanisms, even in polarized conditions, intergroup trust as perceived legitimacy can be enhanced.

Similarly, researchers found DMPs can help to bolster perceived legitimacy in conditions of less polarization, but also with lower levels of trust overall. Kristinn Már and John Gastil found in studying a CIR that trust increased most among disaffected voters: "We found higher levels of trust among people of color compared to Whites, poor compared to rich, and young compared to old."[71] This is positive because part of the goal of a DMP is to include those often not present at the table who can serve as trusted proxies for others. Micha Germann, Sofie Marien, and Lala Muradova found that among nonparticipants who saw the assemblies as acting against their preferences, those with previously low levels of trust who harbored doubts actually developed a better perception of DMP's legitimacy. They write, "This suggests that mini-publics and referendums are able to increase fairness perceptions and decision acceptance precisely among those citizens who are most skeptical about the functioning of current democracies."[72] The researchers also postulate that "involvement of a mini-public in democratic decision-making can, at least in principle, increase perceptions of fair decision-making and the acceptance of negative political decisions."[73] However, they qualify this by saying that "the beneficial effects of mini-publics are largely limited to situations in which the non-binding recommendations of mini-publics are honored, a finding that applied even when the decision to overturn a mini-public recommendation was made by referendum."[74] As they explain: “Citizens at large perceive a decision-making process as fairer and are more willing to accept a negative decision if a mini-public is involved in the process; however, these beneficial effects are largely limited to situations in which mini-public recommendations are honored.”[75] In other words, yet again, giving DMPs authority has instrumental benefits, including perceived legitimacy.

In addition to DMP experiments, scholars use modeling approaches to support these findings. Daan Jacobs and Wesley Kaufmann confirmed that in “much of the existing literature . . . [a] decision-making process that includes a deliberative mini-public is considered to be more legitimate than a decision-making process that does not.”[76] However, they did not find that random selection was perceived as more legitimate than self-selection. Micha Germann found further that DMPs that ended up “reproducing the same, or similar, biases as elections and referendums” were not perceived as more legitimate due to “representative biases.”[77] Germann concludes that “the gains in legitimacy perceptions decrease substantially if mini-publics are not fully representative of the population they are meant to represent, and can even vanish entirely when the deviations are large.”[78]

Others have similarly found that if DMPs do not have the right kind of representativeness or fair selection mechanism, then their perceived legitimacy suffers. People who were invited but declined to participate in a citizens’ panel called “Make Your Brussels—Mobility” held positive views of both the people in the panel and the results of the panel. However, they did not find that diversity was “sufficient to ensure trust in the minipublic and its recommendations,” nor did they have significant interest in the output of the minipublic.[79] This finding is related to concerns, discussed in the previous chapters, about whether diversity will truly have an impact on powerful institutions, extremism, and dominating voices in the room and thus be able to push against the power hierarchies in institutions. The procedures for how the DMP is convened, then, matter as much as the very existence of the DMP, and the moderator plays a crucial role in cultivating intergroup trust and thus perceived legitimacy.

Perceived legitimacy is, as such, an important component of trust, but it is also only one formulation of legitimacy. I argue that perceived legitimacy combined with the instrumental outcomes of substantive benefit and non-domination effects make DMPs legitimate democratic sources of authority. This view, however, can be criticized largely from proceduralist accounts of legitimacy. Proceduralist accounts of democratic legitimacy require specific mechanisms for attaining equality or representativeness, whether they are viewed as such by the public. Maria Paula Saffon and Nadia Urbinati, for example, argue that if one cares about procedure-independent results, such as substantive outcomes like knowledge gains, then we would end up with an epistocracy, a kind of technocracy in which the most knowledgeable rule.[80] For them, this is undesirable because it undermines political equality. As such, they conclude, procedure-independent accounts may be incompatible with democratic equality and thus illegitimate.[81] Similarly, Thomas Christiano argues that the problem with procedure-independent accounts is that if we found a state of inequality beneficial, then democracy would be undermined.[82] Thus, what matters is whether the procedure is valid, not whether it produced certain outcomes (like trust, knowledge, or justice).

My response is threefold. First, this view does not accurately take up the epistemic merits of political equality. Political equality is, contrary to these critiques, *more* conducive to good outcomes, and that itself serves as one valid justification of political equality (alongside other formulations such as inherent rights and the benefits of living in an equal society). Second, even intrinsic equality defenses require some degree of hierarchy to distribute goods and create well-functioning governing institutions. The goal of democracy is to equalize conditions in order to create better (i.e., more just and equitable) distribution of rights, goods, and services, which requires institutional authority to facilitate. As such, democracies are never perfectly equal, and yet there are many legitimate versions of well-justified inequality of power. Third, DMPs create more equalized conditions precisely because they give power to more people than would otherwise have had it and thus may satisfy the proceduralist accounts of legitimacy as well.[83]

In another proceduralist critique of DMP legitimacy, Cristina Lafont argues that DMPs operate as "shortcuts" around previously established fair procedures of representation and thus act as "unauthorized" modes of resolving deep disagreement.[84] Lafont reminds us that the logic of representative governance is that the people authorize the laws that govern them by selecting the people who rule on their behalf.[85] Thus, lacking proper selection mechanisms, where does the authorization for DMPs then come from? For Lafont, *participatory* democracy, as opposed to DMPs, offers fairer mechanisms of inclusion because it is more widely available.[86]

My response to this objection is twofold. First, regulatory agencies, though they were granted authority by Congress, suffer from a democratic deficit, which Lafont similarly laments in DMPs. Given that much of the work of regulation should still be done by these agencies (we do not want to say they are illegitimate, just that there is a deficit that should be filled), the solution is not to return their power to Congress but rather to infuse their system with more democratic accountability and to equalize political representation. This is not a "shortcut" because only some people will be selected for that DMP; rather, it is a way to make administrative agencies more accountable to the wider public. Second, participatory democratic mechanisms already exist in the administrative realm through the comment period, which I have shown to be insufficient because of the problems of self-selection and their consultative-only capacity. It is not clear, on Lafont's account, that participatory modes are less illegitimate than DMPs precisely because of preselection problems and the unfairness of time, resources, and capacity to comment.

Thus, my argument is that the DMP's perceived legitimacy is closely related to its trust-enhancing capacity, which makes possible the non-dominating and epistemic merits of DMP outcomes by shifting sites of authority to the people. This is procedurally legitimate insofar as it uses fair selection mechanisms that equalize and diversify those who hold power, and it is also legitimate insofar as the procedures are justified by their outcomes of non-domination and co-production of knowledge.

Trust, Power, and Knowledge

DMPs can, as Charles Levine writes, help to close the "gap between disaffected citizens and their government."[87] In other words, they can aid in mediating a proper relationship between authority figures and the populous through careful examination of where we locate power and knowledge, who holds power, and whose knowledge we take up. This is a different political mission than *just* the examination of evidence and narratives: it is the recognition of who holds the authority to determine proper evidence and narrative truths.

When a DMP is trusted, it reflects a willingness to accept the DMP as a mitigator of risk on behalf of the public. To trust someone relieves us of the burden of knowing and of performing complicated and painstaking research and assessment of risk. As Niklas Luhmann writes, trust "presupposes a relationship of risk."[88] Studies on risk perception have found that the "lay public relies on social trust when making judgments of risks and benefits when personal knowledge about a hazard is lacking."[89] Regulatory agencies need public trust for their own work of protecting the public interest by mediating risks associated with drugs and food, for instance, and so that parents vaccinate their kids against measles.[90] Perceived risk and benefit are essential for determining the "acceptance of biotechnologies," which explains a significant concern that regulatory institutions and public health communities have expressed about the role of "public acceptance," such as discussed in chapter 2, in whether we treat CRISPR technologies as "GMO 2.0" or as unique innovations.[91] The worry is that even if a regulation is epistemically well founded and with even (let's assume) the best of intentions, preexisting distrust can erode its effective implementation due to (dis)trust-based opposition. In an increasingly uncertain and complicated world, deciding whom to trust is critical.[92]

The relationship between civic engagement, trust, and the healthy functioning of democracy is well recorded by medical institutions as well. As Rishi Manchanda and Jean Silver-Isenbstadt write, "Perhaps in response to this level of civic disengagement and the public's eroding trust in the medical profession, several medical organizations, notably the American Medical Association and the American Board of Internal Medicine, recently began describing civic engagement as an important aspect of medical professionalism."[93] The National Institutes of Health, in conjunction, issued a report in 2001 on "Preserving Public Trust," stipulating its importance as a goal of its institutes.[94] In 2020, the National Academies of Sciences, Engineering, and Medicine published "Earning Trust in the Age of the Pandemic," declaring the need for public acceptance of COVID-19 vaccine(s) by creating "earned trust" or trustworthiness.[95]

Often when scientists and officials call for "public trust," however, they are often calling for compliance.[96] This was seen during the COVID-19 pandemic, when scientists argued that "science" would "save us."[97] We should all trust experts, the argument goes, because they are *epistemic agents*: They offer us knowledge in the form of justified or objective research that we could not have otherwise acquired for ourselves. They, we are told, have more experience in what they are doing, and they have earned authority or a reputation on that basis, so trusting them would be the "rational" thing to do.

There are, of course, many good reasons to trust scientists: It is true that they know far more about epidemiology, virology, and other unique and specialized knowledge sets that *do* make them qualified to give medical advice and direct policy. However, for the scientific community to earn trust from lay participants, as Naomi Scheman writes, "The solution cannot and should not be seen as a matter of image manipulation: Possibly the only effective but certainly the only proper way of producing such confidence would be to work to make the institutions in question—and, generally the broader society within which they exist—more genuinely just, more truly worthy of everyone's trust."[98] In other words, experts and scientists should not take trust for granted: their occupation often demands a dynamic back-and-forth relationship, rather than a top-down one, with the broader public.

In addition to offloading risk assessment, trust is also then a belief in the veracity of another's commitments, or as Annette Baier writes, as a "reliance on another's good will . . . [or] others' competence and willingness to look after, rather than harm, things one cares about which are entrusted to their care."[99] Matthew Cleary and Susan Stokes similarly write that trust presupposes or may be confirmed by the other's beneficence: "If A and B are actors and X is an action that is in A's interest, then if A believes that B will do X, A trusts B."[100] This can be scaled up to explain many elements of group-cohesion. As Rogers Smith writes, trust is the "belief that a community's leaders and members have the capacity to succeed in advancing some of one's important values or interest, if they should try to do so."[101]

Trust, then, exposes our vulnerabilities, but it can also, as Warren writes, help to resolve certain collective-action problems by allowing for "individuals to presume the good will of others with respect to shared interests as well as the divisions of knowledge necessary to make use of explicit rules for collective action."[102] Thus, trust allows us to act in conditions of great uncertainty as well as anxiety.[103] Public trust in scientists generally emerges as an extension of experiencing and benefiting from sciences' achievements, rather than their costs, and accepting the role of uncertainty.[104] When people do not trust their institutions, it may be due to experiences of repeatedly being unable to have their welfare addressed and their wrongs righted, or from finding themselves on the losing end of the risk game.[105]

Before continuing, I offer a couple clarifying definitions: *Untrust* is the general state of lack of trust until evidence or enough belief in the verifiability of the trustee has been attained. Thus, it is not quite the opposite of trust, but it is not enough to create cooperation. *Mistrust* is the result of a betrayal of trust. It occurs when there has been a positive attribution of trust, but then the trustee acted in a way that contradicted how they proclaimed they would or falsely predicted that something would work out that did not ultimately, resulting in betrayal.[106] Many people who experience mistrust have had a negative experience at a doctor visit, for example, but have yet to attribute this to the entire medical profession or health care services.

Distrust is a systemic lack of trust. Distrust is the belief that the distrusted will "actively work against them in a given situation."[107] It is an "active" construct in the sense that it requires information to persist and creates feedback cycles by rearticulating the reasons for distrust. In this way it often features or is based on a lack of information and the need for evidence or verification to affirm that distrust is warranted. Distrust is often considered the "negative mirror-image of trust," and it creates a "'confident negative expectation regarding another's action' in a situation entailing risk to the trusting party."[108]

There are many readily available examples of distrust in the medical context, and they are often related to problems of capture. In the realm of GMOs and vaccination, many distrust vaccines because of the perceived financial-based motives of pharmaceutical companies and the corrupted interests of regulators.[109] During the Ebola outbreak in Congo in 2019 and in 2014–16 in West Africa, for example, distrust of medical authorities resulted in resistance to behaviors that prevent transmission, leading to spread of the disease.[110] Fatality rates were not due to the incredibly dangerous nature of the disease, which is indeed highly contagious and fatal; rather, it was more attributable to failures of "technologies of trust" that are essential to "the development of a framework of practical techniques to improve the acceptance and effectiveness of . . . emergency response strategies."[111] During the COVID-19 pandemic, distrust rose among those who refused to mask, vaccinate, social distance, or even believe that COVID-19 was real.[112] Distrust, as well as trust, then, emerges in moments of vulnerability and uncertainty, and it operates as a mode of taking back power and authorizing certain individuals as sources of knowledge: both for the good and for the deleterious.

The presence of trust in a relationship establishes a hierarchical power dynamic by placing the trustee in a dependent relationship with the trusted.[113] As Baier further writes, "Trust is an acceptance of vulnerability to harm that others could inflict, but which we judge that they will not in fact inflict."[114] The trustee is vulnerable because of their deference to the decision-making authority of the trusted. The trusted's power over the trustee is not, as we saw before with domination, necessarily a kind of interference; rather, it restricts freedom by constraining the likelihood of what that person may believe and their potential for action. In describing doctor-patient relationships, Karen Cook and coauthors write, "Because power alters the relationship between power holders and those who serve them, there are additional reasons why trust—or, more precisely, the problem of assessing others' trustworthiness—may be fraught with difficulty."[115]

After all, trusting those perceived as extremely knowledgeable entails its own risks. To trust a health care provider or geneticist is to create a relationship with them in which they have more information, knowledge, and aligned risk analysis.[116] This places the researcher or doctor in a position of power over the patient, creating opportunities for improving health and well-being, but also conditions for abuse. This can, then, generate a system in which abuse by doctors, like USA Gymnastics team doctor Larry Nassar and Ohio State University doctor Richard Strauss (among many others), is rarely believed, given that doctors are widely thought of as acting in the best interest of their patients.[117] This is often due to the machinations of their position, such as degree, pedigree, and even gender and age. Cults and other "high control groups" are often able to gain power and prominence due to the perception of their leader as a "genius" or "enlightened," allowing them to perform egregious rights abuses on their members who yearn for their precious attention and praise.[118] Trust of this kind is more of an infatuation or coerced obedience, something that asks victims to ignore their moral intuitions and personal needs, and certainly not to pay heed to what I refer to as "healthy skepticism": the thoughtful awareness of valid sources of knowledge and risk assessment balanced with personal intuition and self-knowledge of one's needs.

Even in nonextreme (such as outright abusive) circumstances, because to trust features offloading our risk analysis onto others, this leaves us in a place of susceptibility to a risk analysis gone wrong, resulting in harm. People become vulnerable to domination by trusting that systems are working in their interest, when in fact they are experiencing arbitrary impositions of power, such as in the case of algorithmic determinations of bail in the criminal justice system, or when there is significant bias in the datasets that AI diagnostic tools are trained on in diagnosing skin conditions for nonwhite patients.[119] The opioid crisis was also borne out of an environment of patient trust that doctors were acting in their best interests, and, in turn, where doctors trusted the pharmaceutical companies and sales reps that told them the drugs were nonaddictive.[120] Each level of the ladder is vulnerable to the next in the hierarchy, built by knowledge conditions that create (the potential for) abuses. They can all come crumbling down in moments of betrayal and awareness that the system may not actually be working in the interest of patients, as opposed to say, the industry, as was the case with Purdue Pharma, which hid OxyContin's addictiveness from regulators.[121]

In democratic systems, participation alone is not enough to create trust. There must be sincere listening alongside other efforts, including "higher transparency . . . more involvement of the media, and communal administrative ventures that bring citizens closer to the daily administrative process."[122] As Nancy Roberts warns, this is a fine line to strike, for if "direct involvement is used or seen to be used as a tool for the purposes of manipulation or cooptation, then the levels of dissatisfaction among participants, as well as observers, are likely to increase, setting up a cycle of instability and distrust between the government and its citizens."[123] Thus, institutions require a finely tuned system of engagement or participation to balance the competing ends of the trust-distrust spectrum and to situate ourselves solidly in the middle for the purpose of CRISPR oversight, governance, and risk management.

DMPs can help to cultivate and manage healthy skepticism and root out both the problematic and valid sources of distrust by facilitating accountability and transparency. DMPs can play helpful social roles in managing trusting relationships within and outside their convening institutions in ways that amplify and allow for the co-production of knowledge because the success of the deliberative process requires recognition of lay participants as knowers of their own lived conditions, as thoughtful interlocutors with experts, and as members of communities that have experienced historic injustices. As with epistemic injustice, discussed in chapter 4, Miranda Fricker uses the phrase "epistemic trust" to refer to treating an agent as a source of social information or as part of a "credibility economy."[124] This was seen in the case of pain—to treat a patient when they are in pain is to trust that they are knowers of their pain. It is baked into the framework of shared decision-making and patient autonomy: Consent requires trusting patients as sources of knowing their own well-being.[125]

Epistemic trust refers to the willingness to recognize nontraditional groups as trustworthy sources of knowledge.[126] *Epistemic mistrust*, in turn, is the marginalization of populations seeking to account for their experiences in languages that are disregarded because they do not comport with the dominant narrative or communicative method.[127] To trust someone is to have a positive judgment of their validity, their dependability, and their knowledge-bearing agency. It is one thing to say that we should take seriously women's experiences of sexual assault as knowers of their sexual assault, and another for this disposition to become standardized in public, in courts, and in workplaces. It is not just a matter of whether we trust women's accounts, but how we treat women as trustworthy agents.[128] Often this too reflects population-level power dynamics rather than actually accurate distinctions between those with and without knowledge.

Although trusting others may seem to signal epistemic weakness (for if we trust others, it is because we lack knowledge) in actuality, as John Hardwig argues, *knowing* who to trust is itself an epistemic stance, for identifying the "untrustworthiness of members of epistemic communities is the ultimate foundation for much of our knowledge."[129] As Steven Shapin notes, in "securing our knowledge we rely upon others, and we cannot dispense with that reliance. . . . [Thus] the identification of trustworthy agents is necessary to the constitution of any body of knowledge."[130] Everyone at some point relies on another's knowledge as a source of information, for we cannot all be knowledgeable of everything at all times.[131] This includes those who are "highly knowledgeable" (such as experts) for "the whole fabric of research is trust."[132]

It thus takes wisdom of its own kind to recognize sites of epistemic authority, in both experts and in nonexperts with lived experience. We often choose whom to trust based on proximity, relying on those our already trusted people trust. This leads us to have vastly different sources, until we find ourselves becoming more isolated from different perspectives.[133] Consider that trust in the scientific community can be lost through a variety of complicated factors, including trusting someone more familiar, with greater exposure to in-group shared knowledge, and also amid the scientific community's systemic failure to "share" with "lay communities . . . particularly in the case of marginalized communities."[134] The integration of individuals into the spheres of influence wherein science is communicated creates better conditions for trust, just as it does in other institutional environments and interpersonal relationships.[135]

The kind of trust that forms in DMPs allows for a co-production of knowledge through recognition that others who were previously discounted as knowers are now recognized as sources of knowledge in ways that matter both for the substantive outcome and for identifying true sources of capture. Rather than the silos we have formed for ourselves today, DMPs bring us into closer proximity with those we may disagree with and who have different life narratives, allowing for a greater understanding of trustworthy sources of information than we previously had.

Trustworthy Regulation and CRISPR

In the case of regulating gene editing, the aspiration is that DMP deliberation can help the public and scientists to navigate their thorny trust relationship, particularly when it comes to defining safety, fair access, and equity. Convening a DMP can help to further the relationship between institutions and the public by creating not just a more compliant public but a trustworthy institution where the people act as trusted experiential and informed proxies. Cultivating a healthy skepticism allows for DMP participants to engage in robust learning with experts while having the freedom to interrogate ideas and evidence. In regulating gene editing generally, trust plays a crucial role in the regulatory process because the cultivation of trustworthy regulatory institutions helps to facilitate effective rulemaking that protects the public interest, can be well informed and substantively robust, and may mitigate illicit uses of power and domination. The regulatory agency, as such, has an interest in upholding the authority of DMPs by taking seriously their contributions to rulemaking.

In the case of SCD, and future CRISPR-based gene therapies for historically dominated and vulnerable patient populations, trustworthy guidance from agencies can help patients to better evaluate the risk of participating in clinical trials, navigate language around "treatment" and "cure," and manage pain expectations after treatment. This will hopefully lead to greater patient welfare and effective interventions in populations that have experienced historic medical maltreatment, such as by being used as guinea pigs in research and dismissed as sources of knowledge of their own lived experiences.

In the case of the Deaf community and hereditary hearing loss, where there is greater ambiguity about whether the intervention is needed at all and the stakes of community fissure are potentially higher, transparent procedures and deliberation that reveal careful analysis and inclusion of affected communities are critical if the Deaf community is to trust that the resulting regulation serves their interests in mitigating risk and evaluating safety and also respects their identity. And because there may be disagreement about whether interventions are even needed, including DMP deliberation in rulemaking can enhance the trustworthiness of the institution by showing that it had an inclusive deliberation process, even when the outcome is one that parts of the community may disagree with.

Finally, in the realm of gene drives, as in the case of Florida's modified mosquitoes, much of the opposition was due to distrust of the company and the agency that authorized the project. Gene drive deliberations involve balancing competing risks of current disease and climate crisis impacts with potential future risks. The importance of having trustworthy institutions, researchers, and regulators in this process is crucial for creating interventions that match the needs of communities facing the climate crisis, while also not being taken over by the loudest and most extreme voices in those communities or by ambitious companies developing the gene drive technology. Further, the risk-based categories described at the end of chapter 5, as the goals of deliberation, would help chart a trustworthy path forward for understanding under what conditions gene drives should be developed. The hope is that through this process of weighing values, experiences, evidence, and future worlds, our regulatory agencies, and simultaneously our commitments to democratic values of political equality and well-being, will be strengthened.

Conclusion

From Justice to Knowledge to Trustworthiness and Back Again

In chapter 6, I offered a theory of trust as a social force that binds some together and tears others apart. In offloading information and risk analysis onto others we believe have our interests at heart, trust creates vulnerability, and abusing this vulnerability creates distrust. Trust then plays a vital, though perhaps paradoxical, role in democracies. In many ways, democratic systems emerged from experiences of distrust of oligarchs and aristocrats and are perpetuated by (ideally) a free media, checks and balances, and separation of powers. However, democracies become destabilized when there is too much distrust, as institutions cannot effectively protect the public good. Too much trust, on the other hand, can also signal a breakdown of democratic stability, as it may be a sign of "high" social control or mere compliance. Today, we are witnessing the crises that can arise when democratic institutions and regulatory agencies lack sufficient trust: the erosion from within caused by capitalists' corrosive program of deregulation.

Deliberative minipublics (DMPs), I have suggested, offer a promising way to deepen the democraticity of our regulatory institutions and their perceived legitimacy by infusing the system with greater diversity and discursive representation. This is marked by those in charge truly incorporating the considered judgments of those they govern and delivering on their institutional promises, such as private interest oversight and risk reduction. By serving as a proxy for the people and mitigating captured influence through the scrutiny of experts, DMPs can signal to the broader public that they are more reliable, accountable, and thus trustworthy.

Trustworthiness, then, reinforces the aims of mitigating illicit uses of power and promoting epistemic integrity: When individuals have good reasons to trust the government, they are more willing and able to weigh in and engage with democratic institutions to prevent domination and other power abuses by engendering greater transparency, accountability, and therefore trustworthiness. Because the DMPs may be viewed as more legitimate, the public may be more willing to participate, allowing for fewer barriers to entry and problems of preselection and thus leading to greater epistemic benefits, which in turn bolsters healthy skeptical trust and confidence. In other words, this creates a (non-vicious) cycle that operates in service of democracy's instrumental aims and thus its legitimacy.

Regulatory DMPs will be most effective in a *systemic* and ideally *global* approach to deliberative democracy. While many of the design issues discussed in chapter 5 contribute to the ability of DMPs to thrive, there are further factors that will influence their success. DMPs require significant political capital and resources to become viable tools for public engagement. Because DMPs are a new and likely unfamiliar modification to governance, it may take time and proper messaging for them to be regarded as serious endeavors and incorporated into the everyday fabric of governance. Further, social acceptance and the cultivation of a civic culture centered around deliberation and participation may also take some time and require clear messaging.

Just as the early twentieth-century progressives and pragmatists called for experimentation in democracy, so too will DMPs need continuous modification. And just as the last several decades have witnessed an evolution of DMPs, so too will at least the next several. In many ways, we do not yet know what we do not know about DMPs in regulatory spaces. Thus, the introduction, deployment, and development of DMPs in regulatory agencies may have a steep learning curve that requires innovation along the way, such as in reducing barriers to entry, selecting for diverse viewpoints, mitigating domination by more powerful voices within the DMP itself, and minimizing opportunities for co-option, or even capture. The most important step, though, is to start, lest the "perfect be the enemy of the good."

The ultimate hope, more broadly, is that DMPs will offer a path forward for the public to have good reasons to have confidence in the institutions that are supposed to safeguard them against the private interests of capital. The aim is that pharmaceutical companies and governments will better ensure that vulnerable populations are not exploited in research, and that research is prioritized not necessarily according to what is most profitable but according to the public interest, the protection of group-based identities, and a sensitivity toward historical injustices, such as discrimination against those with disabilities and sickle cell disease. These are ends in themselves and can also benefit the missions of the DMPs by enhancing trustworthiness. One sign would be participants who are more likely to sign up for clinical trials or who are more likely to protect vulnerable populations against the impending tide of climate change. These are the factors that not only strengthen health and welfare but also mitigate domination and reflect the epistemic merits of local, indigenous, and experiential wisdom.

Gene editing is a prime opportunity to implement DMPs within regulatory agencies as modes of public participation because it is not (yet) a topic that is overly polarizing, partisan, technical, or unrelatable. But there are many other applications in which a DMP would support the work of regulatory oversight, including prenatal genetic testing, artificial wombs, and artificial intelligence in clinical care and drug development. These are areas of biotechnological advancement with significant rights-based and uncertainty considerations that have yet to be fully regulated at the federal level. Outside of the biotech realm, myriad other regulatory questions, such as environmental protections and housing, demand careful attention from a diverse body of people. Accordingly, while my analysis has focused on three regulatory agencies specifically, other agencies could benefit from the inclusion of DMPs in their rulemaking process. For example, the Federal Communications Commission, which has also been susceptible to capture, benefits from hearing from affected communities when regulating the internet.[1]

There are, of course, less useful applications of DMPs, such as those focused on much more fact-based identification or ones in which only one community group is affected as opposed to a broader swath of people. For example, I do not think that we need a DMP to assess whether climate change is human caused and an urgent matter, whether vaccinating children is an important public health goal, or whether abortion access is essential medicine. The suggestion is not that every idea or fact is up for debate, but rather that the overall impact of regulation on society requires anti-capture modifications and public accountability. There are many issues between the hypertechnical and the morally extreme that leave room for diverse groups to reasonably disagree.

DMPs are also important to situate in conversation with the global context. Given, for example, that gene drives can cross national boundaries, international bodies, such as the World Health Organization, should also convene DMPs. Further, DMPs largely exist as a Western deliberative concept. This has encountered pushback from those seeking to "decolonize deliberative democracy" to ensure that its deliberative practices do not simply mimic Western views of the world but rather truly take seriously indigenous and local ways of communicating, narrating experiences, and understanding their history.[2] Yet again, we will undoubtedly find ourselves experimenting and learning from mistakes as we proceed.

With CRISPR, our crises of today—regulation, democracy, climate, health care, and trust—can create opportunities for deliberation and resilience. We can decide that by learning from mistakes in the past, by learning about the negative ramifications wrought from excluding certain individuals from the table and the danger of dismissing scientific authority within captured systems, diverse bodies and voices are essential for moral and medical progress. The aim is neither to sacrifice the role of the expert or the regulatory agency, nor to claim that this particular mode of inclusivity will solve every moral or political question. However, if we want to learn from the errors of using oppressive technological tools on marginalized communities—whom the health care system has not only failed but specifically targeted and exploited—it is time to infuse the system with the voices of those who would otherwise not be heard. Or, to create regulation by lot.

Notes

Introduction

1. "Elon Musk," *Forbes*, August 1, 2025; Bobby Allyn and Lauren Frayer, "100 Days of DOGE: What Elon Musk Has Achieved and What Comes Next," *NPR*, April 27, 2025; Jo Yurcaba, "Domestic Violence and Sexual Assault Organizations Sue Trump Administration over Funding Restrictions," *NBC News*, June 16, 2025.

2. Kalyn Belsha and Erica Meltzer, "Teens with Disabilities Were Getting Help with Life After High School. Then DOGE Started Cutting," *Chalkbeat*, February 25, 2025.

3. Cavalcanti et al., "Evaluating the Impact of Two Decades of USAID Interventions."

4. Rhitu Chatterjee, "Trump Administration Ends 988 Lifeline's Special Service for LGBTQ+ Young People," *NPR*, July 19, 2025.

5. Nat Malkus, "Running Down DOGE's Department of Education Receipts," *AEIdeas*, February 21, 2025; Sareen Habeshian, "Musk's xAI Announces $200 Million Contract with Pentagon," *Axios*, July 14, 2025.

6. US Department of Health and Human Services, "About HHS," HHS.gov, accessed July 25, 2025, www.hhs.gov/about/index.html.

7. US Department of Health and Human Services, "HHS Announces Transformation to Make America Healthy Again," March 27, 2025; Helene Branswell, Chelsea Cirruzzo, and Daniel Payne, "Health Secretary RFK Jr. Abruptly Fires CDC Vaccine Advisory Panel," *STAT News*, June 9, 2025.

8. Martin Enserink and Meredith Wadman, "The Members of RFK Jr.'s New Vaccine Committee Have Published Little on Vaccines," *Science*, June 13, 2025; Meslé et al., "Estimated Number of Lives Directly Saved by COVID-19 Vaccination Programmes in the WHO European Region."

9. "U.S. Measles Cases Hit Highest Level Since Declared Eliminated in 2000," *Johns Hopkins Bloomberg School of Public Health*, July 7, 2025.

10. Rob Stein, "RFK Jr. Says COVID Shots No Longer Recommended for Kids, Pregnant Women," *NPR*, May 27, 2025; Walter et al., "Evaluation of the BNT162b2 Covid-19 Vaccine in Children 5 to 11 Years of Age."; Fernández-García et al., "Effectiveness and Safety of COVID-19 Vaccines on Maternal and Perinatal Outcomes."

11. "Congress Must Support Robust Federal Funding for the NIH," *AAMC*, February 4, 2025; Cleary et al., "Comparison of Research Spending on New Drug Approvals by the National Institutes of Health."

12. Marcelo Jauregui-Volpe, "New Brief Finds NIH Has Canceled $1.9 Billion in Grants," *Association of Americans Universities*, May 9, 2025.

13. Annie Waldman, Asia Fields, and Ashley Clarke, "The Trump Administration Said Their Work Was 'Antithetical to the Scientific Inquiry,'" *ProPublica*, June 12, 2025.

14. Yuki Noguchi, "How Safe Is the Food Supply After Federal Cutbacks? Experts Are Worried," *NPR*, May 29, 2025.

15. Mark Doré, "The Risks of Trump's Proposal to Speed Up Drug Approvals," *Media Freedom and Information Access Clinic*, February 20, 2017. This is ironic, given that COVID vaccine skepticism in part originated because of how fast they were able to produce the mRNA vaccines. S. K. Lee et al., "Misinformation of COVID-19 Vaccines."

16. Sarah Fortinsky, "'FDA as We've Known It Is Finished': Former Commissioner," *The Hill*, April 1, 2025.

17. Lisa Friedman and Hiroko Tabuchi, "The E.P.A. Shifts Its Mission," *New York Times*, March 13, 2025.

18. Anna Clark, "Trump's First EPA Promised to Crack Down on Forever Chemicals. His Second EPA Is Pulling Back," *ProPublica*, July 2, 2025.

19. Jared Hayes, "'Forever Chemicals' in Sludge May Taint Nearly 70 Million Farmland Acres," *Environmental Working Group*, January 14, 2025.

20. Sharon Lerner, "Toxic Gaslighting: How 3M Executives Convinced a Scientist the Forever Chimcals She Found in Human Blood Were Safe," *ProPublica*, May 20, 2024.

21. US Environmental Protection Agency, "EPA Releases Proposal to Rescind Obama-Era Endangerment Finding, Regulations That Paved the Way for Electric Vehicle Mandates," July 29, 2025.

22. Gabric, "Climate Change Crisis." The EPA's announcement also came just weeks after catastrophic flooding in Texas killed at least 135 people, many of whom were children at a summer camp, several months after devastating wildfires in Los Angeles, and almost a year after a disastrous hurricane season that caused hundreds of deaths and billions of dollars in property damage in North Carolina and Florida, with predictions of an "above-normal" hurricane season for 2025. Accompanying these events, the Trump administration had plans to eradicate the Federal Emergency Management Agency (FEMA), tasked with emergency response, an act that also requires congressional approval, and it cut the budget of the National Oceanic Atmospheric Administration (NOAA) by 27 percent, impeding NOAA's ability to research and track "tornadoes, hail, flash flooding, hurricanes, wildfires." See also J. David Goodman and Danny Hakim, "FEMA's Director Defends Texas Flood Response During Hearing with Lawmakers," *New York Times*, July 24, 2025; Institute for Applied Economics, "Impact of 2025: Los Angeles Wildfires and Comparative Study," February 2025; Jonathan Belles, "Atlantic Hurricane Season 2024: Florida Walloped, North Carolina Sees Catastrophe," *Weather Channel*, November 27, 2024; National Oceanic Atmospheric Administration, "NOAA Predicts Above-Normal 2025 Atlantic Hurricane Season," May 22, 2025; Rebecca Hersher, "The Trump Administration Says It Wants to Eliminate FEMA. Here's What We Know," *NPR*, June 26, 2025; "New NOAA Budget Details Plans to Eliminate Most Weather and Climate Research," *KCCI*, July 7, 2025.

23. Julia Moskin, "What Are SNAP Benefits and How Will They Change?" *New York Times*, June 1, 2025.

24. Congressional Budget Office, *Dynamic Analysis of Changes to the Supplemental Nutrition Assistance Program (SNAP) Under H.R. 1*; Tony Romm, "Millions Could Lose Food Stamp Benefits Under Republican Bill, Analysis Finds," *New York Times*, May 22, 2025.

25. University of Pennsylvania Leonard Davis Institute of Health Economics and Yale School of Public Health Center for Infectious Disease Modeling and Analysis, "Letter to the Honorable Ron Wyden," June 3, 2025.

26. Sophia Spadafore, "Medicaid Cuts Will Be a Disaster for ERs," *Atlantic*, July 22, 2025; National Rural Health Association, "Estimated Impact on Medicaid Enrollment and Hospital Expenditures in Rural Communities," July 20, 2025.

27. Grace Sparks et al., "Americans' Challenges with Health Care Costs," *KFF*, July 11, 2025.

28. Stefano M. Bertozzi, "U.S. Withdrawal from WHO Could Bring Tragedy at Home and Abroad," *Berkeley Public Health*, University of California, Berkeley, January 24, 2025; Youri Benadjaoud and Georgia Davis, "Former CDC Officials Warn Proposed Budget Cuts Cost American Lives in 2026," *ABC News*, July 2, 2025.

29. US Department of Health and Human Services, "Celebrating Big Wins of the Trump Administration," www.hhs.gov/hhs-big-wins-maha/index.html, accessed July 25, 2025.

30. Julie Corliss, "Seeding Doubt: The Truth About Cooking Oils," *Harvard Heart Letter*, March 1, 2025.

31. Tallie and Popkin, "Promise and Pitfalls"; Lavazza and Farina, "Role of Experts in the Covid-19 Pandemic."

32. Valgarðsson, "Crisis of Political Trust?"; Sarah McCammon, "A Look at Trump's Executive Order That the Government Will Recognize Only Two Sexes," *NPR*, January 22, 2025.

33. "R.F.K. Jr. Says U.S. Will Stop Funding Global Vaccine Agency," *New York Times*, June 25, 2025.

34. Jake Scott, "By Remaking CDC's Vaccine Advisory Panel, RFK Jr. Is Destroying Public Trust," *STAT News*, June 11, 2025; Alex Montero et al., "KFF Tracking Poll on Health Information and Trust: Tylenol-Autism Link and Vaccine Policies," *KFF*, October 9, 2025.

35. Kellner, "Power, Autocracy, and the Crisis of Democracy."

36. Domenico Montanaro, "Poll: Most Feel Democracy Is Threatened and Political Violence Is a Major Problem," *NPR*, July 1, 2025.

37. "New Battleground Civility Poll: Post-Midterms, New Poll Shows Voters Have Deep Concerns over Threat to Democracy, but for Very Different Reasons," news release, Georgetown University Institute of Politics and Public Service, November 30, 2022.

38. Knight Foundation, *Crisis in Democracy*; Kelly Mena et al., "Georgia Republicans Speed Sweeping Elections Bill Restricting Voting Access into Law," *CNN*, March 26, 2021.

39. Isaac Chotiner, "Why 'Constitutional Crisis' Fails to Capture Trump's Attack on the Rule of Law," *New Yorker*, March 20, 2025; Joe Walsh, "Trump's Birthright Citizenship Order Is Unconstitutional, Appeals Court Says," *CBS News*, July 25, 2025.

40. "New Battleground Civility Poll."

41. Formisano, *American Oligarchy*; Carpenter and Moss, *Preventing Regulatory Capture.*

42. Jasanoff, *Fifth Branch*, 9.

43. M. Brown, *Science in Democracy.*

44. Achen and Bartels, *Democracy for Realists*; Brennan, *Against Democracy.*

45. McKee and Bohannon, "Exploring the Reasons Behind Parental Refusal of Vaccines."

46. Pausé, "Die Another Day"; Wiegand, "'It's Probably an STI Because You're Gay.'"

47. Frank et al., "Predictors of Interoperative Pain."

48. Julie Snyder and Ben Phelan, producers, "The Guidelines," episode 3, *The Retrievals*, season 2, podcast series, 2025, www.nytimes.com/2025/07/10/podcasts/c-sections-pain-retrievals-women.html?; Washington, *Medical Apartheid.*

49. Mendell, *Reclaiming Democracy*; see also Clements, *Corporations Are Not People*; Shrader-Frechette, *Environmental Justice.*

50. There is some debate in the literature about whether this was truly deliberative in the way we refer to deliberation today or whether it was too chaotic for that (Cammack, "Deliberation and Discussion in Classical Athens"). However, there is near consensus that this was an original form of "direct democracy" (Arvanitidis and Kyriazis, "Qualities and Workings of Direct Democracy").

51. Elster, *Deliberative Democracy*, 1.

52. Gutmann and Thompson, *Why Deliberative Democracy?* 8.

53. Gustafson, *Imagining Deliberative Democracy in the Early American Republic.*

54. Tocqueville, *Democracy in America*; Hall, "Deliberative Democracy."

55. Leonard, *Illiberal Reformers.*

56. Plotkin, *History of Vaccine Development.*

57. Rothbard, *Progressive Era.*

58. Addams, *Democracy and Social Ethics*, 5.

59. Addams, *Democracy and Social Ethics*, 7. Addams also argued passionately for women's experiences to be taken seriously in understanding the lived conditions of vulnerable people, something that is explored more in chapter 4.

60. Dewey, *Public and Its Problems*, 143.

61. Dewey, *Democracy and Education*, 99.

62. Fung, "Varieties of Participation in Complex Governance," 76.

63. Organisation for Economic Co-operation and Development, "Innovative Citizen Participation and New Democratic Institutions." See also Paulis et al., "POLITICIZE Dataset," 538.

64. Gastil et al., "Convening a Minipublic During a Pandemic."

65. Devaney et al., "Ireland's Citizens' Assembly on Climate Change."

66. Habermas, *Between Facts and Norms*, 306.

67. Abizadeh, "Representation, Bicameralism, Political Equality, and Sortition," 792; Curato et al., *Deliberative Mini-Publics*, 11.

68. Vergne, "Brief Survey of the Literature of Sortition," 69; Heller and Rao, *Deliberation and Development*, 2.

69. Owen and Smith, "Survey Article," 213; Gamson, *Talking Politics.*

70. Gastil and Wright, eds., *Legislature by Lot.*

71. Niemeyer and Jennstål, "Scaling Up Deliberative Effects," 330.

72. Observation of Public Sector Innovation, "Irish Citizens' Assembly," 2016, https://oecd-opsi.org/innovations/the-irish-citizens-assembly/.

73. P. Tucker, *Unelected Power.*

74. Dahl, *Preface to Democratic Theory*, 2.

75. See Pamuk, *Politics and Expertise*; M. Brown, *Science in Democracy.*

76. Nicol et al., "Australian Citizens' Jury."

77. See Nabatchi, "Addressing the Citizenship and Democratic Deficits."

78. Landemore, "Beyond the Fact of Disagreement?" 290; Some authors derive legitimate authority from other sources. For example, for some social contract theorists such as John Locke, "consent to be governed" is the source of legitimate rule (*Two Treatises of Government*). Locke's natural rights theory has been heavily scrutinized, both by his contemporaries such as David Hume, who argued in *Treatise on Human Nature* that Locke's brand of "tacit consent" failed to prove enough about an individual's willingness to be ruled by a government, and by contemporary authors who argue natural rights have little traction in pluralist or secular societies (Fuller, *Morality of Law*). Other writers articulate legitimate authority as deriving from an alternative ("reasonable") consensus to be governed (Rawls, *Theory of Justice*); the moral obligations we have to each other and the government to us (Dworkin, *Law's Empire*); the intrinsic moral well-being for others' contractualism (Scanlon, *What We Owe to Each Other*); or the reasonableness of procedures (Raz, *Authority of Law*).

79. The consequentialist defense of legitimate authority is also referred to as a "procedure independent" account of democracy. That is, the procedures are defensible (or legitimate) insofar as they engender good outcomes, rather than because of the nature of the procedures themselves. This means that if an electoral system, the Electoral College, or the practice of gerrymandering produces inadequately representative outcomes, then the procedure would need to be changed. It also means that some nondemocratic organizations such as schools and hospitals, whose leaders are not selected through procedures involving those they govern, may hold legitimate authority insofar as they produce good educational or life-saving outcomes. If democracy is valuable because it protects human rights, for example, then we should measure its legitimacy by the extent of its human rights protections. See Buchanan, "Political Legitimacy and Democracy."

80. Grönlund, Setälä, and Herne, "Deliberation and Civic Virtue," 95; Scudder, *Beyond Empathy and Inclusion.*

81. Matthew Taylor, "Citizen Deliberation Is the Gateway to a Better Politics," *Economist*, March 11, 2019; Daniel Kopp, "Deliberative Democracy Makes Citizens Happy," *International Politics and Society*, May 20, 2019; Fung, "Varieties of Participation in Complex Governance," 74.

82. Fang et al., "Gene Editing in Regenerative Medicine," 741.

83. “Swiss Army Knife of Gene Editing Gets New Control,” *Mirage News*, September 5, 2019; Fang et al., “Gene Editing in Regenerative Medicine,” 741.

84. While CRISPR-Cas9 together seem to be a powerful tool to edit DNA, other enzymes in the “Cas system” also promise efficient transmission of CRISPR to cells (van Beljouw et al., “RNA-Targeting CRISPR–Cas Systems”).

85. “Over $1bn Raised in Venture Capital Funding for Gene Editing in 2021,” *Pharmaceutical Technology*, January 27, 2022; Eric Niler, “Why Gene Editing Is the Next Food Revolution,” *National Geographic*, August 10, 2018.

86. Cong et al., “Multiplex Genome Engineering.”

87. Adli, “CRISPR Tool Kit,” 911; Liu et al., “Knowledge Mapping and Current Trends of Global Research,” 11.

88. Richard Gray, “Why Gene Editing Could Create So Many Jobs,” *BBC*, October 14, 2018; Timothé Cynober, “CRISPR: One Patent to Rule Them All,” *Labiotech*, November 2, 2019.

89. “CRISPR Market Snapshot from 2024 to 2034,” Future Market Insight, 2024.

90. Collins et al., “Frugal CRISPR Kit”; Stephanie M. Lee, “This Guy Says He’s the First Person to Attempt Editing His DNA with CRISPR,” *Buzzfeed News*, October 14, 2017; Lana Schwartz, “I Edited Human DNA at Home with a DIY CRISPR Kit,” *Vice*, January 23, 2023.

91. It was later revealed that another baby was born as well. Alice Klein and Michael Le Page, “The Gene Editing Revelation That Shocked the World,” *New Scientist*, December 5, 2018; Sharon Begley and Andrew Joseph, “The CRISPR Shocker: How Genome-Editing Scientist He Jiankui Rose from Obscurity to Stun the World,” *STAT News*, December 17, 2018. To protect the twin girls against HIV infection, He used CRISPR to disable the CCR5 gene that plays a role in enabling HIV to enter immune cells. By deleting a part of the gene, he disabled it, “mimicking a naturally occurring mutation . . . that confers resistance to HIV” (Sara Reardon, “Gene Edits to ‘CRISPR Babies’ Might Have Shortened Their Life Expectancy,” *Nature News*, June 3, 2019).

92. Notable exceptions to the near-universal condemnation include Harvard geneticist George Church, who also played a “pioneering role in developing CRISPR.” Church held a self-described “middle” ground position, believing that Dr. He engaged in something worthwhile because the lack of an HIV vaccination and cure presents an “unmet medical need,” and the animal stages of development showed no “evidence of negative consequences” (Jon Cohen, “‘I Feel an Obligation to Be Balanced.’ Noted Biologist Comes to Defense of Gene Editing Babies,” *Science*, November 28, 2018).

93. Ed Yong, “The CRISPR Baby Scandal Gets Worse by the Day,” *Atlantic*, December 3, 2018.

94. Austin Ramzy and Sui-Lee Wee, “Scientist Who Edited Babies’ Genes Is Likely to Face Charges in China,” *New York Times*, January 21, 2019.

95. Ramzy and Wee, “Scientist Who Edited Babies’ Genes”; Smriti Mallapaty, “Disgraced CRISPR-Baby Scientist’s ‘Publicity Stunt’ Frustrates Researchers,” *Nature News*, February 13, 2023.

96. Preetika Rana, "How a Chinese Scientist Broke the Rules to Create the First Gene-Edited Babies," *Wall Street Journal*, May 10, 2019; Yong, "CRISPR Baby Scandal Gets Worse by the Day"; Wang and Yang, "Gene-Edited Babies." Eventually, he served a several-year prison sentence in China (Jon Cohen, "As Creator of 'CRISPR Babies' Nears Release from Prison, Where Does Embryo Editing Stand?" *Science*, March 21, 2022).

97. Hank Greely, "How Should Science Respond to CRISPR'd Babies?" *Issues in Science and Technology* 35 (2019): 34, 36.

98. Rob Schmitz, "Gene-Editing Scientist's 'Actions Are a Product of Modern China,'" *NPR*, February 5, 2019.

99. Landon J. Getz and Graham Dellaire, "CRISPR Gene Editing: Why We Need Slow Science," *Phys Org*, March 19, 2019; Nina Frahm and Tess Doezema, "Are Scientists' Reactions to 'CRISPR Babies' About Ethics or Self-Governance?" *STAT News*, January 28, 2019.

100. Lander et al., "Adopt a Moratorium on Heritable Genome Editing."

101. Lander et al., "Adopt a Moratorium," 165.

102. Lander et al., "Adopt a Moratorium," 166.

103. Lander et al., "Adopt a Moratorium," 167.

104. See, for example, Zhai et al., "Lessons from the HE JIANKUI Incident."

105. Cohen et al., "Handle with Care," 10.

106. WHO Expert Advisory Committee on Developing Global Standards for Governance and Oversight of Human Genome Editing, "Report of the Second Meeting," 1; WHO Expert Advisory Committee on Developing Global Standards for Governance and Oversight of Human Genome Editing, "Human Genome Editing: Recommendations." Further, the moral consensus in the medical and philosophical community is largely that modifications to a germline (reproductive cells, such as ova and embryos) are, at least currently, morally impermissible because germline changes would be heritable, meaning they are transmitted to the next generation. Others have disputed whether there is such a significant moral difference between germline and somatic cell editing because both have widespread implications (Lewis, "Is Germline Gene Editing Exceptional?").

107. International Commission on the Clinical Use of Human Germline Genome Editing, *Heritable Human Genome Editing*, 1.

108. Heidi Ledford, "World's First Personalized CRISPR Therapy Given to Baby with Genetic Disease," *Nature News*, May 15, 2025.

109. Nick Stockton, "How CRISPR Could Snip Away Humanity's Worst Diseases," *Wired*, May 5, 2017.

110. Isaacson, *Code Breaker.*

111. E. A. Posner, *Governance of Dual-Use Technologies*, 1; Orlikowski, "Duality of Technology," 301.

112. Fregulia, *Rich and Tantalizing Brew*; French, *When They Hid Fire.* Mary Shelley's 1818 novel *Frankenstein; or, The Modern Prometheus* warns of the ambitious and reckless rogue scientist's use of electricity to reanimate life. It has also been embraced as an allegory for CRISPR, particularly after He's actions (see Nowlin, "200 Years After Frankenstein").

113. Lander, "Adopt a Moratorium," 167.

114. WHO Expert Advisory Committee, "Human Genome Editing: Recommendations," 16.

115. WHO Expert Advisory Committee, "Heritable Human Genome Editing," 21.

116. Scheinerman, "Public Engagement Through Inclusive Deliberation."

117. Kweon and Choi, "Fueling Conspiracy Beliefs."

118. National Academies of Sciences, Engineering, and Medicine, *Gene Drives on the Horizon*, 79.

119. Nuffield Council on Bioethics, *Novel Techniques*; Nuffield Council on Bioethics, *Genome Editing and Farmed Animal Breeding.*

120. Nuffield Council on Bioethics, *Genome Editing and Human Reproduction*, 128.

Chapter 1

1. Hornblum, *Acres of Skin*; Elliott, *Occasional Human Sacrifice.*

2. Musunuru et al., "Patient-Specific In Vivo Gene Editing," 1.

3. James Moore, "CRISPR Genome Editing Treatment Still Effective After One Year," *BioNews*, March 7, 2022.

4. Meeri Kim, "With CRISPR Gene Editing, Unique Treatments Begin to Take Off for Rare Diseases," *Washington Post*, February 5, 2022; "Intellia Therapeutics Announces FDA Clearance of Investigational New Drug (IND) Application to Initiate a Pivotal Phase 3 Trial of NTLA-2001 for the Treatment of Transthyretin (ATTR) Amyloidosis with Cardiomyopathy," *Intellia Therapeutics*, October 18, 2023.

5. Mark Iskowitz, "Regeneron, Intellia Expand CRISPR-Based Research Effort," *Medical Marketing and Media*, October 4, 2023.

6. Shi et al., "CRISPR/Cas9"; Thapar et al., "Application of CRISPR/Cas9"; Alkanli et al., "CRISPR/Cas9 Mediated Therapeutic Approach."

7. McColgan and Tabrizi, "Huntington's Disease."

8. Hussein et al., "CRISPR-Cas Cure for HIV/AIDS"; Hu et al., "Application and Progression of CRISPR/Cas9 Technology"; Padmanabhan et al., "Beyond Genome-Wide Scans"; Horie et al., "VERVE-101"; Eslahi et al., "New Advancements in CRISPR-Based Gene Therapy"; Azangou-Khyavy et al., "CRISPR/Cas"; Begagic et al., "CRISPR/Cas9-Mediated Gene Therapy"; Liu et al., "Cryo-Shocked Tumor Cells Deliver CRISPR-Cas9"; Asrar et al., "Role of CRISPR/Cas9 Based Therapy in Breast Cancer."

9. Lofti et al., "Application of CRISPR-Cas9 Technology in Diabetes Research."

10. Abavisani et al., "CRISPR-Cas System as a Promising Player"; Zahedipour et al., "Harnessing CRISPR Technology for Viral Therapeutics."

11. Paul Biegler, "CRISPR Hope for Vax-Resistant Disease Treatment," *Cosmos Magazine*, May 20, 2019.

12. Holly Evarts, "New COVID-19 Rapid-Test Technology Performs PCR Faster Than Similar Tests on the Market," *Columbia Engineering*, July 25, 2022.

13. Evarts, "New COVID-19 Rapid-Test Technology."

14. Jayachandran et al., "Genetic Advancements in Obesity Management"; Vlashi et al., "Potential Therapeutic Strategies"; Sharp et al., "Inactivation of

Phosphodiesterase-4B Gene in Rat Nucleus Accumbens Shell by CRISPR/Cas9"; Bekaert et al., "Various Repair Events."

15. Andrea Park, "Astronauts Use CRISPR in 1st Gene Editing Experiment in Space," *HEALTH IT*, May 24, 2019.

16. Yuan et al., "Complement Networks in Gene-Edited Pig Xenotransplantation."

17. "Doctors Transplant a Genetically Modified Pig Heart into a Human for the 1st Time," *NPR*, January 10, 2022.

18. "A Man Who Got the 1st Pig Heart Transplant Has Died After 2 Months," *NPR*, March 9, 2022.

19. "Keep Off-Target Effects in Focus."

20. Aquino-Jarquin, "Current Advances in Overcoming Obstacles," 84.

21. Herai, "Avoiding the Off-Target Effects," 176.

22. Park et al., "Comprehensive Analysis and Accurate Quantification," 1.

23. Herai, "Avoiding the Off-Target Effects," 176; see also Anderson et al., "CRISPR Off-Target Analysis in Genetically Engineered Rats and Mice"; Lin and Wong, "Off-Target Predictions in CRISPR-Cas9 Gene Editing Using Deep Learning."

24. Lopes and Prasad, "Beyond the Promise," 6.

25. Lopes and Prasad, "Beyond the Promise," 8.

26. Rob Stein, "Scientists Plan to Start Human Trials Testing CRISPR Soon," *NPR*, April 16, 2019; Robert Sanders, "FDA Approves First Test of CRISPR to Correct Genetic Defect Causing Sickle Cell Disease," *FDA News*, December 8, 2023.

27. Ajeet et al., "Revolutionary Breakthrough."

28. Byrnes et al., "Process and Strategies for Patient Engagement."

29. "Birth Prevalence of Sickle Cell Disease and County-Level Social Vulnerability—Sickle Cell Data Collection Program, 11 States, 2016–2020," *CDC Morbidity and Mortality Report*, March 28, 2024.

30. Williams, "Sickle Cell Disease in Sub-Saharan Africa."

31. Washington, *Medical Apartheid*.

32. Royal et al., "Living with Sickle Cell Disease."

33. Hollister et al., "Perspectives of Sickle Cell Disease Stakeholders on Heritable Genome Editing," 447.

34. Wailoo, "Sickle Cell Disease."

35. Al Adawi et al., "Lived Experience of Sickle Cell Patients."

36. Dyal et al., "Reflections of Healthcare Experiences of African Americans with Sickle Cell Disease or Cancer," E57.

37. Phillips et al., "Perspectives of Individuals with Sickle Cell Disease on Barriers to Care."

38. Ezenwa et al., "Differences in Sensory Pain, Expectation, and Satisfaction."

39. Haywood et al., "Association of Provider Communication with Trust Among Adults with Sickle Cell Disease."

40. Robin Fields and Maya Miller, "Insurance Executives Refused to Pay for the Cancer Treatment That Could Have Saved Him," *ProPublica*, November 15, 2023.

41. Persaud et al., "A CRISPR Focus on Attitudes and Beliefs Toward Somatic Genome Editing," 1730. From a scientific standpoint, SCD was a promising candidate

for gene-modifying or replacement therapies because it had a "well-characterized monogenic defect," meaning there was a single mutation that was easily identifiable (Ajeet et al., "Revolutionary Breakthrough," 4557).

42. Farrell et al., "End Points for Sickle Cell Disease Clinical Trials," 4003.

43. Rob Stein, "First Sickle Cell Patient Treated with CRISPR Gene-Editing Still Thriving," *NPR*, December 31, 2021.

44. Salcedo et al., "Cost-Effectiveness of a Hypothetical Cell or Gene Therapy Cure for Sickle Cell Disease," 10838.

45. Heidi Ledford, "Doctors Cured Her Sickle-Cell Disease. So Why Is She Still in Pain?" *Nature News*, September 17, 2024.

46. Farooq et al., "CRISPR/Cas9."

47. "A Parent's Guide to Hearing Loss," *Centers for Disease Control and Prevention*, May 15, 2024.

48. He et al., "Update on Gene Therapy in the Treatment of Hereditary Hearing Loss," e70004; see also Gao et al., "Treatment of Autosomal Dominant Hearing Loss by In Vivo Delivery of Genome Editing Agents."

49. Tao et al., "Treatment of Monogenic and Digenic Dominant Genetic Hearing Loss by CRISPR-Cas9 Ribonucleoprotein Delivery In Vivo."

50. Scharp et al., "Conceptualizing Other-Resilience."

51. Nassiri et al., "Current Estimates of Cochlear Implant Utilization in the United States."

52. Sparrow, "Defending Deaf Culture," 136; see also Bernicchia-Freeman, "Blurring of Identity."

53. Chapman and Dammeyer, "Significance of Deaf Identity for Psychological Well-Being."

54. Bradfield, "Hearing Parents' Voices," 143.

55. Cooper, "Can It Be a Good Thing to Be Deaf?"

56. Barnes, "Valuing Disability, Causing Disability."

57. Wallis, "Is It Ever Morally Permissible to Select for Deafness in One's Child?"

58. Barnes, "Valuing Disability, Causing Disability."

59. Arnos, "Genetics and Deafness," 165.

60. Padden and Humphriest, "Who Goes First?"

61. Hoffman-Andrews et al., "Attitudes of People with Inherited Retinal Conditions Toward Gene Editing Technology."

62. Howard, "Political Value of Letting Hopes Die."

63. Haque et al., "Application of CRISPR/Cas9 Genome Editing Technology for the Improvement of Crops," 8.

64. Johann Tasker, "Biotech Crops Can Help Mitigate Climate Change, Say Experts," *Farmers Weekly*, July 16, 2019.

65. Paul et al., "Plant and Fungal Genome Editing to Enhance Plant Disease Resistance."

66. Celik et al., "Crispr/Cas Genome Editing in Fruit Crops"; Priti et al., "Nutritional Enhancement in Horticultural Crops by CRISPR/Cas9."

67. Zhang et al., "Emerging and Uncultivated Potential of CRISPR Technology in Plant Science."

68. Tripathi et al., "CRISPR/Cas9-Mediated Editing of DMR6 Orthologue in Banana."

69. Tiwari et al., "CRISPR/Cas Genome Editing in Tomato Improvement."

70. Hou et al., "CRISPR/Cas Genome Editing System and Its Application in Potato," 14.

71. Rahman et al., "Improvement of Soybean."

72. Shaheen et al., "CRISPR-Cas System"; Karavolioas et al., "CRISPR/Cas9 Knockout of EPFL10 Reduces Stomatal Density"; Claeys et al., "Coordinated Gene Upregulation in Maize"; Ma et al., "Would Consumers Accept CRISPR Fruit Crops If the Benefit Has Health Implications?"

73. Kenny Walter, "Brewers Use CRISPR to Make Hopeless Beer," *Drug Discovery and Development*, March 21, 2018.

74. Shanmugamab et al., "Advanced CRISPR/Cas-based Genome Editing Tools for Microbial Biofuels Production."

75. "FDA Makes Low-Risk Determination for Marketing of Products from Genome-Edited Beef Cattle After Safety Review," *PR Newswire*, March 7, 2022.

76. Jennifer Marston, "BREAKING: SCiFi Foods Bags $22m in a 16z-led Round to Bring Hybrid Alt Beef Burgers to Market Faster," *AgFunder News*, June 8, 2022.

77. Jeremy Hsu, "Supersize My Seabream," *Hakai Magazine*, June 18, 2019.

78. "The Influence of CRISPR-Cas on Animal Breeding," *Technology Networks*, July 2, 2019.

79. Cleves et al., "CRISPR/Caso-Mediated Genome Editing in a Reef-Building Coral."

80. Christie Hemm Klok, "A More Humane Livestock Industry, Brought to You by CRISPR," *Wired*, March 19, 2019.

81. Allyson Mann, "Ecologist Dissects the Science of De-Extinction," *UGA Today*, April 19, 2019; B. Shapiro, *How to Clone a Mammoth*.

82. Kaebnick and Jennings, "De-Extinction and Conservation."

83. Conklin, "On the Road to a Gene Drive in Mammals"; Emerson et al., "Principles for Gene Drive Research"; Esvelt et al., "Concerning RNA-Guided Gene Drives for the Alteration of Wild Populations"; Noble, "Current CRISPR Gene Drive Systems Are Likely to Be Highly Invasive in Wild Populations."

84. National Academies of Sciences, Engineering, and Medicine, *Gene Drives on the Horizon*.

85. Max Kozlov, "Resistant Mosquito Threatens Africa's Fight Against Malaria," *Nature News*, November 2, 2022.

86. Kyrou et al., "A CRISPR–Cas9 Gene Drive Targeting Doublesex"; Kistler et al., "Genome Engineering with CRISPR-Cas9 in the Mosquito Aedes Aegypti."

87. World Health Organization, "World Malaria Report," 2024.

88. Dánica Coto, "Dengue Cases Set a New Record in the Americas This Year as Deaths Also Surge," *AP News*, December 10, 2024.

89. "Current Dengue Outbreak," *Centers for Disease Control and Prevention*, July 29, 2025; Haider et al., "Global Landmark."

90. Timothy C. Winegard, "The Mosquitoes Are Coming for Us," *New York Times*, July 27, 2019.

91. Rebecca Roberts, "CRISPR Gene Drives: Eradicating Malaria, Controlling Pests, and More," *Synthego*, August 9, 2022.

92. Tina Hesman Saey, "Who Decides Whether to Use Gene Drives Against Malaria-Carrying Mosquitoes?" *Science News*, June 3, 2022.

93. Donavyn Coffey, "First Genetically Modified Mosquitoes Released in U.S. Are Hatching Now," *Scientific American*, May 14, 2021.

94. "Say No to Genetically Modified Mosquitoes Release in the Florida Keys," Florida Keys Environmental Coalition, accessed July 25, 2025, https://fkec.org/gmo-mosquitos.

95. Thomas et al., "Public Communication and Outreach by Mosquito Programs in the United States."

96. Sandee LaMotte, "750 Million Genetically Engineered Mosquitoes Approved for Release in Florida Keys," *CNN*, August 20, 2020.

97. "Lyme Disease Surveillance and Data," *Centers for Disease Control and Prevention*, March 14, 2024.

98. Ed Yong, "One Man's Plan to Make Sure Gene Editing Doesn't Go Haywire," *Atlantic*, July 11, 2017.

99. Jimmy Day, "In an Effort to Curb Lyme Disease, Scientists Hope to Release Thousands of Genetically Altered Mice on Nantucket," *MIT Media Lab*, April 21, 2022; Hannah Thomasy, "CRISPR Gene Drives and the Future of Evolution," *Scientist*, March 15, 2024.

100. Jason Graziadei, "Proposal for Genetically Engineered Mice to Fight Lyme Disease on Nantucket Gets Another Look," *Nantucket Current*, October 24, 2024.

101. Gusmano et al., "Public Deliberation in the Wild."

102. Esvelt et al., "Concerning RNA-Guided Gene Drives for the Alteration of Wild Populations."

103. Ed Yong, "New Zealand's War on Rats Could Change the World," *Atlantic*, November 16, 2017.

104. Karen Weintraub, "Despite Controversy, Human Studies of CRISPR Move Forward in the U.S.," *Scientific American*, August 13, 2019.

105. Wagner-Coello et al., "Incorporating Citizen Science Engagement."

106. Cunha et al., "Citizen Science Participation in Research in the Environmental Sciences."

107. Intergovernmental Panel on Climate Change, *IPCC Sixth Assessment Report*.

108. Sara Reardon, "NIH Reiterates Ban on Editing Human Embryo DNA," *Nature News*, April 29, 2015.

109. Macintosh, "Regulation of Human Germline Genome Modification in the United States."

110. Timothé Cynober, "CRISPR: One Patent to Rule Them All," *Labiotech*, November 2, 2019. As I have written with Jordana Goodman, "Just as we cannot discuss the problem of homelessness without first diagnosing the problem of the housing market generally (including rising prices, rent control, and lower-income housing and subsidies), we cannot have a complete conversation about the ethical use of gene editing without first exploring patent law, which sets the stage for how

market forces may shape its future. The manner in which patents are awarded is highly consequential to the way in which gene editing is capitalized on, developed, applied and distributed, all of which strongly influence the aforementioned moral concerns of health, safety, equality and well-being" (Jordana Goodman and Naomi Scheinerman, "Why CRISPR Patents Matter to a World Worried About Designer Babies, Rogue Scientists and Curing Disease," *Genetic Literacy Project*, July 2, 2019).

111. Contreras, *Genome Defense*.

112. Christopher Schwaiger, "Ongoing CRISPR Patent Dispute Complicates Licensing but Hasn't Deterred Gene-Editing Investment," *BioSpace*, July 17, 2024.

113. Scheinerman and Sherkow, "Governance Choices of Genome Editing Patents."

114. Megan Molteni, "In the Wake of Big Patent Decision, It's Business as Usual for CRISPR Therapy Developers," *STAT News*, March 31, 2022.

115. Quoted in Carl Zimmer, "CRISPR, 10 Years On: Learning to Rewrite the Code of Life," *New York Times*, June 27, 2022.

116. Heidi Ledford, "Major CRISPR Patent Decision Won't End Tangled Dispute," *Nature News*, March 9, 2022.

117. Antonio Regaldo, "Two Nobel Prize Winners Want to Cancel Their Own CRISPR Patents in Europe," *MIT Technology Review*, September 25, 2025.

118. "Healthcare Spending Drives Our Debt"; "Technology Revolution Happening Around Us."

119. Jocelyn Kaiser, "Update: House Spending Panel Restores U.S. Ban on Gene-Edited Babies," *Science*, June 4, 2019.

120. US Congress, Senate, Foreign Relations Committee, *S. Res.275—A Resolution Calling for International Ethical Standards in Genome Editing Research*.

121. US House, Agriculture Committee, *H. Res.427—Encouraging the Department of Agriculture to Lead Efforts*.

122. US House, Appropriations Committee, *H.R. 4346—CHIPS and Science Act*; US House, Appropriations Committee, *H.R. 8772—Legislative Branch Appropriations Act, 2025*.

123. Taeihagh et al., "Assessing the Regulatory Challenges of Emerging Disruptive Technologies"; Friedrichs et al., "An Overview of Regulatory Approaches to Genome Editing in Agriculture"; Menz et al., "Genome Edited Crops Touch the Market."

124. Asquer and Krachkovskaya, "Uncertainty, Institutions and Regulatory Responses to Emerging Technologies," 1117.

125. Scott J. Schweikart, "Regulation of Human Genome Editing in the Dawn of the CRISPR Era," *Petrie-Flom Center Blog*, May 8, 2019.

126. US Food and Drug Administration, *21 U.S. Code § 393* (1997).

127. "FDA 101: Clinical Trials and Institutional Review Boards," *EIN Presswire*, February 8, 2021.

128. W. Green, *Contraceptive Risk*.

129. Hutt, "Historical Themes and Developments at FDA over the Past Fifty Years."

130. Harding, "Rehabilitation or Enhancement?"; Ishii, "Germline Genome-Editing Research and Its Socioethical Implications."

131. US Food and Drug Administration, *Human Gene Therapy Products Incorporating Human Genome Editing.*

132. US Food and Drug Administration, *Human Gene Therapy Products.*

133. US Food and Drug Administration, *CVM GFI #187A Heritable Intentional Genomic Alterations in Animals.*

134. Menz et al., "Genome Edited Crops Touch the Market."

135. US Food and Drug Administration, *CVM GFI #236 Clarification of FDA and EPA Jurisdiction over Mosquito-Related Products.*

136. Meghani, "Regulation of Genetically Engineered (GE) Mosquitoes as a Public Health Tool," 6.

137. "FDA Approves Genomically Altered Line of Pigs for Potential Therapeutic Uses," *PharmTech*, January 4, 2021.

138. US Department of Agriculture, "Regulation of the Movement of Animals Modified or Developed by Genetic Engineering."

139. "FDA Makes Low-Risk Determination."

140. "FDA Makes Low-Risk Determination."

141. "About USDA," US Department of Agriculture, accessed July 28, 2025. www.usda.gov/about-usda/general-information/our-agency.

142. Globus and Qimron, "A Technological and Regulatory Outlook on CRISPR Crop Editing."

143. Kuzma, "Policy"; The EU Court of Justice decided on process, but Kuzma is skeptical whether this distinction is robustly scientific or more political; see also Ahmad et al., "Regulatory, Ethical, and Social Aspects of CRISPR Crops."

144. Nxumalo et al., "Patents, Ethics, Biosafety and Regulation Using CRISPR Technology."

145. Duarte Sagawa et al., "Regulatory Framework of CRISPR-Edited Crops," 185.

146. Duarte Sagawa et al., "Regulatory Framework of CRISPR-Edited Crops," 185–86.

147. Duarte Sagawa et al., "Regulatory Framework of CRISPR-Edited Crops," 186.

148. Rich Haridy, "USDA Confirms It Won't Regulate CRISPR Gene-Edited Plants Like It Does GMOs," *New Atlas*, April 3, 2018.

149. US Department of Agriculture, "Movement of Certain Genetically Engineered Organisms."

150. N. E. Hoffman, "Revisions to USDA Biotechnology Regulations," 2021.

151. Selfa et al., "Governing Gene Editing in Agriculture and Food in the United States."

152. Erik Stokstad, "United States Relaxes Rules for Biotech Crops," *Science*, May 18, 2020.

153. Emily Waltz, "CRISPR-Edited Crops Free to Enter the Market, Skip Regulation," *Nature News*, June 6, 2016.

154. Menz et al., "Genome Edited Crops Touch the Market," 9.

155. Environmental Protection Agency, "Pesticides."

Chapter 2

1. Mukherjee, *The Gene.*

2. Woolfson, *Life Without Genes*; D. Roberts, *Fatal Invention*; Campbell, *Contours of Ableism.*

3. I use the words "maternal" and "mother" here fully realizing that they are problematic, as many transgender, nonbinary, and genderqueer individuals have wombs but may "father" or "parent" the pregnancies they carry. "Maternal" is used here as an imperfect placeholder term for those who have procreative and reproductive capacities to gestate and deliver and to acknowledge a long history of gendered expectations of women's bodies and problems of invisible labor.

4. Jamie Metzl, *Hacking Darwin.*

5. Ladd-Taylor, "Saving Babies and Sterilizing Mothers."

6. Megan Molteni, "Buffalo Shooting Ignites a Debate over the Role of Genetics Research in White Supremacist Ideology," *STAT News*, May 23, 2022.

7. Saini, *Superior.*

8. Even respected geneticists have become more confident recently in declaring genetic differences among races, including Harvard geneticist David Reich, who in 2018 "caused a considerable stir . . . by insisting that there are indeed variations among 'races' (his quotes)." Pete Shanks, "Book Review: *Superior—The Return of Race Science*, by Angela Saini," *Biopolitical Times* (blog), Center for Genetics and Society, July 11, 2019; David Reich, "How Genetics Is Changing Our Understanding of 'Race,'" *New York Times*, March 23, 2018.

9. See Torres, "Race, Rare Genetic Variants, and the Science of Human Difference in the Post-Genomic Age."

10. Vinci and Robert, "Aristotle and Modern Genetics"; Miller and Lazcano, "Formation of the Building Blocks of Life."

11. Strathern, *Brief History of Medicine.*

12. This analogy is apt not only because of the ways in which people record tales of genetics and democracy, but also because some CRISPR scientists themselves have made the direct comparison. According to Jim Kozubek, on the fifth floor of the Broad Institute (one of the most powerful players in the industry), sits a poster that "depicts the Acropolis in Greece with Broad Institute members' heads cropped in the bodies of Greek philosophers and heroes" (Kozubek, *Modern Prometheus*, xxv).

13. According to Mukherjee (*The Gene*, 51), Mendel used the word "form," and "allele" was coined later in the 1900s.

14. Mukherjee, *The Gene*, 52.

15. Mukherjee, *The Gene*, 63.

16. Quoted in Mukherjee, *The Gene*, 63.

17. As Rosemary D. Harvey writes, "Bateson had a knack of looking into the future and making pronouncements which seemed at the time rather extreme, but which eventually proved to be true" ("Pioneers of Genetics," 116).

18. Quoted in Mukherjee, *The Gene*, 73.

19. Mukherjee, *The Gene*, 74.

20. Mukherjee, *The Gene*, 75.

21. Hansen and King, *Sterilized by the State*.

22. Bush, *Imperialism, Race, and Resistance*.

23. Bernasconi, "Kant as an Unfamiliar Source of Racism."

24. Yudell et al., "Taking Race Out of Human Genetics." Other countries have experienced their own sterilization campaigns, including India, China, Russia, Japan, and elsewhere. Dowbiggin, *Sterilization Movement and Global Fertility in the Twentieth Century*.

25. Biesold, *Crying Hands*; Kuhl, *Nazi Connection*.

26. Jones, *Bad Blood*.

27. Tapper, *In the Blood*.

28. Dowbiggin, *Sterilization Movement and Global Fertility in the Twentieth Century*.

29. Sanger's project was referred to as "The Negro Project" (D. Roberts, *Fatal Invention*).

30. Kevles, *In the Name of Eugenics*.

31. Lombardo, *Three Generations No Imbeciles*.

32. *Buck v. Bell*, 274 U.S. 200 (1927).

33. A. Cohen, *Supreme Court, American Eugenics, and the Sterilization of Carrie Buck*, 7.

34. Jake Taylor, "Trump Says People Crossing the Border Bring 'Very Contagious Disease' with Them," *NBC News*, May 22, 2024; Zolan Kannos-Youngs and Hamed Aleaziz, "Inside Trump's Search for a Health Threat to Justify His Immigration Crackdown," *New York Times*, January 9, 2025.

35. Rapoport, "Rosalind Franklin."

36. Mukherjee, *The Gene*, 158.

37. Mukherjee, *The Gene*, 163.

38. Mukherjee, *The Gene*, 179.

39. Mukherjee, *The Gene*, 184.

40. Kanaya and Ceci, "Are All IQ Scores Created Equal?"

41. Beit-Hallahmi, "Science, Ideology, and Ideals"; Conley and Fletcher, *Genome Factor*.

42. Porter, "Measurement and Meritocracy."

43. Watson also had pithy things to say about gender differences, again unsurprisingly finding men to be intellectually superior. Gillborn, "Softly, Softly."

44. Herrnstein and Murray, *Bell Curve*.

45. W. H. Tucker, *"Bell Curve" in Perspective*.

46. D. Roberts, *Fatal Invention*.

47. This idea has largely been debunked as "myth" (MacNamee, *Meritocracy Myth*) and the tendency to assume its accuracy as a "trap" (Markovits, *Meritocracy Trap*).

48. Ogbu, *Cultural Amplifiers of Intelligence*.

49. Centerwall, and Centerwall, "Discovery of Phenylketonuria."

50. Jennifer Henderson, "HHS Scraps Advisory Committee on Newborn Screening," *MedPage Today*, April 10, 2025; Health Resources and Services Administration, "Recommended Uniform Screening Panel," HRSA.org, January 2023.

51. Alper et al., *Double-Edged Helix*. Note, however, this was *not* the rationale behind the dissolving of the advisory board.

52. Conley and Fletcher, *Genome Factor*.

53. Mukherjee, *The Gene*.

54. E. Green, "Sequencing the Human Genome."

55. Sheryl Gay Stolberg, "The Biotech Death of Jesse Gelsinger," *New York Times*, November 28, 1999.

56. Moreno, *Body Politic*.

57. "FDA Approves First Gene Therapy for Adults with Severe Hemophilia A," *FDA News*, June 29, 2023.

58. Wong et al., "Estimated Annual Financial Impact of Gene Therapy in the United States"; Aiuti et al., "Ensuring a Future for Gene Therapy for Rare Diseases."

59. Rueda et al., "Affordable Pricing of CRISPR Treatments Is a Pressing Ethical Imperative."

60. Davies, *Editing Humanity*.

61. Amy Harmon, "Indian Tribe Wins Fight to Limit Research of Its DNA," *New York Times*, April 21, 2025.

62. Orr et al., "American Indian Genes in the Media," 67.

63. Garrison, "Informed Consent and the Ethics of IRB Research."

64. Beauchamp and Faden, "Informed Consent." The tribe sued and, in 2010, was rewarded with $700,000, the return of their blood, and assistance for scholarships to help in obtaining federal funding for a health clinic (Credo and Ingram, "Perspective Developing Successful Collaborative Research Partnerships with AI/AN Communities").

65. Garrison et al., "Genomic Research Through an Indigenous Lens."

66. Kuhn et al., "Indigenous Research Ethics Requirements"; Jacobs et al., "Bridging the Divide Between Genomic Science and Indigenous Peoples." In particular, the Center for the Ethics of Indigenous Genomic Research at the University of Oklahoma has been at the forefront of pressing the importance of tribal data sovereignty.

67. National Human Genome Research Institute, "Whole Genome Association Studies," July 15, 2011, www.genome.gov/17516714/2006-release-about-whole-genome-association-studies.

68. Larena et al., "Multiple Migrations to the Philippines During the Last 50,000 Years."

69. Wailoo, "Introduction: Genetic Claims and the Unsettled Past," 3. See also Stranger et al., "Progress and Promise of Genome-Wide Association Studies for Human Complex Trait Genetics."

70. K. Richardson, "GWAS and Cognitive Abilities."

71. Green and Scheinerman, "Against Genetic Determinism of Welfare and Behavior"; Comfort, *Science of Human Perfection*; Herd, "Reconstructing Sociogenomics Research."

72. Fatumo et al., "Roadmap to Increase Diversity in Genomic Studies," 243.

73. Fatumo et al., "Roadmap to Increase Diversity in Genomic Studies," 243.

74. This period also featured the rise of DNA "biobanks" that are "associated with electronic health records in the hope that population genetics can speed the

identification of disease susceptibility genes or diagnostic biomarkers" (Maschke, "Navigating an Ethical Patchwork," 539). Biobanks are rife with the similar large data problems we see elsewhere, namely privacy, population homogeneity, and informed consent problems for those contributing their data to, at times, unknown uses.

75. Gwynne et al., "Genomics and Inclusion of Indigenous Peoples in High Income Countries." As of March 27, 2024, according to the NIH's website, 781,00 participants had consented to join the program and completed all initial steps. National Institutes of Health, *All of Us*, 2025, accessed July 25, 2025, https://allofus.nih.gov/.

76. Baylis, "Human Genome Editing."

77. This also inspired the 1997 movie *Gattaca,* which portrayed a world in which the social order is determined by genetic selection of embryos.

78. De Boeck, "Cystic Fibrosis in the Year 2020."

79. The ability to sequence entire genomes also ushered in "personalized medicine," which involves the use of "genomics-based knowledge and tools [that] promise the ability to approach each patient as the biological individual he or she [or they] is, thereby radically changing our paradigms and improving efficacy." Guttmacher et al., "Educating Health-Care Professionals About Genetics and Genomics," 151; See also Burke and Psaty, "Personalized Medicine in the Era of Genomics."

80. MacKellar and Bechtel, *Ethics of the New Eugenics.*

81. Daar, *New Eugenics.*

82. Lagu et al., "I Am Not the Doctor for You."

83. Labube et al., "Vulnerability and the Ethics of Human Germline Genome Editing."

84. Parens and Asch, *Prenatal Testing and Disability Rights.*

85. J. L. Scully, *Disability Bioethics*, 25; Davis, "Introduction: Normality, Power, and Culture."

86. Shakespeare, "Social Model of Disability."

87. A. Knight, "Gene Editing Technologies, Utopianism, and Disability Politics"; T. Rulli, "Reproductive CRISPR Does Not Cure Disease."

88. Retel Helmrich et al., "Discrepancy Between Disability and Reported Well-Being After Traumatic Brain Injury."

89. Mailhan et al., "Life Satisfaction and Disability After Severe Traumatic Brain Injury."

90. Ubel et al., "Whose Quality of Life?"

91. Shimazono, "Reasons Against the Selection of Life," 306.

92. Kamm, "What Is and Is Not Wrong with Enhancement?" 109, 114.

93. Savulescu, "New Breeds of Humans," 36.

94. Savulescu, "New Breeds of Humans."

95. Molhoek, "Raising the Virtuous Bar"; Persson and Savulescu, *Unfit for the Future.*

96. J. Harris, "Moral Enhancement and Freedom," 102.

97. Rakic, "Genome Editing for Involuntary Moral Enhancement," 46.

98. Cwik, "Moving Beyond 'Therapy' and 'Enhancement' in the Ethics of Gene Editing."

99. Agar, *Truly Human Enhancement*, 137.

100. Agar, *Truly Human Enhancement*, 19.

101. Baylis and Robert, "The Inevitability of Genetic Enhancement Technologies," 1.

102. Fuller and Lipinska, *Proactionary Imperative*.

103. Juengst, "What's Taxonomy Got to Do with It?"

104. Annas, *Bioethics*.

105. C. Thomas, "Disability and Diversity."

106. Karpin, "Choosing Disability," 89.

107. Glover, *Choosing Children*; Camporesi, "Choosing Deafness with Preimplantation Genetic Diagnosis."

108. Parens, *Shaping Our Selves*.

109. Steinbock, "Designer Babies."

110. Johnston and Juengst, "Are Parents Really Obligated to Learn as Much as Possible About Their Children's Genomes?" S14. A similar concern arises with direct-to-consumer testing, which some critique for fueling a kind of obsession with genetics over other health concerns, making people vulnerable to manipulation, anxiety, and further data collection (Bob Thomas, "Genetic Testing: The Next Big Arena for Fraud?" *STAT News*, June 7, 2019).

111. Buchanan, "Human Nature and Enhancement," 145.

112. Sparrow, "A Not-So-New EUGENICS"; Eilers, Grüber, and Rehmann-Sutter, *Human Enhancement Debate and Disability*.

113. Karpin, "Choosing Disability." This is similar to the world described in Aldous Huxley's 1932 dystopian novel *Brave New World*.

114. Agar, "Liberal Eugenics"; Dieke et al., "Disparities in Assisted Reproductive Technology Utilization by Race and Ethnicity"; Seifer et al., "Disparities in ART (DART) Hypothesis."

115. Sandel, *Case Against Perfection*, 16.

116. Collier Meyerson, "Crispr Babies, IVF, and the Ethics of Genetic Class Warfare," *Wired*, December 16, 2018.

117. Røe, "High Cost of New Cancer Therapies"; Cushing et al., "The Haves, the Have-Nots, and the Health of Everyone."

118. Eubanks, *Automating Inequality*.

119. Shelton et al., "The 'Haves' and 'Have-Nots' of Personal Protective Equipment During the COVID-19 Pandemic"; Ali et al., "COVID-19 and Inequality"; Suarez-Alvarez and Lopez-Menendez, "Is COVID-19 Vaccine Inequality Undermining the Recovery from the COVID-19 Pandemic?"; Haelermans et al., "Sharp Increase in Inequality in Education in Times of the COVID-19 Pandemic."

120. Peter Sullivan, "Breakthrough Drugs Raise Worries over Higher Costs," *The Hill*, July 25, 2019.

121. Scheffer and Tessema, "Double-Edged Sword," 703.

122. Scheffer and Tessema, "Double-Edged Sword," 703–4.

123. Centers for Medicare and Medicaid Services, "Cell and Gene Therapy (CGT) Access Model," 2025, accessed July 25, 2025, www.cms.gov/priorities/innovation/innovation-models/cgt.

124. Peter Shin, "Risk of Medicaid Cuts: Millions of Community Health Center Patients Stand to Lose Coverage," *National Association of Community Health Centers*,

May 16, 2025; Department of Health and Human Services, "CMS Expands Access to Lifesaving Gene Therapies Through Innovative State Agreements," CMS.gov, July 15, 2025.

125. Harden, *Genetic Lottery.*

126. In his treatise *A Theory of Justice* (1971), John Rawls argued that because the conditions of our birth—namely, our genetics, socioeconomic status, geographic location, and education—are pure luck, we should equalize lived conditions by distributing resources that maximize those who are the least well off.

127. Kathryn Paige Harden, "Why Progressives Should Embrace the Genetics of Education," *New York Times*, July 24, 2018, www.nytimes.com/2018/07/24/opinion/dna-nature-genetics-education.html.

128. Harden, "Why Progressives Should Embrace the Genetics of Education."

129. D. Roberts, "Can Research on the Genetics of Intelligence Be 'Socially Neutral'?" S50.

130. Martschenko, "Elephant in the Room"; Evans, *White Supremacy.*

131. See Buchanan et al., *From Chance to Choice*; Robertson, *Children of Choice.*

132. Lam, *New Reproductive Technologies and Disembodiment.* Some scholars push back against this, arguing that reproductive technologies are liberating and empowering, creating new possibilities for procreation that would not have existed otherwise. (Lublin, *Pandora's Box.*)

133. Gimenez, *Marx, Women, and Capitalist Social Reproduction.*

134. Rachel Adams, "Gene-Edited Babies Don't Grow in Test Tubes." *Conversation*, May 29, 2019.

135. Adams, "Gene-Edited Babies."

136. Crain et al., *Invisible Labor.*

137. Adams, "Gene-Edited Babies."

138. Adams, "Gene-Edited Babies."

139. Davies, *Editing Humanity.*

140. Gyngell et al., "Ethics of Germline Gene Editing," 506.

141. Knoppers and Kleiderman, "Heritable Genome Editing."

142. See MacAskill, *What We Owe the Future*; M. Scully, *Dominion.*

143. See Kaebnick, *Humans in Nature.*

144. Kofler and Grogan, "Giving Voice to the Voiceless in Environmental Gene Editing," S67.

145. Jamie Metzl, *Hacking Darwin*, 195.

146. A subspecies of GMOs, transgenic organisms, are those that contain added genetic material from a different organism. The field of "synthetic biology" also covers human-create attempts to alter the world around us by reorganizing biological material. The applications are also generally environmental, such as creating organisms that consume toxic chemicals in water (to help with oil spills) or creating phytoplankton to absorb more carbon dioxide than usual. (Heidari et al., "CRISPR and the Rebirth of Synthetic Biology.")

147. Fedoroff et al., "Radically Rethinking Agriculture for the 21st Century."

148. Mukherjee, *The Gene*, 95.

149. Ali et al., "Genetically Modified Foods"; Potrykus, "Golden Rice and Beyond"; Hew et al., "Transgenic Salmon."

150. There were, and are, further concerns of Western dominance and colonialist tactics used to assert their influence on farmers in other parts of the world. (Gbashi et al., "Food Safety.")

151. Eric Lipton, "Food Industry Enlisted Academics in G.M.O. Lobbying War, Emails Show," *New York Times*, September 5, 2015.

152. Phillips and Isaac, "GMO Labeling." Part of this reflects the tactics and optics of corporations like Monsanto with aggressive litigation strategies for going after small farms. This created a systemic distrust of the GMO industry, whose tactics reveal a willingness to poison people to earn a profit. However, as Jamie Metzl points out in *Hacking Darwin*, generally most of the public is either unaware or fine with GMO consumption. Nevertheless, scientists who call for inclusion and discussion are worried that the small, passionate, and vocal percentage of the public that is suspicious of industry will try to block their progress through regulation, particularly labeling requirements.

153. US Department of Agriculture, "National Bioengineered Food Disclosure Standard."

154. Landon J. Getz and Graham Dellaire, "CRISPR Gene Editing: Why We Need Slow Science," *The Conversation*, March 19, 2019.

Chapter 3

1. Bekkers et al., *Governance and the Democratic Deficit.*
2. Hamilton, "Federalist 81."
3. Madison, "Federalist 10."
4. Eleena Shao and Ashley Wu, "The Federal Work Force Cuts So Far, Agency by Agency," *New York Times*, May 12, 2025.
5. Moreno, *Body Politic.*
6. Weber, "Bureaucracy."
7. Burris, *Technocracy at Work*, 10.
8. Burris, *Technocracy at Work*, 151.
9. Burris, "Technocratic Organization and Gender," 447.
10. Steinberger, *Concept of Political Judgment.*
11. Steinberger, *Concept of Political Judgment*, 219.
12. Lohse and Canali, "Follow the Science?"
13. Beiner, *Political Judgment*, 1.
14. Beiner, *Political Judgment*, 1–2.
15. Wilson, *Bureaucracy*, 376; Carpenter, *Forging of Bureaucratic Autonomy.*
16. Fischer, "Organizational Expertise and Bureaucratic Control," 192.
17. Burris, *Technocracy at Work*, 153.
18. Leonard, "American Economic Reform in the Progressive Era."
19. Leonard, *Illiberal Reformers.*
20. Fischer, *Democracy and Expertise*, 25.

21. Burris, *Technocracy at Work*, 179.

22. J. Miller, *Can Democracy Work?* 178.

23. Quoted in Jasanoff, *Fifth Branch*, 9.

24. Cairney, *Politics of Evidence-Based Policy Making*, 6.

25. Law and Libecap, "Progressive Era Reform," 325.

26. Catastrophe, or "tragedy," is used here both to indicate that this is how the thalidomide saga is commonly referred to as well as to mark a critique by disability rights advocates who may lament referring to babies born with limb deformities as a "tragedy" as opposed to, perhaps, a mere difference (see chapter 2). See Meyers and Jutai, "The Disability Experience."

27. Waggoner and Lyerly, "Clinical Trials in Pregnancy and the 'Shadows of Thalidomide.'"

28. Blum, *Poison Squad.*

29. Fischer, *Democracy and Expertise*, 27.

30. This outlook is complicated because it took place amid the eugenics movement (see chapter 2), which passed policies that disfavored the procreation of the poor who were seen as genetically inferior and destined to always be poor due to natural selection, rather than a lack of education, skill training, and social safety networks.

31. Fischer, *Democracy and Expertise*, 4; Dewey, *The Public and Its Problems.*

32. Fischer, *Democracy and Expertise*, 4–5.

33. Wilson, *Bureaucracy*, 377.

34. Fischer, *Democracy and Expertise*, 377.

35. Saini, *Superior*; Markovits, *Meritocracy Trap*; Sandel, *Tyranny of Merit.*

36. United Nations Development Programme, *Global Report on Gender Equality in Public Administration.*

37. "A Profile of the 2023 Federal Workforce," *Partnership for Public Service*, accessed April 20, 2026, https://ourpublicservice.org/fed-figures/a-profile-of-the-2023-federal-workforce/.

38. M. Thompson, "Explanations of the Gender Gap in Philosophy": Cimpian et al., "Understanding Persistent Gender Gaps in STEM."

39. Jasanoff, *Ethics of Invention*, 46.

40. Jasanoff, *Ethics of Invention*, 47.

41. Burris, *Technocracy at Work*, 155.

42. J. Friedman, *Power Without Knowledge.*

43. Burgess and Tansey, "Technology, Democracy, and Ethics"; Maggetti, "Legitimacy and Accountability of Independent Regulatory Agencies."

44. Laffont and Tirole, "Politics of Government Decision-Making"; Wright et al., "Tracking Mask Mandates During the Covid-19 Pandemic."

45. Decker, *Modern Economic Regulation*; Acharya et al., *Regulating Wall Street*; Stiglitz, "Role of the State in Financial Markets."

46. Grossman, *Choose Your Medicine.*

47. Whiteside and Poku, *AIDS and Governance.*

48. Jasanoff, *Ethics of Invention*, 47.

49. Adolph, *Bankers, Bureaucrats, and Central Bank Politics.*

50. Kasperson and Moser, *Risk Conundrums in a Fast and Complex World*, 1.

51. Wiener, "Rhetoric of Precaution."

52. Majone, "Political Institutions and the Principle of Precaution," 411.

53. Resnik, "Environmental Disease, Biomarkers, and the Precautionary Principle," 247.

54. Resnik, "Environmental Disease, Biomarkers, and the Precautionary Principle," 242.

55. Weber and Ancker, "Risk Perceptions and Risk Attitudes in the United States and Europe," 480.

56. Ribaritis et al., "Genome-Edited Plants," 430.

57. Wong-Rieger, "Moving from Patient Advocacy to Partnership."

58. Jensen and McGrath, "Making Rules About Rulemaking."

59. Yackee and Yackee, "Politics of Rulemaking in the United States," 40.

60. US Code, *Administrative Procedure Act*, 5 U.S.C. § 553 (2022).

61. Kerwin and Furlong, *Rulemaking*.

62. Kerwin and Furlong, *Rulemaking*, 40.

63. Garvey, "Brief Overview of Rulemaking and Judicial Review." In addition, courts have generally only intervened in administrative agency's actions to ensure that the agencies adhere to the APA's procedural requirements and not the content of their rulemaking (Rose-Ackerman et al., *Due Process of Lawmaking*).

64. Lyons, "E-Rulemaking and the Politicization of the Comment Process," 2.

65. Shapiro, "Law, Expertise and Rulemaking Legitimacy," 667.

66. Shapiro, "Law, Expertise and Rulemaking Legitimacy," 668.

67. United States Department of Agriculture, Agricultural Marketing Services, "Swiss Cheese, Emmentaler Cheese Grades and Standards," February 22, 2001.

68. Bingham et al., "New Governance"; Fung, "Varieties of Participation in Complex Governance."

69. Gordon and Rashin, "Stakeholder Participation in Policy Making"; see also Dwidar, "Diverse Lobbying Coalitions and Influence in Notice-and-Comment Rulemaking."

70. Golden, "Interest Groups in the Rulemaking Process," 245.

71. Golden, "Interest Groups in the Rulemaking Process," 245.

72. Golden, "Interest Groups in the Rulemaking Process," 261.

73. Yackee and Yackee, "A Bias Towards Business?"

74. Yackee and Yackee, "A Bias Towards Business?"

75. Yackee and Yackee, "A Bias Towards Business?" 128.

76. Yackee and Yackee, "Sweet-Talking the Fourth Branch," 103.

77. Yackee and Yackee, "'Science' of Policy Development During Administrative Rulemaking," 12.

78. Yackee and Yackee, "'Science' of Policy Development During Administrative Rulemaking," 12.

79. Others raise concerns of hacking or "fraudulent comments," such as in the Federal Communications Commission's net neutrality rulemaking in 2017 wherein a "record 22 million comments were submitted, but millions of those were submitted by computers (bots) using fictitious names or the names of real people, living

and dead, who had no connection to the comment" (Herz, "~~Fraudulent~~ Malattributed Comments in Agency Rulemaking," 2).

80. Coffee, "Political Economy of Dodd-Frank," 1021.

81. Coffee, "Political Economy of Dodd-Frank," 1029.

82. Coffee, "Political Economy of Dodd-Frank," 1029.

83. Coffee, "Political Economy of Dodd-Frank," 1029.

84. US Food and Drug Administration, *Human Gene Therapy Products Incorporating Human Genome Editing.*

85. US Food and Drug Administration, *Human Gene Therapy Products,* 1.

86. US Food and Drug Administration, *Human Gene Therapy Products.*

87. US Department of Agriculture, "Movement of Certain Genetically Engineered Organisms," 29790.

88. US Department of Agriculture, "Movement of Certain Genetically Engineered Organisms," 29790.

89. US Department of Agriculture, "Movement of Certain Genetically Engineered Organisms," 29790.

90. Carpenter and Moss, *Preventing Regulatory Capture,* 13.

91. Rex, "Anatomy of Agency Capture."

92. Becker, "Competition and Democracy," 109.

93. R. A. Posner, "Concept of Regulatory Capture," 49.

94. John Beck and Jane Dekleva, "EPA Chemical Industry Collaboration: A New Era," *New York Times,* January 22, 2025.

95. Balleisen and Eisner, "Promise and Pitfalls of Co-Regulation."

96. Little, "Governments as Regulators."

97. Natalie Kitroeff, David Gelles, and Jack Nicas, "The Roots of Boeing's 737 Max Crisis," *New York Times,* July 27, 2019.

98. Nunn, "Grounded," 727.

99. Meghani and Kuzma, "'Revolving Door' Between Regulatory Agencies and Industry," 575.

100. C. W. Thomas, "Maintaining and Restoring Public Trust in Government Agencies."

101. Carpenter and Moss, *Preventing Regulatory Capture,* 5.

102. Geransar, "Banking on Trust," 185. See also Einsiedel and Geransar, "Framing Genetic Risk."

103. Yackee and Yackee, "Regulatory Capture's Self-Serving Application," 866.

104. Hausman, *Anti/Vax*; Hamilton et al., "Trust in Scientists on Climate Change and Vaccines."

105. Tucker Carlson, "RFK Jr. Provides an Update on His Mission to End Skyrocketing Autism and Declassifying Kennedy Files," *YouTube,* June 30, 2025, video, 19:09, www.youtube.com/watch?v=w_fzlwxJZAA.

106. Teddy Rosenbluth, "Kennedy Claims Doctors Profit Off Vaccines. In Fact, Many Lose Money on Them," *New York Times,* July 15, 2025.

107. Akhtar Allana, "Some Americans Were Primed for Vaccine Skepticism After Decades of Mistrust in Big Pharma," *Business Insider,* October 24, 2021.

108. Carpenter and Moss, "Corrosive Capture?" 172.

109. Lee et al., "Hurdles to Herd Immunity."

110. Graham, "Ambiguous Capture."

111. Singleton, "Co-operation or Capture?" 2–3; see also Singleton, *Constructing Cooperation.*

112. Reiss, "Benefits of Capture," 569.

113. Schwarcz, "Preventing Capture Through Consumer Empowerment Groups," 370.

114. Magill, "Courts and Regulatory Capture," 151.

115. Revesz and Livermore, *Retaking Rationality*, 123.

116. Jeannie Suk Gersen, "The Supreme Court Is One Vote Away from Changing How the U.S. Is Governed," *New York Times*, July 3, 2019.

117. Kellough, *Fragility of Merit.*

118. House Committee on Oversight and Accountability, Democratic Staff, *White House for Sale.*

119. Furlong and Kerwin, "Interest Group Participation in Rule Making"; Le Dantec and DiSalvo, "Infrastructuring and the Formation of Publics in Participatory Design."

120. Kerwin and Furlong, *Rulemaking*, 66.

121. Kerwin and Furlong, *Rulemaking*, 62.

122. Natow, "Negotiated Rulemaking for U.S. Higher Education Regulatory Policy."

123. Asamoah and Sharfstein, "Transparency at the Food and Drug Administration"; Bingham et al., "New Governance."

124. See Vermeule, *Mechanisms of Democracy*; Chayes and Handler Chayes, *New Sovereignty.*

125. Grant and Keohane, "Accountability and Abuses of Power in World Politics."

126. Fung, Graham, and Weil, *Full Disclosure.*

127. Daniels et al., "Testing Transparency," 1263.

128. Daniels et al., "Testing Transparency," 1270.

129. Fenster, "Opacity of Transparency," 927.

130. Braun and Busuioc, "Stakeholder Engagement as a Conduit for Regulatory Legitimacy?" 1600.

Chapter 4

1. Jasanoff, *States of Knowledge.*

2. Wartenberg, *Forms of Power*, 8.

3. This is an intentionally broad definition that authors of metaphysics and politics have endeavored to articulate more fully, particularly regarding what it means to have "power over" someone else or groups, to be an "agent" of power, and to be subject to another's power (Wrong, *Power*).

4. Pitkin, *Wittgenstein and Justice*, 276.

5. Arendt, *Human Condition*, 200.

6. Hayward, *De-Facing Power*, 3–4.

7. Mill, *On Liberty*, 9.

8. Foucault, "Ethics of the Concern of the Self," 136.

9. Foucault, *History of Sexuality*, 138.

10. While Foucault may have coined the words "biopower" and "biopolitics," elements of these concepts preexisted the postmodern period of the late twentieth century when Foucault was writing. For example, Mika Ojakangas traces the "roots of biopolitics" to ancient philosophy, such as in Plato's *Republic*, which describes a stable society as composed of stringent class hierarchies that categorize individuals' roles. Similarly, Aristotle's "biopolitical regulation of population" includes directives on procreation, child-rearing, and education. Ojakangas, *On the Greek Origins of Biopolitics.*

11. See also Hont, *Politics in Commercial Society.*

12. A. Smith, *An Inquiry into the Nature and Causes of the Wealth of Nations.*

13. Bentham, "Panopticon, Constitution, Colonies, Codification," 39. Bentham's Panopticon should also be understood in the context of his utilitarian ethics: Government resources should be distributed to maximize utility or pleasure. The prison construct was seen as fulfilling that goal by using minimal resources and thus maximizing efficiency in observation and disciplinary action.

14. Locke, *Two Treatises of Government*; Hobbes, *Leviathan*; Rousseau; *Social Contract.*

15. Foucault, *Security, Territory, and Population.*

16. In the Marxist tradition, power lies with those who have economic control by owning "the means of production." According to Karl Marx (and Friedrich Engels), the solution to the problems of commercial societies and capitalist systems is to eliminate private property (Marx, *Capital*; Marx and Engels, *Communist Manifesto*).

17. Cisney and Morar, "Introduction: Why Biopower? Why Now?," 4.

18. Mensch and Mensch, *IQ Mythology.*

19. Sharpes, *Evolution of the Social Sciences.*

20. V. Harris, "Sex on the Margins."

21. Foucault, "Politics of Health in the Eighteenth Century."

22. Nadesan, *Governmentality, Biopower, and Everyday Life*, 108.

23. Nussbaum, *Hiding from Humanity*; J. M. Metzl, *Why "Against Health"?*

24. "Fat" has become an embraced identity, with many advocates using the term to decrease its stigma. See Gordon, *What We Don't Talk About*; Fruh et al., "Weight Bias and Stigma."

25. Kyle et al., "Regarding Obesity as a Disease."

26. Gordon, *You Just Need to Lose Weight.*

27. Murray et al., "Could Polygenic Risk Scores Be Useful?"; Borsa et al., "New Genetics of Sexuality."

28. Theresa Gaffney, "House Passes Bill to Ban Gender-Affirming Care for Young People," *STAT News*, December 17, 2025; Baker et al., "Hormone Therapy, Mental Health, and Quality of Life"; Surendran et al., "A Scoping Review of the Ethical Issues in Gender-Affirming Care."

29. A. Becker, *Get It Out.*

30. Tasnim, "A Foucauldian Perspective on Covid-19 Response."

31. Nadesan, *Crisis Communication*, 31. The distinction between positive and negative liberty is associated with Isaiah Berlin's 1969 essay, "Two Concepts of Liberty." Berlin distinguishes between "positive liberty" as something requiring the presence to create freedom and "negative liberty," which requires the absence of something to permit freedom. The distinction can also be described as "free to" or "free from" liberty. Positive liberty may be having a protected right to vote, whereas a negative liberty may be the right to be free from discrimination. The discussion of domination will take up this question of whether "free from" is the best way to describe problems of power and violations of freedom.

32. John Rawls's liberalism features an egalitarian claim for the equitable distribution of resources wherein inequalities were only permissible insofar as they advantaged the least advantaged. This is regardless of merit or desert (Rawls, *Justice as Fairness*).

33. Hayek, *Fatal Conceit*; Hayek, *Economic Freedom*; M. Friedman, *Capitalism and Freedom*.

34. Nadesan, *Crisis Communication*, 29–30.

35. Nadesan, *Crisis Communication*, 62.

36. Or, in Marxists terms, the capitalists as the owners of the means of production.

37. W. Brown, *Undoing the Demos*.

38. W. Brown, *Undoing the Demos*, 61–62.

39. Rose, "Government, Authority and Expertise in Advanced Liberalism," 296–97.

40. Patton, "Power and Biopower in Foucault," 114.

41. McKean, *Disorienting Neoliberalism*, 24 (emphasis in original).

42. Zuboff, *Age of Surveillance Capitalism*. See also Paul Mozur et al., "Made in China, Exported to the World: The Surveillance State," *New York Times*, April 24, 2019, www.nytimes.com/2019/04/24/technology/ecuador-surveillance-cameras-police-government.html.

43. See also Lemke, *Biopolitics*.

44. J. E. Cohen, *Between Truth and Power*.

45. Makary, *Price We Pay*; Reinhardt, *Priced Out*.

46. Dutescu and Hillier, "Encouraging the Development of New Antibiotics"; Guidone, "Under the Influence."

47. Roberts, "Social Immorality of Health in the Gene Age," 62.

48. Ronald M. Green, "By Design: Building Baby from the Genes Up," *Washington Post*, April 13, 2008.

49. Roberts, "Collateral Consequences," 567.

50. Esposito, *Biopolitics and Philosophy*.

51. Tomes, *Gospel of Germ*.

52. Murcho et al., "Oral Pre-Exposure Prophylaxis (PrEP) to Prevent HIV."

53. Imran and Marri, "Gene Therapy for Sickle Cell Disease."

54. Best et al., "A Systematic Review of Geographical Inequities for Accessing Clinical Genomic and Genetic Services."

55. Marx, *Capital*.

56. Young, *Justice and the Politics of Difference.*

57. Oppression and marginalization are also important concepts in understanding injustices embedded in CRISPR regulation. For this chapter, I focus on different problems of injustice.

58. Fluit et al., "Social Marginalization."

59. Hamel et al., "Race, Health, and COVID-19"; Price-Haywood et al., "Hospitalization and Mortality Among Black Patients and White Patients with Covid-19."

60. Kashmir Hill, "Another Arrest, and Jail Time, Due to a Bad Facial Recognition Match," *New York Times*, December 29, 2020.

61. Lovett, *A General Theory of Domination and Justice*, 127.

62. McCammon, "Domination."

63. Thus, it may well be the case that to not be dominated, some forms of equality or fairness are required, but they are not necessarily the ultimate motivation or sufficient requirements. Shapiro, *Politics Against Domination*, 21.

64. This inability to "exit" as a basis of domination is also present in the work of many other scholars. See Lovett, "What Counts as Arbitrary Power?"

65. Anderson, *Private Government.*

66. Rahman, *Democracy Against Domination*, 80.

67. Rahman, *Democracy Against Domination*, 2.

68. Rahman, *Democracy Against Domination*, 3.

69. Azariadis and Stachurski, "Poverty Traps."

70. Zimring, *Clean and White*; Benevolenza and DeRigne, "Impact of Climate Change and Natural Disasters on Vulnerable Populations."

71. Lovett, "What Counts as Arbitrary Power?" 150.

72. Michelle Alexander argues that lacking a clear agenda of racist policies, the effects of numerous "race blind" institutions create racist systems that keep Blacks and minorities trapped, marginalized, and dominated. Just as the system of Jim Crow once dominated, the modern system of criminal justice is equally, if not more, dominating by perpetuating the illusion of being race blind when in fact it has extreme racially disparate impacts (Alexander, *New Jim Crow*).

73. Young, *Justice and the Politics of Difference*, 76.

74. Young, *Justice and the Politics of Difference*, 77.

75. Pettit, *Republicanism*, 61.

76. Jennings, "Public Health and Civic Republicanism"; Latham, "Political Theory, Values and Public Health."

77. See, for example, D. Allen, "A New Theory of Justice." Although I do not ultimately apply the Kantian intrinsic right to non-domination formulation, it is worth mentioning that Kant also upheld the values of participation and self-governance through institutions or representation. See Varden, "Self-Governance and Reform in Kant's Liberal Republicanism."

78. Rahman, *Democracy Against Domination*, 13.

79. Rahman, *Democracy Against Domination*, 144–45.

80. Rahman, *Democracy Against Domination*, 14.

81. H. S. Richardson, *Democratic Autonomy*, 17.

82. H. S. Richardson, *Democratic Autonomy*, 18.

83. H. S. Richardson, *Democratic Autonomy*, 19.

84. H. S. Richardson, *Democratic Autonomy*, 220.

85. H. S. Richardson, *Democratic Autonomy*, 220.

86. H. S. Richardson, *Democratic Autonomy*, 222.

87. H. S. Richardson, *Democratic Autonomy*, 222.

88. McKean, *Disorienting Neoliberalism*.

89. Zuboff, *Age of Surveillance Capitalism*.

90. Woodly, *Black Lives Matter*; Weldon, *When Protest Makes Policy*; Corry and Reiner, "Protests and Policies."

91. Heydari Fard, "Transformative Power of Social Movements."

92. Anderson, *Private Government*, 37.

93. Breen, "Freedom, Republicanism, and Workplace Democracy."

94. Dolan and Rosenbloom, *Representative Bureaucracy*, 3, 114.

95. Brummer, "What Do the Data Reveal."

96. Farmbry, *Administration and the Other*, iv.

97. Feinstein et al., "Board Diversity Matters."

98. Opstrup and Villadsen, "The Right Mix?"; Chen et al., "Why Female Board Representation Matters."

99. Epstein and Knight, "How Social Identity and Social Diversity Affect Judging"; Rackley, *Women, Judging and the Judiciary*; Hunter, "More Than Just a Different Face?"

100. Feeney and Camarena, "Gender, Race, and Diversity Values Among Local Government Leaders."

101. White-Lewis et al., "Moving the Science Forward."

102. Mertkan et al., "Diversity and Equality in Academic Publishing."

103. Montalmant and Ettinger, "Racial Disparities in Maternal Mortality."

104. Landor, "Right to Informed Consent, Right to a Doula."

105. Alsan et al., "Does Diversity Matter for Healthy?," 4107.

106. Salsberg et al., "Estimation and Comparison of Current and Future Racial/Ethnic Representation."

107. Chisowa, "Racial Disparities Are Rife in Biotech."

108. Bolger and Ecklund, "Seeing Is Achieving."

109. See Urbinati, *Democracy Disfigured*.

110. Lewandowsky et al., "Beyond Misinformation."

111. Bacon, *Meditationes Sacrae*; Hobbes, *Leviathan*.

112. Bacon, *New Atlantis*. This defense of technocratic power relied on knowledge conveyed through scientific discovery and prowess and has also contributed to the philosophical justifications of administrative power as explored in chapter 3.

113. Lucas, "Bacon's New Atlantis."

114. García, "Scientia Potestas Est," 110.

115. B. Allen, "Foucault's Nominalism," 101–2.

116. Dotson, "Accumulating Epistemic Power."

117. Anwar et al., "Impact of Jury Race in Criminal Trials"; Flanagan, "Race, Gender, and Juries."

118. Leila Ettachfini, "Court Reporters May Be Writing Down Black People's Testimonies Wrong," *Vice*, May 23, 2019.

119. Fricker, *Epistemic Injustice.*

120. Fricker, "Epistemic Oppression and Epistemic Privilege"; Fricker, "Rational Authority and Social Power."

121. Fricker, *Epistemic Injustice*, 1, 6. See also D. Coady, "Two Concepts of Epistemic Injustice."

122. McConkey, "Knowledge and Acknowledgement," 198. See also Tsosiel, "Indigenous Peoples and Epistemic Injustice"; Crenshaw, "Demarginalizing the Intersection of Race and Sex"; Dotson, "A Cautionary Tale."

123. Medina, "Hermeneutical Injustice and Polyphonic Contextualism"; Medina, *Epistemology of Resistance.*

124. Quashie, *Sovereignty of the Quiet*; de Sousa Santos, *End of the Cognitive Empire.*

125. Bohman, "Domination, Epistemic Injustice and Republican Epistemology," 180.

126. Hänel, "#MeToo and Testimonial Injustice"; Jacques, "'Me Too': Epistemic Injustice."

127. Manne, *Down Girl.*

128. Bergen, "Marital Rape."

129. Kidd and Carel, "Epistemic Injustice and Illness," 172. See also Wardrop, "Medicalization and Epistemic Injustice."

130. Hoffman et al., "Racial Bias in Pain Assessment."

131. Wiggleton-Little, "'Just' a Painful Period."

132. Elizabeth Anderson, "Feminist Epistemology and Philosophy of Science," *Stanford Encyclopedia of Philosophy*, August 1, 2024, https://plato.stanford.edu/archives/fall2024/entries/feminism-epistemology.

133. Addams, *Democracy and Social Ethics*, 9.

134. Gauckler, "Expert Identification for Ethics Expertise."

135. Code, "Ignorance, Injustice and the Politics of Knowledge," 148.

136. Code, "Ignorance, Injustice and the Politics of Knowledge," 148.

137. Dotson, "Conceptualizing Epistemic Oppression," 115.

138. Apffel-Marglin and Marglin, *Decolonizing Knowledge.*

139. Janack, "Standpoint Epistemology," 130.

140. Collins, *Artificial Experts*, 270.

141. Klein et al., "Opinion: Sex Inclusion in Basic Research."

142. Perez, *Invisible Women.*

143. A. Ryan et al., "Impact of Sex."

144. Linder and Svensson, "Road Safety," 142.

145. Terrier and Terrier, "Efficacy and Safety of Flibanserin"; ChoGlueck, "Still No Pill for Men?"; Segal, "Sex, Drugs, and Rhetoric."

146. See Comen, *All in Her Head.*

147. Shansky, "Are Hormones a 'Female Problem'?"

148. Claudia López Lloreda, "Exclusive: NIH Appears to Archive Policy Requiring Female Animals in Studies," *Transmitter*, February 24, 2025.

149. Polanyi, *The Tacit Dimension.*

150. Borkman, "Experiential Knowledge," 446–47.

151. Andrew Joseph, "Following 'CRISPR Babies' Scandal, Senators Call for International Gene Editing Guidelines," *STAT News*, July 15, 2019.

152. Kettle, "Knowledge Co-Production in Contested Spaces."

153. Buckwalter and Turi, "Knowledge, Adequacy, and Approximate Truth."

154. C. A. J. Coady, *Testimony*, 125.

155. McKinnon, "Epistemic Injustice," 438.

156. Kitcher, *Science, Truth, and Democracy*, 439–41.

157. Frey, "The 'Diversity Explosion,'" 16.

158. Lewis and Cantor, "Introduction: The Value of Diversity," 5; Carnevale and Smith, "Economic Value of Diversity," 106; Tienda, "Diversity as a Strategic Advantage," 195.

159. Callon et al., *Acting in an Uncertain World.*

160. Graham and Jones, *Expert Systems, Knowledge, Uncertainty and Decision*, 84.

161. Townley, *A Defense of Ignorance*, ix.

162. Soyer and Hogarth, "Illusion of Predictability"; Cooke, "Five Arguments."

163. Lindley, *Uncertainty*, 2.

164. Sloman and Fernbach, *Knowledge Illusion.*

165. Van Asselt, *Perspectives on Uncertainty and Risk*, 1.

166. Van Asselt, *Perspectives on Uncertainty and Risk*, 1.

167. Landemore, "Deliberation, Cognitive Diversity, and Democratic Inclusiveness," 1210.

168. Conor Friedersdorf, "Can Neurodiversity Defeat Doublethink?" *Atlantic*, June 25, 2019. Self-described "autistic blogger" Jonathan Mitchell critiques the endorsement of "neurodiversity" for problematically perpetuating a disability condition by depicting autism as a benefit rather than a hinderance that needs treatment. His concern is that by comparing neurodiversity to the kinds of experiential knowledge offered by other social groups, they are equating it as an identity worth preserving (Mitchell, "The Dangers of 'Neurodiversity': Why Do People Want to Stop a Cure for Autism Being Found?" *Spectator*, January 19, 2012). The argument is subtly different, however: Being neurodiverse provides helpful knowledge. Being Black provides helpful knowledge of lived conditions of Black people, but it is not to say that we want to keep those lived conditions static (i.e., in states of domination).

169. Friedersdorf, "Can Neurodiversity Defeat Doublethink?"

170. Anderson, *Imperative of Integration.*

171. Anderson, "Epistemic Justice as a Virtue of Social Institutions," 163.

172. See Goldman, "Experts."

173. Ostrom, *Understanding Institutional Diversity.*

174. Hong and Page, "Groups of Diverse Problem Solvers," 16389.

175. Hong and Page, "Groups of Diverse Problem Solvers," 16389.

176. Hong and Page, "Groups of Diverse Problem Solvers," 16389.

177. Page, *The Difference*, 257, 10 (emphasis in original).

178. Page, *The Difference*, 5.

179. Page, *The Difference*, 9 (emphasis in original).

180. Hong and Page, "Groups of Diverse Problem Solvers," 16385; Page, *The Diversity Bonus*, 160 (emphasis added). Page is not always clear about what he means by "identity." The diversity literature often conflates identity and lived experiences, however here he is showing that cognitive diversity may also be impacted by identity diversity in important ways.

181. Hong and Page, "Groups of Diverse Problem Solvers," 16389.

182. Sahay et al., *Financial Inclusion.*

183. Calkin, "Tapping Women for Post-Crisis Capitalism," 612.

184. See, for example, Haldar et al., "Tokenism or Realism?"

185. Haidt, *Righteous Mind*, 92.

186. Du Bois, *Darkwater*, 143–44.

187. Landemore, "Democratic Reason," 251.

188. Bohman, "Deliberative Democracy," 175. See also Landemore, "Beyond the Fact of Disagreement?"; Goodin and Spiekermann, *An Epistemic Theory of Democracy*; List and Goodin, "Epistemic Democracy"; Anderson, "Epistemology of Democracy."

189. Bovens and Rabinowicz, "Democracy and Argument"; Holst and Molander, "Epistemic Democracy and the Role of Experts"; Landemore, "Democratic Reason."

190. Moore, *Critical Elitism.*

191. Misak, "A Culture of Justification."

192. Räber, *Knowing Democracy.*

193. Landemore, "Collective Wisdom," 5; Saxton et al., "Rules of Crowdsourcing"; Estellés-Arolas and González-Ladrón-de-Guevara, "Towards an Integrated Crowdsourcing Definition."

194. Surowiecki, *Wisdom of Crowds.*

195. Joanne Chen, "Do Not Trust That Stranger's 5-Star Review," *New York Times*, May 25, 2019; Sullivan et al., "Can Real Social Epistemic Networks Deliver?"

196. Sorensen, *Mob Rule.*

197. Xenos, "Citizens Make Sense of Science Issues"; Simpson, "Evaluating Google." This is similar to Ian Shapiro's argument against deliberation (Shapiro and Hacker-Cordón, "Promises and Disappointments").

198. Christiano, *Rule of the Many*, 264.

199. Gamson, *Talking Politics.*

200. Chambers, "Balancing Epistemic Quality and Equal Participation," 273.

201. Chambers, "Balancing Epistemic Quality and Equal Participation," 274.

202. Gamson, *Talking Politics.*

203. Fung, "Minipublics," 167.

204. Habermas, "Political Communication in Media Society"; Chambers, "Balancing Epistemic Quality and Equal Participation," 270, 269, 271.

205. Bagg, "Sortition as Anti-Corruption."

206. Science and technology studies (STS) scholars endorse democracy for similar epistemic virtues wrought from diverse collaboration. See Hurlbut, "Limits of Responsibility." Further, Anderson's argument in *The Imperative of Integration*

features a case for democratic equality through considered "exercise of practical intelligence in testing value judgments" (97). See also Dzur, *Punishment, Participatory Democracy, and the Jury.*

207. Brennan, *Against Democracy,* 30.

208. Brennan, *Against Democracy,* 28.

209. Brennan, *Against Democracy,* 47.

210. Landemore "Yes, We Can."

211. Mansbridge, "Conflict and Self-Interest in Deliberation."

212. Landemore "Yes, We Can," 189.

213. Grönlund et al., "Does Enclave Deliberation Polarize Opinions?"

214. Fishkin, *Democracy When the People Are Thinking*; Ding and Pivato, "Deliberation and Epistemic Democracy."

215. Dryzek, "Theory, Evidence, and the Tasks of Deliberation," 249.

216. Kahneman, *Thinking, Fast and Slow,* 5, 4.

217. Collins, *Artificial Experts*; Fishkin et al., "Deliberative Polling"; Tetlock, *Expert Political Judgment.*

218. Fischer, *Democracy and Expertise*; Grundmann, "Problem of Expertise"; Viehoff, "Authority and Expertise."

219. Shapiro, "Collusion in Restraint of Democracy."

220. Gastil, "Lessons and Limitations of Experiments in Democratic Deliberation."

221. Archon Fung, "COVID-19 Requires More Democracy, Not Less," *Boston Review,* April 23, 2020.

222. Marzinke et al., "Limited Evidence for Use of a Black Race Modifier."

223. Organ Procurement and Transplantation Network, "Understanding Race and eGFR," 2025, https://www.hrsa.gov/optn/patients/resources/kidney/understanding-race-egfr.

224. Usha Lee McFarling and Katie Palmer, "Inside the Bruising Battle to Purge Race from a Kidney Disease Calculator," *STAT News,* September 5, 2024.

225. Velasco and Snodgrass, "Use of Race in eGFR."

226. McFarling and Palmer, "Inside the Bruising Battle."

Chapter 5

1. Sintomer, "From Deliberative to Radial Democracy," 57.

2. MacKenzie and O'Doherty, "Deliberating Future Issues," 5.

3. Fung, "Survey Article: Recipes for Public Spheres"; Goodin and Dryzek, "Deliberative Impacts."

4. Curato et al., *Deliberative Mini-Publics,* 3.

5. O'Flynn, *Deliberative Democracy,* 49.

6. O'Flynn, *Deliberative Democracy,* 50.

7. Blok, "Experts on Public Trial," 176.

8. "Tracking Danish-Style, Citizen-Based Deliberative Consensus Conferences Worldwide," The Loka Institute, accessed August 1, 2024, www.loka.org/TrackingConsensus.html.

9. Renn, "Stakeholder and Public Involvement in Risk Governance," 16.

10. Joss and Durant, *Public Participation in Science*. The exact numbers, procedures, and timing of the conferences, particularly as the model has expanded to other countries, is subject to (slight) changes.

11. Curato et al., *Deliberative Mini-Publics*, 8.

12. This included "a non-U.S. Department of Health and Human Services, non-advocate 15-member panel representing the fields of obstetrics and gynecology, maternal-fetal medicine, pediatrics, diabetic research, biostatistics, women's health issues, health services research, decision analysis, health management and policy, health economics, epidemiology, and community engagement. In addition, 16 experts from pertinent fields presented data to the panel and conference audience" (Vandorsten et al., "NIH Consensus Development Conference," 1).

13. Olszewski, "Between Bench and Bedside."

14. PytlikZillig et al., *Deliberative Public Engagement with Science*, 26.

15. See, for example, Rudow et al., "Consensus Conference on Best Practices"; Taner et al., "Summary of a Consensus Conference"; Kasiske et al., "Report of a Consensus Conference."

16. Andersen and Jæger, "Scenario Workshops and Consensus Conferences."

17. Powell and Kleinman, "Building Citizen Capacities," 329.

18. Calvello et al., "Emergency Care in Sub-Saharan Africa," 42.

19. Wortman et al., "Do Consensus Conferences Work?" 495.

20. Fung, "Varieties of Participation in Complex Governance," 67.

21. Fung, "Survey Article: Recipes for Public Spheres," 342.

22. Crosby, "Citizens Juries," 157.

23. Hans et al., "Deliberative Democracy and the American Civil Jury."

24. Smith and Wales, "Citizen Juries and Deliberative Democracy," 58, 53.

25. Nicol et al., "Genome Editing," 20.

26. Nicol et al., "Genome Editing," 23.

27. Nicol et al., "Genome Editing," 23.

28. Nicol et al., "Genome Editing," 25.

29. Australian Government National Health and Medical Research Council, *Mitochondrial Donation Community Consultation Citizens' Panel Statement*.

30. Scuffham et al., "Engaging the Public in Healthcare Decision-Making"; Oswald and Laverty, "Data Sharing in a Pandemic"; Ballantyne et al., "Sharing Precision Medicine Data."

31. Street et al., "Use of Citizens' Juries in Health Policy Decision-Making," 2.

32. Dryzek et al., "Promethean Elites Encounter Precautionary Publics," 270.

33. Dryzek et al., "Promethean Elites Encounter Precautionary Publics."

34. Aasen and Vatn, "Deliberation on GMOs," 464.

35. Pimbert and Wakeford, "Prajateerpu," 2786.

36. Rusike, "Exploring Food and Farming Futures."

37. Newell, "Trade and Biotechnology in Latin America," 365.

38. Pimbert and Barry, "Let the People Decide," 1097–98.

39. Nuffield Council on Bioethics, "Public Dialogue on Genome Editing in Farmed Animals," 4.

40. Mansbridge, "Deliberative Polling"; Fishkin, *Democracy When the People Are Thinking*, 75.

41. "What Is Deliberative Polling®?" Deliberative Democracy Lab, https://cdd.stanford.edu/what-is-deliberative-polling/.

42. Ackerman and Fishkin, *Deliberation Day.*

43. Fishkin and Luskin, "Experimenting with a Democratic Ideal," 288.

44. Fishkin, *Democracy When the People Are Thinking*, 75.

45. Fishkin, "Deliberative Democracy in Context," 28.

46. Fishkin, "Deliberative Democracy in Context," 31–32.

47. Fishkin, *Democracy When the People Are Thinking*, 113.

48. Fishkin, *Democracy When the People Are Thinking*, 79.

49. Fishkin, *Democracy When the People Are Thinking*, 105.

50. Taiwan Ministry of Digital Affairs, "Utilizing AI to Enhance Information Integrity," June 5, 2024, https://moda.gov.tw/en/major-policies/alignment-assemblies/2024-deliberative-assembly/1521.

51. Kuo, "Public Policy in Taiwan"; Fan and Siu, "Effects of Public Deliberation."

52. "Deliberative Polling on Utilizing AI to Enhance Information Integrity," *Participedia*, n.d., https://participedia.net/case/deliberative-polling-on-utilizing-ai-to-enhance-information-integrity.

53. Fang-Yu Chen and Chia Ying Wang, "Using AI to Strengthen Democracy: Audrey Tang on Taiwan's Global Role," *Taiwan 20*, October 21, 2024.

54. Taiwan Ministry of Digital Affairs, "Utilizing AI."

55. Taiwan Ministry of Digital Affairs, "Utilizing AI."

56. Chen and Wang, "Using AI."

57. Paolo Spada, "Social Movements Demanding Minipublics: A New Phenomenon?" *Public Agenda*, May 25, 2021.

58. Thompson, "Who Should Govern Who Governs?" 26.

59. Warren and Pearse, *Designing Deliberative Democracy*, 6.

60. James, Pearse, and Warren, "Descriptive Representation," 106.

61. Fournier et al., *When Citizens Decide*; Herath, *Real Power.*

62. Fournier et al., *When Citizens Decide.*

63. Fournier et al., *When Citizens Decide*, 26.

64. Landemore, "Inclusive Constitution-Making," 167; see also Ólafsson, "Constituent Assembly."

65. O'Grady, "Power to the People"; Johnson, *Deliberation Democracy.*

66. Sarah Allan, "Parliament Sends 30,000 Invitations for Citizens' Assembly on Climate Change," *Involve*, November 2, 2019.

67. Giraudet et al., "'Co-construction' in Deliberative Democracy," 2.

68. Shannon Osaka, "Can 'The People' Solve Climate Change?: France Decided to Find Out." *Grist*, November 15, 2021.

69. Aurelien Breeden, "France Passes Climate Law, but Critics Say It Falls Short," *New York Times*, July 20, 2021; Germann, Marien, and Muradova, "Scaling Up?" 5.

70. "French Citizens' Assembly on Assisted Dying." *Participedia*, accessed August 25, 2024, https://participedia.net/case/french-citizens-assembly-on-assisted-dying.

71. Pauline Rouquette, "'Aid in Dying': What's in the New French Bill on Assisted Suicide," *France 24*, November 11, 2024.

72. Jessica Jones, "'We Have Never Been So Close': Why France May Soon Legalise Assisted Dying," *Connexion France*, April 25, 2025.

73. Elizabeth Pineau, "French Lawmakers Approve Assisted Dying Bill, Paving the Way for Approval," *Reuters*, May 27, 2025.

74. Dryzek et al., "Global Citizen Deliberation on Genome Editing," 1436.

75. Some researchers dispute the second claim. For example, in studying the Northern Ireland Assembly, James Pow found that the method of selection did not matter on the public's perceptions of it, suggesting that "mini-public organizers have a degree of flexibility over design choices without sacrificing perceptions of legitimacy from the perspective of the wider public, even in deeply divided political settings." Pow, "Mini-Publics and the Wider Public," 2.

76. Pitkin, *Concept of Representation.*

77. Fung, "Varieties of Participation in Complex Governance," 67–68.

78. Hendriks, "Coupling Citizens and Elites in Deliberative Systems."

79. Pimbert and Barry, "Let the People Decide."

80. Caluwaerts and Reuchamps, *Legitimacy of Citizen-led Deliberative Democracy.*

81. Stone, *Luck of the Draw*; Flanigan et al., "Fair Algorithms for Selecting Citizens' Assemblies," 552.

82. Gül, "Representation in Minipublics," 37; Mansbridge, "Should Blacks Represent Blacks?" 637.

83. See Jonathan Jackson's interview about diversity in clinical drug trials in Nicholas St. Fleur, "Listen: Diverse Clinical Trials: Why Aren't We There Yet?" *STAT News*, June 13, 2022.

84. See Jackson, "Methodological Framework for Investigating Diversity."

85. Dryzek and Niemeyer, "Discursive Representation," 481.

86. Sénit, "Transforming Our World?" 412.

87. Mansbridge, "Should Blacks Represent Blacks?" 628; Hendriks, *Politics of Public Deliberation.*

88. Smith and Wales, *Democratic Innovations.*

89. Longstaff and Burgess, "Recruiting for Representation," 219.

90. Longstaff and Burgess, "Recruiting for Representation," 220.

91. Nicol et al., "Australian Citizens' Jury," 15.

92. Nicol et al., "Australian Citizens' Jury," 14.

93. Ryfe, "Does Deliberative Democracy Work?" 53.

94. Caluwaerts and Reuchamps, "The G1000," 17; Goodin and Dryzek, "Deliberative Impacts," 221.

95. Devillers et al., "Looking in from the Outside."

96. Curato et al., *Deliberative Mini-Publics*, 23.

97. World Health Organization, *Citizen Engagement in Evidence-Informed Policy-Making.*

98. O'Doherty et al., "Sequencing the Salmon Genome."

99. Though Alexis de Tocqueville (*Democracy in America*, 1835) and John Stuart Mill (*On Liberty*, 1859) were early users of the phrase "tyranny of the majority," John Adams also used the concept to justify multiple branches of government ("mixed government") and its systems of checks and balances as protective systems in arguing for the ratification of the Constitution (*In Defense of the Constitutions of Government of the United States of America*, 1788). Thus, mixed government using a variety of mechanisms has often been invoked as protective against dominant groups.

100. Melilo, "Gene Editing"; Mendz and Cook, "Transhumanist Genetic Enhancement"; Mizrahi, *Playing God*.

101. Nacos and Bloch-Elkon, "How Far-Right Extremism Changed American Body Politic."

102. Bächtiger and Dryzek, *Deliberative Democracy*, 127.

103. Bächtiger and Dryzek, *Deliberative Democracy*, 128.

104. Bächtiger and Dryzek, *Deliberative Democracy*, 128.

105. Bächtiger and Dryzek, *Deliberative Democracy*, 128.

106. Bächtiger and Dryzek, *Deliberative Democracy*, 131.

107. Bächtiger and Dryzek, *Deliberative Democracy*, 141.

108. Bächtiger and Dryzek, *Deliberative Democracy*, 144.

109. Bächtiger and Dryzek, *Deliberative Democracy*, 146.

110. Carpini et al., "Public Deliberation," 325.

111. O'Flynn, *Deliberative Democracy*, 61.

112. Landemore, "Deliberation, Cognitive Diversity, and Democratic Inclusiveness," 1229.

113. Farrell and Stone, "Sortition and Mini-Publics."

114. Boswell, "Why and How Narrative Matters."

115. Morrell, "Participant Bias and Success."

116. Curato et al., *Deliberative Mini-Publics*, 42.

117. Curato et al., *Deliberative Mini-Publics*, 42.

118. "How Are the Participants Selected?" Convention Citoyenne pour le Climate, accessed July 25, 2025, www.conventioncitoyennepourleclimat.fr/en/how-are-the-participants-selected/.

119. Curato et al., *Deliberative Mini-Publics*, 43.

120. Curato et al., *Deliberative Mini-Publics*, 43.

121. There are other important details to keep in mind, including preparing for people to drop out and oversampling marginalized populations as participation will be hardest for those communities.

122. Curato et al., *Deliberative Mini-Publics*, 44.

123. Caluwaerts and Reuchamps, "Does Inter-group Deliberation Foster Inter-Group Appreciation?" 43.

124. Ackerman and Fishkin, *Deliberation Day*.

125. Hans et al., "Deliberative Democracy and the American Civil Jury," 1328.

126. See, for example, Macklin, "'Due' and 'Undue' Inducements"; Lynch et al., "Promoting Ethical Payment"; T. Phillips, "Exploitation in Payments."

127. Organisation for Economic Co-operation and Development, *Innovative Citizen Participation and New Democratic Institutions*, 83.

128. Fishkin and Farrar, "Deliberative Polling," 74.

129. Giraudet et al., "'Co-construction' in Deliberative Democracy."

130. Talpin, "When Democratic Innovations Let the People Decide"; Fiket and Memoli, "Improving the Quality of Democracy."

131. Curato et al., *Deliberative Mini-Publics*, 62.

132. Strandberg and Grönlund, "Online Deliberation."

133. Elstub et al., "Resilience of Pandemic Digital Deliberation"; Willis et al., "Taking Deliberative Research Online."

134. Niessen and Reuchamps, "Designing a Permanent Deliberative Citizens' Assembly."

135. Organisation for Economic Co-operation and Development, *Innovative Citizen Participation and New Democratic Institutions*, 14.

136. Junius, "What About Elite Manipulation?"

137. Bagg, "Sortition as Anti-Corruption."

138. World Health Organization, *Citizen Engagement in Evidence-Informed Policy-Making*.

139. This concern featured strongly in the case of many European CAs. See Naomi O'Leary, "The Myth of the Citizens' Assembly," *Politico*, June 18, 2019.

140. Longstaff and Burgess, "Recruiting for Representation."

141. S. Shapiro, "Does the Amount of Participation Matter?"

142. Naughton et al., "Understanding Commenter Influence," 261.

143. Jacquet and van der Does, "Consequences of Deliberative Minipublics"; Howlett and Giest, "Policy-Making Process."

144. Parkinson, *Deliberating in the Real World*.

145. O'Flynn, *Deliberative Democracy*, 29.

146. Robert Dahl was particularly interested in agenda-setting power also as a hallmark of democratic legitimacy. See Dahl, *On Democracy*; Bau and Escobar, "Participatory-Deliberative Processes."

147. Beauvais, "Deliberation and Equality," 145.

148. W. Smith, "Transnational and Global Deliberation," 865.

149. Goodin and Dryzek, "Deliberative Impacts."

150. Scheinerman, "Public Engagement Through Inclusive Deliberation."

151. Hartz-Karp and Carson, "Putting the People into Politics"; Curato and Boker, "Linking Mini-Publics to the Deliberative System."

152. Longstaff and Secko, "Assessing the Quality of a Deliberative Democracy Mini-Public Event," 258.

153. Nicol et al., "Genome Editing," 72.

154. Fishkin, *Democracy When the People Are Thinking*.

155. Fan, "Indigenous Participation and Knowledge Justice in Deliberative Systems," 1495.

156. Muradova et al., "Climate Change Communication and Public Engagement," 1322.

157. Hébert, "Indigenous Spheres of Deliberation."

158. Druckman and Nelson, "Framing and Deliberation," 729.
159. Calvert and Warren, "Deliberative Democracy," 203.
160. Leino et al., "Expert Hearings in Mini-Publics."
161. Brown, "Expertise and Deliberative Democracy."
162. Leino et al., "Expert Hearings in Mini-Publics," 442.
163. Drury et al., "Deliberative Quality and Expertise," 34.
164. J. Roberts et al., "Experts and Evidence in Deliberation," 17.
165. Chambers, "Balancing Epistemic Quality and Equal Participation," 273.
166. Batt et al., "Pharmaceutical Ethics and Grassroots Activism," 51.
167. Batt et al., "Pharmaceutical Ethics and Grassroots Activism," 52.
168. Grossman, "FDA and the Rise of the Empowered Consumer," 646.
169. Batt et al., "Pharmaceutical Ethics and Grassroots Activism," 53.
170. Batt et al., "Pharmaceutical Ethics and Grassroots Activism," 53.
171. Batt et al., "Pharmaceutical Ethics and Grassroots Activism," 53.
172. Batt et al., "Pharmaceutical Ethics and Grassroots Activism," 53.
173. Batt et al., "Pharmaceutical Ethics and Grassroots Activism," 54
174. McCoy et al., "Conflicts of Interest."
175. Bhat et al., "Medical Product Industry Ties," 1164–65.
176. Fabbri et al., "Industry Funding."
177. Taylor and Denegri, "Industry Links with Patient Organisations"; Bruno and Rose, "Patient Organizations."
178. Batt et al., "Pharmaceutical Ethics and Grassroots Activism," 55.
179. Hennessy, Fahey, and Larkin, "Commercial Influences."
180. Moynihan and Bero, "Toward a Healthier Patient Voice," 351.
181. Singh, *Multiple Autisms.*
182. Roche, Adams, and Clark, "Research Priorities of the Autism Community," 344.
183. Roche, Adams, and Clark, "Research Priorities of the Autism Community," 346.
184. Rosenblatt, "Autism, Advocacy Organizations, and Past Injustice."
185. Villaneuva and Majumder, "Hashtag Who's Missing?"
186. "Autism Speaks' MSSNG Project Reaches 10,000 Genome Milestone," *Autism Speaks*, May 6, 2019.
187. Villaneuva and Majumder, "Hashtag Who's Missing?" 2.
188. Villaneuva and Majumder, "Hashtag Who's Missing?" 3.
189. Hobson et al., "Toward Reproducible and Respectful Autism Research."
190. See Kuyper and Wolkenstein, "Complementing and Correcting Representative Institutions"; Hendriks, *Politics of Public Deliberation.*
191. Doyle and Walsh, "Constitutional Amendment," 420.
192. Extinction Rebellion, "Citizens' Assembly," accessed July 28, 2025, https://extinctionrebellion.uk/decide-together/citizens-assembly/.
193. Curato et al., *Deliberative Mini-Publics*, 62.
194. Landwehr, "Facilitating Deliberation."
195. Curato et al., *Deliberative Mini-Publics*, 62.
196. Doerr, *Political Translation*, 13.

197. Sanders, *Against Deliberation*, 349.

198. Curato and Böker, "Twelve Key Findings," 30.

199. Young, *Justice and the Politics of Difference*; Pateman, *Participation and Democratic Theory*; Young, "Rawlsian Reasonableness"; Forrester, *In the Shadow of Justice*.

200. Curato and Boker, "Twelve Key Findings," 30.

201. See Landwehr, "How Can We Agree on Democratic Procedures?"

202. Curato et al., *Deliberative Mini-Publics*, 63.

203. Bächtiger and Goldberg, "Towards a More Robust, but Limited and Contingent, Defence," 38.

204. Curato et al., *Deliberative Mini-Publics*.

205. Popescu and Loveland, "Judging Deliberation."

206. Curato et al., *Deliberative Mini-Publics*, 76.

207. Gutmann and Thompson, *Spirit of Compromise*. See also Joshua Cohen, "Epistemic Conception of Democracy"; Elster, "Market and the Forum."

208. Fuerstein, "Democratic Consensus," 282.

209. Hysse, "Consensus and Power," 268.

210. Silver, Scott, and Kazepov, "Participation in Urban Contention"; Karpowitz and Mansbridge, "Disagreement and Consensus."

211. Schkade et al., "What Happened on Deliberation Day?" 915.

212. Kadlec and Friedman, "Deliberative Democracy."

213. Hébert, "Indigenous Spheres of Deliberation," 106.

214. Hébert, "Indigenous Spheres of Deliberation," 106.

215. See Fuerstein, "Democratic Consensus."

216. Fuerstein, "Democratic Consensus."

217. There are other details to consider, such as "if the deliberation will aim at unanimity, then it would be better to have a lower percentage of women in each group; if there is majority rule, however, then women need to be a larger presence in each group" (Morrell, "Participant Bias and Success," 168). Sometimes the decision-making procedure then determines the kind of deliberation that should occur, rather than the other way around.

218. Paulis et al., "POLITICIZE Dataset."

219. Breeden, "France Passes Climate Law."

220. Ólafsson, "Constituent Assembly."

221. Johnson, *Democratic Illusion*.

222. Farrell and Stone, "Sortition and Mini-Publics," 239.

223. Curato et al., *Deliberative Mini-Publics*, 31.

224. Goodin, "If Deliberation Is Everything."

225. Thompson, "Deliberative Democratic Theory," 503.

226. Joshua Cohen, "Deliberative Democracy," 234; Rosenberg, *Deliberation, Participation and Democracy*, 133.

227. Strandberg and Grönlund, "Online Deliberation," 108. See also Coleman and Moss, "Under Construction."

228. Van Crombrugge, "Derailed Promise."

229. Jacquet and van der Does, "Consequences of Deliberative Minipublics."

230. This evokes a similar "constituent power" problem: When the DMP consists of actors situated downstream from their hosting institution that appoints them, can they truly contravene upon the powers of the upstream actor? See Wenman, *Agonistic*; Negri, *Insurgencies*.

231. Risse, "Arguing and Deliberation."

232. Grimmelikhuijsen et al., "Can Decision Transparency Increase Citizen Trust?"

233. Curato et al., *Deliberative Mini-Publics*, 54.

234. Fishkin and Luskin, "Broadcasts of Deliberative Polls."

235. Curato et al., *Deliberative Mini-Publics*, 54.

236. Parkinson, "Rickety Bridges," 175.

237. Naurin and Reh, "Deliberative Negotiation," 735.

238. Even then, drug decisions, though they demand extensive clinical knowledge and specialized skill, still require rigorous risk-based analyses based on the needs and vulnerabilities of different patient populations and "surrogate endpoints." See Fleming, "Surrogate Endpoints."

239. US Food and Drug Administration, *CVM GFI #187A Heritable Intentional Genomic Alterations in Animals*.

Chapter 6

1. "Public Trust in Government: 1958–2024," *Pew Research Center*, June 24, 2024; Wolfensberger and Wrigley, *Distrust, Fear, and Science-Denial in Medicine and Healthcare*; Sandra J. Sucher and Shalene Gupta, "Trust Crisis," *Harvard Business Review*, July 16, 2019.

2. Bok, *Lying*, 31.

3. Kirk, "Landscape of Distrust," 28.

4. Niessen, "When Citizen Deliberation Enters Real Politics," 482.

5. Jamal and Nooruddin, "Democratic Utility of Trust"; Tilly, *Trust and Rule*; Putnam, "What Makes Democracy Work?"; Uslaner, "Democracy and Social Capital"; Boix and Posner, "Social Capital."

6. Kenneth Arrow also referred to trust as an "invisible institution" ("Difficulty in the Concept of Social Welfare").

7. Fishkin, "Democratic Practice."

8. See Carey, *The Federalist*.

9. Rosanvallon, *Counter-Democracy*, 8.

10. Schmidthuber et al., "Government Openness and Public Trust."

11. Knight Foundation, *Crisis in Democracy*.

12. Rosanvallon, *Counter-Democracy*, 9.

13. Keefe, *Empire of Pain*; Carreyrou, *Bad Blood*; G. Harris, *No More Tears*; Elmore, *Seed Money*.

14. Scheufele et al., "What We Know About Effective Public Engagement."

15. Mill, *On Liberty*.

16. Mill, *On Liberty*, 92–93.

17. Urbinati, *Mill on Democracy*, 83.

18. Quine, "Epistemology Naturalized."

19. King et al., "Question of Participation," 388.

20. Warren, "Democratic Theory and Trust," 340.

21. Warren, "Democratic Theory and Trust," 340 (emphasis in original).

22. Warren, "Democratic Theory and Trust," 341.

23. Warren, "Democratic Theory and Trust," 341.

24. Warren, "Democratic Theory and Trust," 340 (emphasis in original). Others show that social trust is rather a prerequisite for deliberation (see Mclaverty and Halpin, "Deliberative Drift").

25. In my articulation of deliberation's instrumental aims, this would mean it failed to accomplish one of the other two goals—namely, non-domination and epistemic gain—though Warren's metrics are different.

26. Gastil and Richards, "Making Direct Democracy Deliberative," 260.

27. Warren, "Democratic Theory and Trust," 343. This is also connected to the Deweyan pragmatist vision of democracy in which we change the culture (and individuals) more than a particular institutional practice (although in my case, I think institutional innovations are essential as well).

28. Setälä, "Role of Deliberative Mini-Publics," 208.

29. Niemeyer and Jennstål, "Scaling Up Deliberative Effects."

30. Niemeyer and Jennstål, "Scaling Up Deliberative Effects," 337.

31. Kübler et al., "Who Supports Deliberative Mini-Publics?" 1.

32. Kübler et al., "Who Supports Deliberative Mini-Publics?" 4.

33. See Knight and Johnson, *Priority of Democracy.*

34. MacKenzie and Warren, "Two Trust-Based Uses of Minipublics," 96. See also Burgess, "From 'Trust Us' to Participatory Governance"; Nabatchi, "Addressing the Citizenship and Democratic Deficits"; and Martinez, "Realizing the Value of Public Input."

35. Suiter et al., "Scaling up Deliberation," 256.

36. Gastil and Richards, "Making Direct Democracy Deliberative."

37. Escobar and Elstub, "Forms of Mini-Publics"; *Democracy in Motion*; Grönlund et al., "Deliberation and Civic Value"; Halvorsen, "Assessing the Effects of Public Participation"; Luskin et al., "Considered Opinions"; Strandberg and Grönlund, "Online Deliberation"; Boulianne, "Mini-Publics and Public Opinion"; Jacobs and Kaufmann, "Right Kind of Participation?"; Werner and Marien, "Process vs. Outcome?"

38. Grönlund et al., "Deliberation and Civic Virtue."

39. Grönlund et al., "Deliberation and Civic Virtue," 95.

40. Grönlund et al., "Deliberation and Civic Virtue," 96.

41. Karlsson et al., "Democratic Innovation," 128. The researchers qualified this finding by showing that political trust by people who already distrusted the government did not increase.

42. List and Dryzek, "Social Choice Theory," 12.

43. Fishkin and Luskin, "Experimenting with a Democratic Ideal," 287; see also Schweigert, "Strengthening Citizenship."

44. Fishkin and Luskin, "Experimenting with a Democratic Ideal," 339.

45. Niemeyer and Jennstål, "Scaling Up Deliberative Effects," 336.

46. MacKenzie and Warren, "Two Trust-Based Uses of Minipublics," 96. Of note, MacKenzie and Warren describe the public this way in reference to legislative

contexts, but much of my work here is to show how legislative contexts can translate to regulatory ones; see also Beauvais and Warren, "What Can Deliberative Mini-Publics Contribute?"

47. James Fishkin (*Democracy When the People Are Thinking*) has similarly argued that it is important for society at large to accept deliberative efforts as trustworthy and not just the participants of the deliberation.

48. Gastil, "Lessons and Limitations."

49. Setälä et al., "Beyond Polarization," 484.

50. Setälä et al., "Beyond Polarization," 485.

51. Grönlund et al., "Deliberation and Civic Value."

52. Ryan and Smith, "Defining Mini-Publics."

53. Fournier et al., *When Citizens Decide.*

54. Suiter et al., "Scaling Up Deliberation," 256.

55. Goodin and Dryzek, "Deliberative Impacts."

56. Geisler, "Public Trust in Citizens' Juries," 3.

57. Geisler, "Public Trust in Citizens' Juries," 17.

58. MacKenzie and Warren, "Two Trust-Based Uses of Minipublics," 98.

59. Pilet et al., "Public Support for Deliberative Citizens' Assemblies," 885.

60. Pilet et al., "Public Support for Deliberative Citizens' Assemblies," 887.

61. Pilet et al., "Public Support for Deliberative Citizens' Assemblies," 882.

62. Garry et al., "Perception of the Legitimacy of Citizens' Assemblies?" 532. They also examine politicians' perceptions of the ability of these assemblies to help deeply conflicted areas break deadlock. Contrary to evidence cited elsewhere in this book, they found that politicians are more likely to support the use of such assemblies when they provide recommendations as opposed to holding real power.

63. Crosby, "Citizens Juries."

64. Werner and Marien, "Process vs. Outcome?" See also Gibson et al., "Measuring Attitudes"; A. G. Miller et al., "Prediction and Perception"; Sunshine and Tyler, "Role of Procedural Justice and Legitimacy"; Joshua Cohen, "Procedure and Substance in Deliberative Democracy"; Goodin and Dryzek, "Deliberative Impacts."

65. Gaus et al., "Defending Democracy," 345.

66. Font and Blanco, "Procedural Legitimacy and Political Trust," 559.

67. Font and Blanco, "Procedural Legitimacy and Political Trust," 584.

68. Van Dijk et al., "Perceived Legitimacy of Deliberative Minipublics," 2.

69. Van Dijk et al., "Perceived Legitimacy of Deliberative Minipublics," 5.

70. Van Dijk et al., "Perceived Legitimacy of Deliberative Minipublics," 11.

71. Már and Gastil, "Do Voters Trust Deliberative Minipublics?," 975.

72. Germann et al., "Scaling Up?" 694.

73. Germann et al., "Scaling Up?" 682.

74. Germann et al., "Scaling Up?" 694.

75. Germann et al., "Scaling Up?" 679.

76. Jacobs and Kaufmann, "Right Kind of Participation?" 104.

77. Germann, *Mini-Publics*, 2–3.

78. Germann, *Mini-Publics*, 8.

79. Devillers et al., "Looking in from the Outside," 153.

80. Saffon and Urbinati, "Procedural Democracy."

81. Saffon and Urbinati, "Procedural Democracy."

82. Christiano, "Authority of Democracy"; Christiano, *Rule of the Many.*

83. Just as in the case of equality and representation, legitimacy exists on a spectrum: Some procedures are better than others insofar as some consequences are more desirable than others. It follows, then, that by tracking some procedures as *less* instrumentally valuable, they could be said to suffer a democratic deficit for failing to enact benefit society. For example, one defense of workplace democracy (see chapter 4) is that it better protects worker rights in more robust ways than, say, just unionizing. However, if unionizing was the more effective route to worker rights, then that would also be preferable.

84. Lafont, *Democracy Without Shortcuts*, 7.

85. Cristina Lafont and Nadia Urbinati make a similar argument in their 2024 book *The Lottocratic Mentality: Defending Democracy Against Lottocracy.*

86. See also Mutz, *Hearing the Other Side.*

87. Levine, "Citizenship and Service Delivery," 81.

88. Luhmann, "Familiarity, Confidence, Trust," 97–98.

89. Siegrist and Cvetkovich, "Perceptions of Hazards," 713.

90. National Academies for Science, Engineering, and Medicine, *Strategies for Public Engagement to Combat Mistrust and Build COVID-19 Vaccine Confidence.*

91. Siegrist, "Influence of Trust," 195.

92. Harré, "Trust and Its Surrogates."

93. Manchanda and Silver-Isenbstadt, "Prescription for a Healthier Democracy," 50.

94. National Academy of Sciences, *Preserving Public Trust.*

95. "Earning Trust in the Age of the Pandemic," *National Academies of Science*, September 16, 2020, www.nationalacademies.org/event/09-16-2020/earning-trust-in-the-age-of-the-pandemic. See also Opel et al., "Building Trust."

96. Doug Hubley, "Commencement 2019: Scientists Must 'Rebuilt Public Trust' in Their Work, Says CRISPR Pioneer," *Bates News*, May 26, 2019.

97. Diya Chacko, "Coronavirus Today: Science Will Save Us," *LA Times*, April 13, 2020.

98. Scheman, "Epistemology Resuscitated," 38. Scheman continues that "one could concede all this and still argue that the results of scientific research, conducted according to internal norms, are in fact, objectively, true (or, more reasonably, leading toward the truth), and justifiably believed by anyone who is in a position—as the lay public lamentably is not—to see how it is that the practices of trust-grounding work, despite the injustices of the situation in which they are housed" (38).

99. Baier, "Trust and Antitrust," 234, 259.

100. Cleary and Stokes, *Democracy and the Culture of Skepticism*, 11.

101. R. Smith, *Stories of Peoplehood*, 59.

102. Bianco, "Uncertainty, Appraisal, and Common Interest"; Warren, *Democracy and Trust*, 14.

103. Albertson and Gadarian, *Anxious Politics*, 96. Researchers found that in comparing public health with immigration studies that, during moments of health

crisis, "anxiety increased trust in potentially relevant government actors (e.g., Centers for Disease Control and Prevention, HHS) and generally not in less relevant government agencies (e.g., the Chair of the Federal Reserve Board)" (98–99).

104. Luhmann, "Familiarity, Confidence, Trust."

105. Shapin, *Social History of Truth.*

106. Marsh and Dibben, "Trust, Untrust, Distrust and Mistrust."

107. Marsh and Dibben, "Trust, Untrust, Distrust and Mistrust," 20.

108. Marsh and Dibben, "Trust, Untrust, Distrust and Mistrust," 19.

109. Yaqub et al., "Attitudes to Vaccination"; Calvao, "GMOs."

110. Tom Grady, "Ebola Outbreak in Congo Is Declared a Global Health Emergency," *New York Times,* July 17, 2019; Earnshaw et al., "Medical Mistrust"; Vinck et al., "Institutional Trust and Misinformation"; Trapido, "Ebola."

111. M. J. Ryan et al., "Technologies of Trust," 1.

112. Pagliaro et al., "Trust Predicts COVID-19 Prescribed and Discretionary Behavioral Intentions."

113. Luhmann, *Trust and Power.*

114. Baier, *Moral Prejudices,* 152.

115. Kramer and Gavrieli, "Power, Uncertainty, and the Amplification of Doubt," 344.

116. Cook et al., "Trust and Distrust."

117. Abrams and Potts, "Language of Harm"; Wertheim, *Surviving Ohio State.*

118. Berman, *Don't Call It a Cult.*

119. Tschandl, "Risk of Bias."

120. Sally Satel, "Truth About Painkiller Addiction," *Atlantic,* August 7, 2019.

121. Vertinsky, "Pharmaceutical (Re)Capture."

122. Vigoda, "From Responsive to Collaboration," 485. See also Scudder, *Beyond Empathy and Inclusion.*

123. D. Roberts, "Public Deliberation," 343–44.

124. Fricker, *Epistemic Injustice,* 2, 30.

125. Buchman et al., "Investigating Trust, Expertise, and Epistemic Injustice."

126. Origgi, "Epistemic Injustice and Epistemic Trust."

127. Knox, "Epistemic Mistrust."

128. Koenig and Harris, "Basis of Epistemic Trust."

129. Hardwig, "Role of Trust in Knowledge," 694.

130. Shapin, *Social History of Truth,* xxv–xxvi.

131. Webb, "Epistemology of Trust."

132. Quoted in Hardwig, "Role of Trust in Knowledge," 694.

133. Uslaner, *Segregation and Mistrust.*

134. Grasswick, "Scientific and Lay Communities," 387.

135. See Hardin, "Do We Want Trust?"

Conclusion

1. Rosenberg, "Environmental Procedures at the FCC."

2. Mendonça and Asenbaum, "Decolonizing Deliberative Democracy."

Bibliography

Periodicals

AAMC
ABC News
AEIdeas
AgFunder News
AP News
Association of American Universities
Atlantic
Autism Speaks
Axios
Bates News
BBC
BioNews
Biopolitical Times
BioSpace
Boston Review
Business Insider
Buzzfeed News
CBS News
Centers for Disease Control and Prevention
Centers for Medicare and Medicaid Services
Chalkbeat
CNN
Columbia Engineering
Connexion France
Conversation
Cosmos Magazine
Drug Discovery and Development
EIN Presswire
Environmental Working Group
Farmers Weekly
FDA News
Florida Keys Environmental Coalition
Forbes
France 24
Future Market Insight
Genetic Literacy Project
Grist
Hakai Magazine
Harvard Business Review
Harvard Heart Letter
HEALTH IT
Hill
Intellia Therapeutics
International Politics and Society
Issues in Science and Technology
Johns Hopkins Bloomberg School of Public Health
Kaiser Foundation
KCCI
KFF
Labiotech
Lancet Infectious Diseases
LA Times
Loka Institute
Media Freedom and Information Access Clinic
Medical Marketing and Media
MedPage Today
Mirage News
MIT Media Lab
MIT Technology Review
Nantucket Current
National Association of Community Health Centers
National Geographic
National Public Radio (NPR)

National Rural Health Association
Nature News
NBC News
New Atlas
New Yorker
New York Times
Participedia
Partnership for Public Service
Petrie-Flom Center Blog
Pew Research Center
Pharmaceutical Technology
PharmTech
Phys Org
Politico
PR Newswire
ProPublica
Public Agenda
Reuters
Science Direct
Science News
Scientific American
Scientist
Scotus Blog
Spectator
Stanford University
STAT News
Synthego
Taiwan 20
Technology Networks
Transmitter
UGA Today
Vice
Wall Street Journal
Washington Post
Weather Channel
Wired

Primary Sources

US Reports and Regulations

Attorney General of New York. *Background on Trump Organization Properties and Fraudulent Schemes.* 2023.

Congressional Budget Office. *Dynamic Analysis of Changes to the Supplemental Nutrition Assistance Program (SNAP) Under H.R. 1.* CBO Publication No. 61505. Washington, DC: Congressional Budget Office, June 1, 2025.

Environmental Protection Agency. "Pesticides; Exemptions of Certain Plant-Incorporated Protectants (PIPs) Derived from Newer Technologies." *Federal Register* 88 (May 31, 2023): 34756–79.

House Committee on Oversight and Accountability, Democratic Staff. *White House for Sale: How Princes, Prime Ministers, and Premiers Paid Off President Trump.* January 4, 2024.

National Academy of Sciences. *Preserving Public Trust: Accreditation and Human Research Participant Protection Programs.* Washington, DC: National Academies Press, 2001.

National Academies of Sciences, Engineering, and Medicine. *Gene Drives on the Horizon: Advancing Science, Navigating Uncertainty, and Aligning Research with Public Values.* Washington, DC: National Academies Press, 2016.

——. *Strategies for Public Engagement to Combat Mistrust and Build COVID-19 Vaccine Confidence.* Washington DC: National Academies Press, 2021.

US Department of Agriculture. Agricultural Marketing Services. "Swiss Cheese, Emmentaler Cheese Grades and Standards." February 22, 2001.

——. Animal and Plant Health Inspection Service. Food Safety and Inspection Services. "Regulation of the Movement of Animals Modified or Developed by Genetic Engineering." *Federal Register* 85 (December 28, 2020): 84269–75.
——. Animal and Plant Health Inspection Service. "Movement of Certain Genetically Engineered Organisms." *Federal Register* 85 (May 18, 2020): 29790–838.
——. "National Bioengineered Food Disclosure Standard: List of Bioengineered Foods." November 29, 2023.
US Food and Drug Administration. *CVM GFI #187A Heritable Intentional Genomic Alterations in Animals: Risk-Based Approach.* May 2024.
——. *CVM GFI #236 Clarification of FDA and EPA Jurisdiction over Mosquito-Related Products.* Guidance Document. October 2017.
——. *Human Gene Therapy Products Incorporating Human Genome Editing: Guidance for Industry.* January 2024.
White House. "Withdrawing the United States from the World Health Organization." January 20, 2025. www.whitehouse.gov/presidential-actions/2025/01/withdrawing-the-united-states-from-the-worldhealth-organization/.

Court Cases

Buck v. Bell, 274 US 200 (1927).
Dobbs v. Jackson Women's Health Organization, 597 US 215 (2022).
Planned Parenthood of Southeastern Pennsylvania v. Casey, 505 US 833 (1992).
Roe v. Wade, 410 US 113 (1973).
Whole Woman's Health v. Hellerstedt, 579 US 582 (2016).

Legislative Actions

US Code. *Administrative Procedure Act*, 5 USC § 553 (2022).
US Food and Drug Administration. 21 USC § 393 (1997).
US House. Agriculture Committee. *H. Res.427—Encouraging the Department of Agriculture to Lead Efforts, in Partnerships with States, Localities, Universities, and Various Farming and Ranching Groups, to Prepare for the Disruptive Transformation of the Entire Agriculture Industry.* 117th Cong., 2nd sess., May 26, 2021.
US House. Appropriations Committee. *H.R. 4346—CHIPS and Science Act.* 117th Cong., 2021–22.
——. *H.R. 8772—Legislative Branch Appropriations Act, 2025.* 118th Cong., July 11, 2024.
US House. *Healthcare Spending Drives Our Debt.* 117th Cong., 2nd sess., July 6, 2025. www.congress.gov/congressional-record/volume-168/issue-119/house-section/article/H6869-5.
US House of Representatives. *Articles of Impeachment Against Donald John Trump.* 116th Cong., 1st sess., December 18, 2019.
US Senate. Foreign Relations Committee. *S. Res.275—A Resolution Calling for International Ethical Standards in Genome Editing Research.* 116th Cong., July 15, 2019.

International Reports

Australian Government National Health and Medical Research Council. *Mitochondrial Donation Community Consultation Citizens' Panel Statement.* June 27, 2018.

Intergovernmental Panel on Climate Change. *IPCC Sixth Assessment Report: Mitigation of Climate Change*. New York: United Nations, 2022.

International Commission on the Clinical Use of Human Germline Genome Editing. *Heritable Human Genome Editing.* Washington, DC: National Academies Press, 2020.

Nuffield Council on Bioethics. *Genome Editing and Farmed Animal Breeding: Social and Ethical Issues.* London: Nuffield Council on Bioethics, 2021.

———. *Genome Editing and Human Reproduction: Social and Ethical Issues.* London: Nuffield Council on Bioethics, 2018.

———. *Novel Techniques for the Prevention of Mitochondrial DNA Disorders: An Ethical Review.* London: Nuffield Council on Bioethics, 2013.

———. "Public Dialogue on Genome Editing in Farmed Animals," October 12, 2022.

United Nations Development Programme. *Global Report on Gender Equality in Public Administration.* New York: UNDP, 2021.

WHO Expert Advisory Committee on Developing Global Standards for Governance and Oversight of Human Genome Editing. "Human Genome Editing: Recommendations," 2021.

———. "Report of the Second Meeting," Geneva, Switzerland, August 26–28, 2019.

World Health Organization. *Citizen Engagement in Evidence-Informed Policy-Making: A Guide to Mini-Publics: A Review of the Empirical Literature.* Geneva: WHO, 2024.

———. "Statement on Governance and Oversight of Human Genome Editing," July 26, 2019. www.who.int/news/item/26-07-2019-statement-on-governance-and-oversight-of-human-genome-editing.

———. "World Malaria Report." December 11, 2024. www.who.int/teams/global-malaria-programme/reports/world-malaria-report-2024.

Videos

Carlson, Tucker. "RFK Jr. Provides an Update on His Mission to End Skyrocketing Autism and Declassifying Kennedy Files." *YouTube.* June 30, 2025, video, 19:09. www.youtube.com/watch?v=w_fzlwxJZAA.

"Earning Trust in the Age of the Pandemic." *National Academies of Science,* September 16, 2020. www.nationalacademies.org/event/9-16-2020/earning-trust-in-the-age-of-the-pandemic.

Niccol, Andrew, director. *Gattaca.* Culver City, CA: Columbia Pictures, 1997.

Wertheim, Jon, writer. *Surviving Ohio State.* Directed by Eva Orner. HBO, 2025.

Secondary Sources

Aasen, Marianne, and Arild Vatn. "Deliberation on GMOs: A Study of How a Citizens' Jury Affects the Citizens' Attitudes." *Environmental Values* 22 (2013): 461–81.

Abavisani, Mohammad, et al. "CRISPR-Cas System as a Promising Player Against Bacterial Infection and Antibiotic Resistance." *Drug Resistance Updates* 68 (2023): 100948.

Abizadeh, Kian. "Representation, Bicameralism, Political Equality, and Sortition." *Philosophy & Public Affairs* 46 (2018): 99–132.

Abrams, Jamie R., and Amanda Potts. "The Language of Harm: What the Nasser Victim Impact Statements Reveal About Abuse and Accountability." *University of Pittsburgh Law Review* 82 (2020): 71–134.

Acharya, Viral V., et al. *Regulating Wall Street: The Dodd-Frank Act and the New Architecture of Global Finance.* New York University Stern School of Business, 2011.

Achen, Christopher H., and Larry M. Bartels. *Democracy for Realists: Why Elections Do Not Produce Responsive Government.* Princeton University Press, 2017.

Ackerman, Bruce, and James S. Fishkin. *Deliberation Day.* Yale University Press, 2004.

Adams, John. *In Defense of the Constitutions of Government of the United States of America.* Poultry, 1787.

Addams, Jane. *Democracy and Social Ethics.* University of Illinois Press, 2002. Originally published in 1902.

Adolph, Christopher. *Bankers, Bureaucrats, and Central Bank Politics: The Myth of Neutrality.* Cambridge University Press, 2013.

Agar, Nicholas. *Liberal Eugenics: In Defence of Human Enhancement.* Wiley-Blackwell, 2004.

———. *Truly Human Enhancement: A Philosophical Defense of Limits.* MIT Press, 2014.

Ahmad, Aftab, et al. "Regulatory, Ethical, and Social Aspects of CRISPR Crops." In *CRISPRized Horticulture Crops*, edited by A. Ahmad et al., 261–87. Academic Press, 2021.

Aiuti, Alessandro, et al. "Ensuring a Future for Gene Therapy for Rare Diseases." *Nature* 28 (2022): 1985–88.

Ajeet, Singh, et al. "Revolutionary Breakthrough: FDA Approves CASGEVY, the First CRISPR/Cas9 Gene Therapy for Sickle Cell Disease." *Annals of Medicine & Surgery* 86 (2024): 4555–59.

Al Adawi, Mudhar Mohammed, et al. "Lived Experience of Sickle Cell Patients During and After Crisis." *Open Journal of Nursing* 11 (2021): 219–29.

Albertson, Bethany, and Shana Kushner Gadarian. *Anxious Politics: Democratic Citizenship in a Threatening World.* Cambridge University Press, 2015.

Alexander, Michelle. *The New Jim Crow: Mass Incarceration in the Age of Colorblindness.* New Press, 2010.

Ali, Arfan, et al. "Genetically Modified Foods: Engineered Tomato with Extra Advantages." *Advancements in Life Sciences* 1 (2023): 139–52.

Ali, Shehzad, et al. "COVID-19 and Inequality: Are We All in This Together?" *Canadian Journal of Public Health* 111 (2020): 415–16.

Alkanli, Suleyman Serdar, et al. "CRISPR/Cas9 Mediated Therapeutic Approach in Huntington's Disease." *Molecular Neurobiology* 60 (2023): 1486–98.

Allen, Barry. "Foucault's Nominalism." In *Foucault and the Government of Disability*, edited by Shelley Tremain, 93–107. University of Michigan Press, 2005.

Allen, Danielle. "A New Theory of Justice." In *Difference Without Domination: Pursuing Justice in Diverse Democracies*, edited by Danielle Allen and Rohini Somanathan, 27–58. University of Chicago Press, 2020.

Alper, Joseph S., et al. *The Double-Edged Helix: Social Implications of Genetics in a Diverse Society*. Johns Hopkins University Press, 2002.

Alsan, Marcella, et al. "Does Diversity Matter for Health? Experimental Evidence from Oakland." *National Bureau of Economic Research* 109 (2019): 4071–111.

Andersen, Ida-Elisabeth, and Birgit Jæger. "Scenario Workshops and Consensus Conferences: Towards More Democratic Decision-Making." *Science and Public Policy* 26 (1999): 331–40.

Anderson, Elizabeth. "Epistemic Justice as a Virtue of Social Institutions." *Social Epistemology* 26 (2012): 163–73.

——. "The Epistemology of Democracy." *Episteme* 3 (2006): 8–22.

——. *The Imperative of Integration*. Princeton University Press, 2010.

——. *Private Government: How Employers Rule Our Lives (and Why We Don't Talk About It)*. Princeton University Press, 2017.

Anderson, K. R., et al. "CRISPR Off-Target Analysis in Genetically Engineered Rats and Mice." *Nature Methods* 15 (2018): 512–14.

Annas, George. *Bioethics: Crossing Human Rights and Health Law Boundaries*. Oxford University Press, 2005.

Anwar, Shamena, et al. "The Impact of Jury Race in Criminal Trials." *Quarterly Journal of Economics* 127 (2012): 1017–55.

Apffel-Marglin, Frédérique, and Stephen A. Marglin. *Decolonizing Knowledge: From Development to Dialogue*. Oxford University Press, 1996.

Aquino-Jarquin, Guillermo. "Current Advances in Overcoming Obstacles of CRISPR/Cas9 Off-target Genome Editing." *Molecular Genetics and Metabolism* 134 (2021): 77–86.

Arendt, Hannah. *The Human Condition*. University of Chicago Press, 1958.

Aristotle. *The Complete Works of Aristotle*. Revised Oxford Translation. Edited by Jonathan Barnes. 2 vols. Princeton University Press, 1984. Reprinted with corrections in 1995.

Arnos, Kathleen S. "Genetics and Deafness: Impacts on the Deaf Community." *Sign Language Studies* 2 (2002): 150–68.

Arrow, Kenneth J. "A Difficulty in the Concept of Social Welfare." *Journal of Political Economy* 58 (1950): 328–46.

Arvanitidis, P. A., and N. C. Kyriazis. "The Qualities and Workings of Direct Democracy: Lessons from Classical Athens." In *Essays in Contemporary Economics*, edited by George C. Bitros and Nicholas C. Kyriazis, 161–74. Springer, 2015.

Asamoah, Afia K., and Jashua M. Sharfstein. "Transparency at the Food and Drug Administration." *New England Journal of Medicine* 362 (2010): 2341–43.

Asquer, Alberto, and Inna Krachkovskaya. "Uncertainty, Institutions and Regulatory Responses to Emerging Technologies: CRISPR Gene Editing in the US and the EU (2012–2019)." *Regulation & Governance* 15 (2021): 1111–27.
Asrar, Asim, et al. "Role of CRISPR/Cas9 Based Therapy in Breast Cancer: A Future Direction." *Nucleus* 67 (2024): 251–61.
Azangou-Khyavy, Mohammadreza, et al. "CRISPR/Cas: From Tumor Gene Editing to T Cell-Based Immunotherapy of Cancer." *Frontiers in Immunology* 11 (2020).
Azariadis, Costas, and John Stachurski. "Poverty Traps." In *Handbook of Economic Growth*, edited by Philippe Aghion and Steven Durleaf, 295–384. North Holland, 2005.
Bächtiger, André, and John Dryzek. *Deliberative Democracy for Diabolical Times: Confronting Populism, Extremism, Denial, and Authoritarianism*. Cambridge, UK, 2024.
Bächtiger, André, and Saskia Goldberg. "Towards a More Robust, but Limited and Contingent Defence of the Political Uses of Deliberative Minipublics." *Journal of Deliberative Democracy* 16 (2020): 33–42.
Bacon, Francis. *Meditationes Sacrae*. Whitefish: Kessinger Publishing, 2010 (1597).
———. *New Atlantis*. Edited by Jerry Weinberger. Harlan Davidson, 1989 (1627).
Bagg, Samuel. "Sortition as Anti-Corruption: Popular Oversight against Elite Capture." *American Journal of Political Science* 68 (2022): 93–105.
Baier, Annette. *Moral Prejudices*. Harvard University Press, 1995.
———. "Trust and Antitrust." *Ethics* 96 (1986): 231–60.
Baker, Kellan E., et al. "Hormone Therapy, Mental Health, and Quality of Life Among Transgender People: A Systematic Review." *Journal of the Endocrine Society* 5 (2021): 1–16.
Ballantyne, Angela, et al. "Sharing Precision Medicine Data with Private Industry: Outcomes of a Citizens' Jury in Singapore." *Big Data & Society* 9 (2022).
Balleisen, Edward J., and Marc Eisner. "The Promise and Pitfalls of Co-Regulation." In *New Perspectives on Regulation*, edited by D. Moss and J. Cisternino, 127–49. Tobin Project, 2009.
Barnes, Elizabeth. "Valuing Disability, Causing Disability." *Ethics* 125 (2014): 88–113.
Batt, Sharon, Judy Butler, Olivia Shannon, and Adriane Fugh-Berman. "Pharmaceutical Ethics and Grassroots Activism in the United States: A Social History Perspective." *Bioethical Inquiry* 17 (2020): 49–60.
Bau, Adrian, and Oliver Escobar. "Participatory-Deliberative Processes and Public Policy Agendas: Lessons for Policy and Practice." *Policy Design and Practice* 1 (2018): 126–40.
Baylis, Françoise. *Altered Inheritance: CRISPR and the Ethics of Human Genome Editing*. Harvard University Press, 2019.
———. "Human Genome Editing: Our Future Belongs to All of Us." *Issues in Science and Technology* 35 (2019): 42–44.

Beauchamp, Tom, and Ruth R. Faden. "Informed Consent." In *Encyclopedia of Bioethics*, edited by Bruce Jennings, 1673–725. Macmillan, 2012.

Beauvais, Edana. "Deliberation and Equality." In *The Oxford Handbook of Deliberative Democracy*, edited by André Bächtiger et al. Oxford University Press, 2018.

Beauvais, Edana, and Mark E. Warren. "What Can Deliberative Mini-Publics Contribute to Democratic Systems?" *European Journal of Political Research* 58 (2019): 893–914.

Becker, Andréa. *Get It Out: On the Politics of Hysterectomy*. Health, Society, and Inequality series. New York University Press, 2025.

Becker, Gary S. "Competition and Democracy." *Journal of Law and Economics* 1 (1958): 105–9.

Begagic, Emir, et al. "CRISPR/Cas9-Mediated Gene Therapy for Glioblastoma: A Scoping Review." *Biomedicines* 12 (2024): 238.

Beiner, Ronald. *Political Judgment*. University of Chicago Press, 1983.

Beit-Hallahmi, Benjamin. "Science, Ideology, and Ideals: The Social History of IQ Testing." *Centennial Review* 38 (1994): 341–60.

Bekaert, B., et al. "Various Repair Events Following CRISPR/Cas9-based Mutational Correction of an Infertility-Related Mutation in Mouse Embryos." *Journal of Assisted Reproduction and Genetics* 41 (2024).

Bekkers, Victor, et al. *Governance and the Democratic Deficit*. Routledge, 2007.

Benevolenza, Mia A., and LeaAnne DeRigne. "The Impact of Climate Change and Natural Disasters on Vulnerable Populations: A Systematic Review of Literature." *Journal of Human Behavior in the Social Environment* 29 (2019): 266–81.

Bentham, Jeremy. "Panopticon, Constitution, Colonies, Codification." In *The Works of Jeremy Bentham*, edited by John Bowring. Liberty Fund, 1843.

Bergen, Raquel Kennedy. "Marital Rape: New Research and Directions." *Applied Research Forum* (2006).

Berlin, Isaiah. "Two Concepts of Liberty." In *Liberty: Incorporating Four Essays on Liberty*, edited by Henry Hardy, 166–217. Oxford University Press, 2002. Originally published in 1969.

Berman, Sarah. *Don't Call It a Cult: The Shocking Story of Keith Raniere and the Women of NXIVM*. Steerforth, 2021.

Bernasconi, Robert. "Kant as an Unfamiliar Source of Racism." In *Philosophers on Race: Critical Essays*, edited by Julie K. Ward and Tommy L. Lott, 145–66. Blackwell Publishing, 2002.

Bernicchia-Freeman, Zoe. "The Blurring of Identity: Cochlear Implants and the Deaf Community." *Journal of Integrative Research & Reflection* 1 (2018): 64–74.

Best, Stephanie, et al. "A Systematic Review of Geographical Inequities for Accessing Clinical Genomic and Genetic Services for Non-Cancer Related Rare Disease." *European Journal of Human Genetics* 30 (2022): 645–52.

Bhat, Shamik, Joseph S. Ross, and Reshma Ramachandran. "Medical Product Industry Ties to Patient Advocacy Organizations' Executive Leadership." *JAMA Internal Medicine* 183 (2023): 1164–65.

Bianco, William T. “Uncertainty, Appraisal, and Common Interest: The Roots of Constituent Trust.” In *Trust and Government*, edited by Valerie Braithwaite and Margaret Levi, 245–66. Russell Sage Foundation, 2005.
Biesold, Horst. *Crying Hands: Eugenics and Deaf People in Nazi Germany.* Translated by William Sayers. Gallaudet University Press, 1988.
Bingham, Lisa Blomgren, Tina Nabatchi, and Rosemary O’Leary. “The New Governance: Practices and Processes for Stakeholder and Citizen Participation in the Work of Government.” *Public Administration Review* 65 (2005): 547–58.
Blok, Anders. “Experts on Public Trial: On Democratizing Expertise Through a Danish Consensus Conference.” *Public Understanding of Science* 16 (2007): 163–82.
Blum, Deborah. *The Poison Squad: One Chemist’s Single-Minded Crusade for Food Safety at the Turn of the Twentieth Century.* Penguin Press, 2018.
Bohman, James. “Deliberative Democracy and the Epistemic Benefits of Diversity.” *Episteme* 3 (2006): 175–91.
——. “Domination, Epistemic Injustice and Republican Epistemology.” *Social Epistemology* 26 (2012): 175–87.
Boix, Carles, and Daniel Posner. “Social Capital: Explaining Its Origins and Effects on Government Performance.” *British Journal of Political Science* 28 (1998): 686–93.
Bok, Sissela. *Lying: Moral Choice in Public and Private Life.* Pantheon Books, 1978.
Bolger, Daniel, and Elaine Howard Ecklund. “Seeing Is Achieving: Religion, Embodiment, and Explanations of Racial Inequality in STEM.” *Ethnic and Racial Studies* 45 (2020): 3–21.
Borkman, Thomasina. “Experiential Knowledge: A New Concept for the Analysis of Self-Help Groups.” *Social Science Review* 50 (1979): 445–56.
Borsa, Alexander, Miriam Miyagi, Kelsey Ichikawa et al. “The New Genetics of Sexuality.” *GLQ: A Journal of Lesbian and Gay Studies* 30, no. 1 (2024): 119–40.
Boschele, Marco. “COVID-19 Science Policy, Experts, and Publics: Why Epistemic Democracy Matters in Ecological Crises.” *Journal of Integrative Biology* 24 (2020): 479–82.
Boswell, John. “Why and How Narrative Matters in Deliberative Systems.” *Political Studies* 61 (2012): 620–36.
Boulianne, Shelley. “Building Faith in Democracy: Deliberative Events, Political Trust and Efficacy.” *Political Studies* 67 (2019): 4–30.
——. “Mini-Publics and Public Opinion: Two Survey-Based Experiments.” *Political Studies* 66 (2018): 119–36.
Bovens, Luc, and Wlodek Rabinowicz. “Democracy and Argument: Tracking Truth in Complex Social Decisions.” In *Deliberation and Decision: Economics, Constitutional Theory, and Deliberative Democracy*, edited by Anne van Aaken et al., 143–57. Routledge, 2004.
Bradfield, Owen M. “Hearing Parents’ Voices: Parental Refusal of Cochlear Implants and the Zone of Parental Discretion.” *Bioethical Inquiry* 19 (2022): 143–50.

Braun, Caelesta, and Madalina Busuioc. "Stakeholder Engagement as a Conduit for Regulatory Legitimacy?" *Journal of European Public Policy* 27 (2020): 1599–611.
Breen, Keith. "Freedom, Republicanism, and Workplace Democracy." *Critical Review of International Social and Political Philosophy* 18 (2015): 470–85.
Brennan, Jason. *Against Democracy*. Princeton University Press, 2016.
Brown, Mark. "Expertise and Deliberative Democracy." In *Deliberative Democracy: Issues and Cases*, edited by Stephen Elstub and Peter McLaverty. Edinburgh University Press, 2014.
——. *Science in Democracy: Expertise, Institutions, and Representation*. MIT Press, 2009.
Brown, Wendy. *Undoing the Demos: Neoliberalism's Stealth Revolution*. Zone Books, 2015.
Brummer, Chris. "What Do the Data Reveal About (the Absence of Black) Financial Regulators?" *Brookings* (2020).
Bruno, Bethany, and Susanna Rose. "Patient Organizations and Conflicts of Interest." *British Medical Journal* 364 (2019): 1129.
Buchanan, Allen. "Human Nature and Enhancement." *Bioethics* 23 (2009): 141–50.
——. "Political Legitimacy and Democracy." *Ethics* 112 (2002): 689–719.
——, et al. *From Chance to Choice: Genetics and Justice*. Cambridge University Press, 2000.
Buchman, Daniel Z., Anita Ho, and Daniel Goldberg. "Investigating Trust, Expertise, and Epistemic Injustice in Chronic Pain." *Bioethical Inquiry* 24 (2017): 31–42.
Buckwalter, Wesley, and John Turri. "Knowledge, Adequacy, and Approximate Truth." *Consciousness and Cognition* 83 (2020): 102950.
Burgess, Michael M. "From 'Trust Us' to Participatory Governance: Deliberative Publics and Science Policy." *Public Understanding of Science* 23 (2014): 48–52.
——, and James Tansey. "Technology, Democracy, and Ethics: Democratic Deficit and the Ethics of Public Engagement." In *Emerging Technologies: From Hindsight to Foresight*, edited by Edna F. Einsiedel, 275–88. UBC Press, 2009.
Burke, Wylie, and Bruce M. Psaty. "Personalized Medicine in the Era of Genomics." *JAMA* 298 (2007): 1682–84.
Burris, Beverly H. *Technocracy at Work*. State University of New York Press, 1993.
——. "Technocratic Organization and Gender." *Forum* 12 (1989): 447–62.
Bush, Barbara. *Imperialism, Race, and Resistance: Africa and Britain*. Routledge, 1999.
Byrnes, C., et al. "Process and Strategies for Patient Engagement and Outreach in the Sickle Cell Disease (SCD) Community to Promote Clinical Trial Participation." *Journal of the National Medical Association* 114 (2022): 211–17.
Cairney, Paul. *The Politics of Evidence-Based Policy Making*. Palgrave Macmillan, 2016.

Calkin, Sydney. "Tapping Women for Post-Crisis Capitalism: Evidence from the 2012 World Development." *International Feminist Journal of Politics* 17 (2015): 611–29.

Callon, Michel, et al. *Acting in an Uncertain World: An Essay on Technical Democracy*. Translated by Graham Burchell. MIT Press, 2001.

Caluwaerts, Didier, and Min Reuchamps. "Does Inter-Group Deliberation Foster Inter-Group Appreciation? Evidence from Two Experiments in Belgium." *Politics* 34 (2014): 101–15.

——. "The G1000." In *The Malaise of Electoral Democracy and What to Do About It*. Rel-Bel e-book, 2014.

——. *The Legitimacy of Citizen-Led Deliberative Democracy: The G1000 in Belgium*. Routledge, 2018.

Calvao, Tania. "GMOs: An Examination of Issues Surrounding GMO Regulations." In *The Intersection of Food and Public Health*, edited by A. Bryce Hoflund et al. Routledge, 2017.

Calvello, Emilie, et al. "Emergency Care in Sub-Saharan Africa: Results of a Consensus Conference." *African Journal of Emergency Medicine* 3 (2013): 42–48.

Calvert, Aubin, and Mark E. Warren. "Deliberative Democracy and Framing Effects: Why Frames Are a Problem and How Deliberative Mini-Publics Might Overcome Them." In *Deliberative Mini-Publics: Involving Citizens in the Democratic Process*, edited by Kimmo Grönlund et al., 203–23. ECPR Press, 2014.

Cammack, Daniela. "Deliberation and Discussion in Classical Athens." *Journal of Political Philosophy* 29 (2021): 135–66.

Campbell, Fiona Kumari. *Contours of Ableism: The Production of Disability and Abledness*. Palgrave Macmillan, 2009.

Camporesi, Silvia. "Choosing Deafness with Preimplantation Genetic Diagnosis: An Ethical Way to Carry on a Cultural Bloodline?" *Cambridge Quarterly of Healthcare Ethics* 19 (2010): 86–96.

Carey, George Wescott. *The Federalist: Design for a Constitutional Republic*. University of Illinois Press, 1994.

Carnevale, Anthony, and Nicole Smith. "The Economic Value of Diversity." In *Our Compelling Interests: The Value of Diversity for Democracy and a Prosperous Society*, edited by Earl Lewis and Nancy Cantor, 106–57. Princeton University Press, 2016.

Carpenter, Daniel. *The Forging of Bureaucratic Autonomy: Reputations, Networks, and Policy Innovation in Executive Agencies, 1862–1928*. Princeton University Press, 2001.

——, and David A. Moss, eds. "Corrosive Capture? The Dueling Forces of Autonomy and Industry Influence in FDA Pharmaceutical Regulation." In *Preventing Regulatory Capture: Special Interest Influence and How to Limit It*, edited by Daniel Carpenter and David A. Moss, 152–72. Cambridge University Press, 2014.

——. *Preventing Regulatory Capture: Special Interest Influence and How to Limit It*. Cambridge University Press, 2014.

Carpini, Michael X. Deli, et al. "Public Deliberation, Discursive Participation, and Citizen Engagement: A Review of the Empirical Literature." *Annual Review of Political Science* 7 (2004): 315–44.

Carreyrou, John. *Bad Blood: Secrets and Lies in a Silicon Valley Startup*. Random House, 2018.

Cavalcanti, Daniella Medeiros, et al. "Evaluating the Impact of Two Decades of USAID Interventions and Projecting the Effects of Defunding on Mortality Up to 2030: A Retrospective Impact Evaluation and Forecasting Analysis." *Lancet* 406 (2025): P283–P294.

Celik, Selime, et al. "CRISPR/Cas Genome Editing in Fruit Crops: Recent Advances, Challenges, and Future Prospects." In *CRISPRized Horticulture Crops: Genome Modified Plants and Microbes in Food and Agriculture*, edited by Kamel A. Abd-Elsalam, Aftab Ahmad, and Baohong Zhang, 349–64. Academic Press, 2024.

Centerwall, Siegried A., and Willard R. Centerwall. "The Discovery of Phenylketonuria." *American Academy of Pediatrics* 105 (2000).

Chambers, Simone. "Balancing Epistemic Quality and Equal Participation in a System Approach to Deliberative Democracy." *Social Epistemology* 31 (2017): 266–76.

Chapman, Madeleine, and Jesper Dammeyer. "The Significance of Deaf Identity for Psychological Well-Being." *Journal of Deaf Studies and Deaf Education* 22 (2017): 70–90.

Chayes, Abram, and Antonia Handler Chayes. *The New Sovereignty: Compliance with International Regulatory Agreements*. Harvard University Press, 1998.

Chen, Jie, et al. "Why Female Board Representation Matters: The Role of Female Directors in Reducing Male CEO Overconfidence." *Journal of Empirical Finance* 53 (2019): 70–90.

Chisowa, Tendai. "Racial Disparities Are Rife in Biotech: We Must Act Now." *BioCentury* 1 (2022): 242–43.

ChoGlueck, Christopher. "Still No Pill for Men? Double Standards & Demarcating Values in Biomedical Research." *Studies in History and Philosophy of Science* 91 (2022): 66–76.

Christiano, Thomas. "The Authority of Democracy." *Journal of Political Philosophy* 12 (2004): 266–90.

——. *The Rule of the Many: Fundamental Issues in Democratic Theory*. Routledge, 1996.

Cimpian, Joseph R., et al. "Understanding Persistent Gender Gaps in STEM." *Science* 368 (2020): 1317–19.

Cisney, Vernon W., and Nicolae Morar. "Introduction: Why Biopower? Why Now?" *Biopower: Foucault and Beyond*, edited by Vernon W. Cisney and Nicolae Morar, 1–25. University of Chicago Press, 2016.

Claeys, Hannes, et al. "Coordinated Gene Upregulation in Maize Through CRISPR/Cas-Mediated Enhancer Insertion." *Plant Biotechnology Journal* 22 (2024): 16–18.

Cleary, Ekaterina Galkina, et al. "Comparison of Research Spending on New Drug Approvals by the National Institutes of Health vs the Pharmaceutical Industry, 2010–2019." *JAMA* 28 (2023): e230511.

Cleary, Matthew R., and Susan C. Stokes. *Democracy and the Culture of Skepticism: Political.* Russell Sage Foundation, 2006.

Cleghorn, Elinor. *Unwell Women.* Dutton, 2021.

Clements, Jeffrey D. *Corporations Are Not People: Reclaiming Democracy from Big Money and Global Corporations.* Berrett-Koehler, 2012.

Cleves, P. A., et al. "CRISPR/Caso-Mediated Genome Editing in a Reef-Building Coral." *Proceedings of the National Academies of Science* 115 (2018): 5235–40.

Coady, C. A. J. *Testimony.* Oxford University Press, 1992.

Coady, David. "Two Concepts of Epistemic Injustice." *Episteme* 7 (2010): 101–13.

Code, Lorraine. "Ignorance, Injustice and the Politics of Knowledge: Feminist Epistemology Now." *Australian Feminist Studies* 29 (2014): 148–60.

Coffee, John C. "The Political Economy of Dodd-Frank: Why Financial Reform Tends to Be Frustrated and Systemic Risk Perpetuated." *Cornell Law Review* 97 (2011): 1019–82.

Cohen, Adam. *The Supreme Court, American Eugenics, and the Sterilization of Carrie Buck.* Penguin, 2016.

Cohen, I. Glenn, et al. "Handle with Care: The WHO Report on Human Genome Editing." *Hastings Center Report* 52 (2022): 10–14.

Cohen, Joshua. "Deliberative Democracy." In *Deliberation, Participation and Democracy: Can the People Govern?*, edited by Shawn W. Rosenberg, 219–36. Palgrave Macmillan, 2007.

———. "An Epistemic Conception of Democracy." *Ethics* 97 (1986): 26–38.

———. "Procedure and Substance in Deliberative Democracy." In *Deliberative Democracy: Essays on Reason and Politics*, edited by James Bohman and William Rehg, 67–92. MIT Press, 1997.

Cohen, Julie E. *Between Truth and Power.* Oxford University Press, 2019.

Coleman, Stephen, and Giles Moss. "Under Construction: The Field of Online Deliberation Research." *Journal of Information Technology & Politics* 9 (2012): 1–15.

Collins, Harry. *Artificial Experts: Social Knowledge and Intelligent Machines.* MIT Press, 1990.

Collins, Marvin, et al. "A Frugal CRISPR Kit for Equitable and Accessible Education in Gene Editing and Synthetic Biology." *Nature* 15 (2024): 6563.

Comen, Elizabeth. *All in Her Head: The Truth and Lies Early Medicine Taught Us About Women's Bodies and Why It Matters Today.* HarperWave, 2024.

Comfort, Nathaniel. *The Science of Human Perfection: How Genes Became the Heart of American Medicine.* Yale University Press, 2012.

Cong, Le, et al. "Multiplex Genome Engineering Using CRISPR/Cas Systems." *Science* 339 (2013): 819–23.

Conklin, Bruce R. "On the Road to a Gene Drive in Mammals." *Nature* 566 (2019): 43–45.

Conley, Dalton, and Jason Fletcher. *The Genome Factor: What the Social Genomics Revolution Reveals About Ourselves, Our History, and the Future.* Princeton University Press, 2017.

Contreras, Jorge L. *The Genome Defense: Inside the Epic Legal Battle to Determine Who Owns Your DNA.* Algonquin Books of Chapel Hill, 2021.

Cook, Karen S., et al. "Trust and Distrust in Patient-Physician Relationships: Perceived Determinants of High- and Low-Trust Relationships in Managed-Care Settings." In *Trust and Distrust in Organizations: Dilemmas and Approaches,* edited by Roderick Kramer, M. Cook, and Karen S. Cook, 65–98. Russell Sage Foundation, 2004.

Cooke, M. "Five Arguments for Deliberative Democracy." *Political Studies* 48 (2000): 947–69.

Cooper, Rachel. "Can It Be a Good Thing to Be Deaf?" *Journal of Medicine and Philosophy* 32 (2007): 563–83.

Corry, Olaf, and David Reiner. "Protests and Policies: How Radical Social Movement Activists Engage with Climate Policy Dilemmas." *Sociology* 55 (2020): 197–217.

Crain, Marion, et al., eds. *Invisible Labor: Hidden Work in the Contemporary World.* University of California Press, 2016.

Credo, Jonathan, and Jani C. Ingram. "Perspective Developing Successful Collaborative Research Partnerships with AI/AN Communities." *International Journal of Environmental Research and Public Health* 18 (2021): 9089.

Crenshaw, Kimberle. "Demarginalizing the Intersection of Race and Sex: A Black Feminist Critique of Antidiscrimination Doctrine, Feminist Theory, and Antiracist Politics." *University of Chicago Legal Forum* (1989): 139–68.

Crosby, Ned. "Citizens Juries: One Solution for Difficult Environmental Questions." In *Fairness and Competence in Citizen Participation,* edited by O. Renn et al., 157–14. Springer, 1995.

Cunha, Davi G. F., et al. "Citizen Science Participation in Research in the Environmental Sciences: Key Factors Related to Projects' Success and Longevity." *Anais da Academia Brasileira de Ciências* 89 (2017): 2229–45.

Curato, Nicole, and Jürg Steiner. "Deliberative Democracy and Comparative Democratization Studies." In *The Oxford Handbook of Deliberative Democracy,* edited by André Bächtiger et al. Oxford University Press, 2018.

Curato, Nicole, and Marit Boker. "Linking Mini-Publics to the Deliberative System: A Research Agenda." *Policy Sciences* 49 (2015): 173–90.

Curato, Nicole, et al. *Deliberative Mini-Publics: Core Design Features.* Bristol: Bristol University Press, 2021.

———. "Twelve Key Findings in Deliberative Democracy Research." *Daedelus* 146 (2017): 28–38.

Cushing, Lara, et al. "The Haves, the Have-Nots, and the Health of Everyone: The Relationship Between Social Inequality and Environmental Quality." *Annual Review of Public Health* 36 (2015): 193–209.

Cwik, Bryan. "Moving Beyond 'Therapy' and 'Enhancement' in the Ethics of Gene Editing." *Cambridge Quarterly of Healthcare Ethics* 28 (2019): 695–707.
Daar, Judith. *The New Eugenics.* Yale University Press, 2017.
Dahl, Robert. *On Democracy.* Yale University Press, 1998.
——. *A Preface to Democratic Theory.* University of Chicago Press, 2006.
Daniels, Brigham, Mark T. Buntaine, and Tanner Bangerter. "Testing Transparency." *Northwestern University Law Review* 114 (2020): 1263–334.
Davies, Kevin. *Editing Humanity: The CRISPR Revolution and the New Era of Genome Editing.* Pegasus Books, 2020.
Davis, Leonnard J. "Introduction: Normality, Power, and Culture." In *The Disability Studies Reader,* edited by Leonnard J. Davis, 1–16. Routledge, 2017.
De Boeck, Kris. "Cystic Fibrosis in the Year 2020: A Disease with a New Face." *Acta Paediatrica* 109 (2020): 893–99.
Decker, Charles. *Modern Economic Regulation: An Introduction to Theory and Practice.* Cambridge University Press, 2010.
De Sousa Santos, Boaventura. *The End of the Cognitive Empire: The Coming Age of Epistemologies of the South.* Duke University Press, 2018.
Devaney, Laura, et al. "Ireland's Citizens' Assembly on Climate Change: Lessons for Deliberative Engagement and Communication." *Environmental Communication* 14 (2020): 141–46.
Devillers, Sophie, et al. "Looking in from the Outside: How Do Invited but Not Selected Citizens Perceive the Legitimacy of a Minipublic?" *Journal of Deliberative Democracy* 17 (2021): 149–59.
Dewey, John. *Democracy and Education: An Introduction to the Philosophy of Education.* Macmillan, 1916.
——. *The Later Works, 1925–1953.* Vol. 8, 1933: *Essays and How We Think, Revised Edition.* Edited by Jo Ann Boydston. Southern Illinois University Press, 1986.
——. *The Public and Its Problems.* Swallow Press, 1927.
Dieke, Ada C., et al. "Disparities in Assisted Reproductive Technology Utilization by Race and Ethnicity, United States, 2014: A Commentary." *Journal of Women's Health* 26 (2017): 605–8.
Ding, Huihui, and Marcus Pivato. "Deliberation and Epistemic Democracy." *Journal of Economic Behavior & Organization* 185 (2021): 138–67.
Doerr, Nicole. *Political Translation: How Social Movement Democracies Survive.* Cambridge University Press, 2018.
Dolan, Julie A., and David H Rosenbloom. *Representative Bureaucracy: Classic Readings and Continuing Controversies.* M. E. Sharpe, 2003.
Dotson, Kristie. "Accumulating Epistemic Power: A Problem with Epistemology." *Philosophical Topics* 46 (2018): 129–54.
——. "A Cautionary Tale: On Limiting Epistemic Oppression." *Frontiers: A Journal of Women Studies* 33 (2012): 24–47.
——. "Conceptualizing Epistemic Oppression." *Social Epistemology* 33 (2012): 24–47.
Dowbiggin, Ian. *The Sterilization Movement and Global Fertility in the Twentieth Century.* Oxford University Press, 2008.

Doyle, Oran, and Rachael Walsh. "Constitutional Amendment and Public Will Formation: Deliberative Mini-publics as a Tool for Consensus Democracy." *International Journal of Constitutional Law* 20 (2022): 398–427.

Dressel, Julia, and Hany Farid. "The Dangers of Risk Prediction in the Criminal Justice System." *Science Advances* 4 (2018).

Druckman, James N., and Kjersten R. Nelson. "Framing and Deliberation: How Citizens' Conversations Limit Elite Influence." *American Journal of Political Science* 47 (2003): 729–45.

Drury, Sara A. Mehltretter, et al. "Deliberative Quality and Expertise: Uses of Evidence in Citizens' Juries on Wind Farms." *Journal of Deliberative Democracy* 7 (2021): 31–44.

Dryzek, John S. *Deliberative Democracy and Beyond: Liberals, Critics, Contestations.* Oxford University Press, 2000.

———. "Theory, Evidence, and the Tasks of Deliberation." In *Deliberation, Participation and Democracy: Can the People Govern?*, edited by Shawn W. Rosenberg, 237–50. Palgrave Macmillan, 2007.

———, and Simon Niemeyer. "Discursive Representation." *American Political Science Review* 102 (2008): 481–93.

———, et al. "Global Citizen Deliberation on Genome Editing." *Science* 369 (2020): 1435–37.

———, et al. "Promethean Elites Encounter Precautionary Publics: The Case of GM Foods." *Science, Technology, & Human Values* 34 (2009): 263–88.

Duarte Sagawa, Cintia H., et al. "Regulatory Framework of CRISPR-Edited Crops in the United States." In *Global Regulatory Outlook for CRISPRized Plants*, edited by Kamel A. Abd-Elsalam and Aftab Ahmad, 167–95. Elsevier, 2024.

Du Bois, W. E. B. *Darkwater: Voices from Within the Veil.* Dover Publications, 1999. Originally published in 1920.

Dutescu, Illinca A., and Sean A. Hillier. "Encouraging the Development of New Antibiotics: Are Financial Incentives the Right Way Forward? A Systematic Review and Case Study." *Infection and Drug Resistance* 14 (2021): 415–34.

Dwidar, Maraam. "Diverse Lobbying Coalitions and Influence in Notice-and-Comment Rulemaking." *Policy Studies Journal* 50 (2021): 199–240.

Dworkin, Ronald. *Law's Empire.* Belknap Press, 1986.

Dyal, Brenda, et al. "Reflections of Healthcare Experiences of African Americans with Sickle Cell Disease or Cancer: A Qualitative Study." *Cancer Nursing* 44 (2021): E53–E61.

Dzur, Albert W. *Punishment, Participatory Democracy, and the Jury.* Oxford University Press, 2012.

Earnshaw, Valerie A., et al. "Medical Mistrust in the Context of Ebola: Implications for Intended Care-Seeking and Quarantine Policy Support in the United States." *Journal of Health Psychology* 24 (2017): 219–28.

Eiler, Miriam, Katherine Grüber, and Christoph Rehmann-Sutter, eds. *The Human Enhancement Debate and Disability: New Bodies for a Better Life.* Palgrave Macmillan, 2014.

Einsiedel, Edna F., and Rose Geransar. "Framing Genetic Risk: Trust and Credibility Markers in Online Direct-to-Consumer Advertising for Genetic Testing." *New Genetics and Society* 28 (2009): 339–62.

Elliott, Carl. *The Occasional Human Sacrifice: Medical Experimentation and the Price of Saying No.* W. W. Norton, 2024.

Elmore, Bartow J. *Seed Money: Monsanto's Past and Our Food Future.* W. W. Norton, 2021.

Elster, Jon. *Deliberative Democracy.* Cambridge University Press, 1998.

——. "The Market and the Forum." In *Foundations of Social Choice Theory,* edited by Jon Elster and A. Hylland, 103–32. Cambridge University Press, 1986.

Elstub, Stephen, et al. "The Resilience of Pandemic Digital Deliberation: An Analysis of Online Synchronous Forums." *Journal of the European Institute for Communication and Culture* 28 (2021): 237–55.

Emerson, Claudia, et al. "Principles for Gene Drive Research." *Science and Society* 358 (2017): 1135–36.

Epstein, Lee, and Jack Knight. "How Social Identity and Social Diversity Affect Judging." *Leiden Journal of International Law* 35 (2022): 897–911.

Escobar, Oliver, and Elstub, Stephen. "Forms of Mini-Publics: An Introduction to Deliberative Innovations in Democratic Practice." New Democracy Foundation, 2017.

Eslahi, Atieh, et al. "New Advancements in CRISPR-Based Gene Therapy of Duchenne Muscular Dystrophy." *Gene* 867 (2023): 147358.

Esposito, Roberto. *Biopolitics and Philosophy.* University of Minnesota Press, 2008.

Estellés-Arolas, Enrique, and Fernando González-Ladrón-de-Guevara. "Towards an Integrated Crowdsourcing Definition." *Journal of Information Science* 38 (2012): 189–200.

Esvelt, Kevin M., et al. "Concerning RNA-Guided Gene Drives for the Alteration of Wild Populations." *eLife Sciences: Emerging Technology* 3 (2014): e03401.

Eubanks, Virginia. *Automating Inequality: How High-Tech Tools Profile, Police, and Punish the Poor.* St. Martin's Press, 2018.

Evans, Gavin. *White Supremacy: From Eugenics to Great Replacement.* Icon Books, 2024.

Ezenwa, Miriam O., et al. "Differences in Sensory Pain, Expectation, and Satisfaction Reported by Outpatients with Cancer or Sickle Cell Disease." *Pain Management Nursing* 19 (2018): 322–32.

Fabbri, Alice, et al. "Industry Funding of Patient and Health Consumer Organisations: A Systematic Review with Meta-Analysis." *British Medical Journal* 368 (2020).

Fan, Mei-Fang. "Indigenous Participation and Knowledge Justice in Deliberative Systems: Flooding and Wild Creek Remediation Controversies in Taiwan." *Environment and Planning C: Politics and Space* 39 (2021): 1492–510.

Fan, Mei-Fang, and Alice Siu. "The Effects of Public Deliberation on Attitude Change Toward Net-Zero Pathway: Generational Differences in Taiwan." *Energy and Climate Change* 6 (2025): 100183.

Fang, Yunlan, et al. "Gene Editing in Regenerative Medicine." In *Principles of Regenerative Medicine (Third Edition)*, edited by Anthony Atala et al., 741–59. Academic Press, 2019.

Farmbry, Kyle. *Administration and the Other: Explorations of Diversity and Marginalization in the Political Administrative State*. Lexington Books, 2009.

Farooq, et al. "CRISPR/Cas9: Targeted Genome Editing for the Treatment of Hereditary Hearing Loss." *Journal of Applied Genetics* 61 (2020): 51–65.

Farrell, Ann T., et al. "End Points for Sickle Cell Disease Clinical Trials: Renal and Cardiopulmonary, Cure, and Low-Resource Settings." *Blood Advances* 3 (2019): 4002–20.

Farrell, David M., and Peter Stone. "Sortition and Mini-Publics: A Different Kind of Representation." In *The Oxford Handbook of Political Representation in Liberal Democracies*, edited by Robert Rohrschneider and Jacques Thomassen, 228–46. Oxford University Press, 2020.

Fatumo, Segun, et al. "A Roadmap to Increase Diversity in Genomic Studies." *Nature Medicine* 28 (2022): 243–50.

Fedoroff, N. V., et al. "Radically Rethinking Agriculture for the 21st Century." *Science* 327 (2010): 833–34.

Feeney, Mary K., and Leonor Camarena. "Gender, Race, and Diversity Values Among Local Government Leaders." *Review of Public Personnel Administration* 41 (2021): 105–31.

Feinstein, Brian D., et al. "Board Diversity Matters: An Empirical Assessment of Community Lending at Federal Reserve-Regulated Banks." *Yale Program on Financial Stability* (2022).

Fenster, Mark. "The Opacity of Transparency." *Iowa Law Review* 91 (2006): 885–950.

Ferejohn, John. "The Citizens' Assembly Model." *Journal of Political Philosophy* 28 (2020): 191–212.

Fernández-García, Silvia, et al. "Effectiveness and Safety of COVID-19 Vaccines on Maternal and Perinatal Outcomes: A Systematic Review and Meta-Analysis." *British Medical Journal* 9 (2024): e014247.

Fiket, Irena, and Vincenzo Memoli. "Improving the Quality of Democracy: The Case of Deliberative Poll Held in 2007 in Turin." In *Participatory Democratic Innovations in Europe: Improving the Quality of Democracy?,* edited by Brigitte Giebel and Marko Joas, 123–42. Barbara Budrich Publishers, 2013.

Fischer, Frank. *Democracy and Expertise: Reorienting Policy Inquiry*. Oxford University Press, 2009.

———. "Organizational Expertise and Bureaucratic Control: Behavioral Science as Managerial Ideology." In *Critical Studies in Organization and Bureaucracy*, edited by Frank Fischer and Carmen Sirianni, 174–95. Temple University Press, 1984.

Fishkin, James S. "Deliberative Democracy in Context: Reflections on Theory and Practice." *Deliberative Mini-Publics: Involving Citizens in the Democratic Process*, edited by Kimmo Grönlund et al., 27–39. ECPR Press, 2014.

——. *Democracy When the People Are Thinking: Revitalizing Our Politics Through Deliberation.* Oxford University Press, 2018.
——. "Democratic Practice: Filter v. Mirror." *New Perspectives Quarterly* 30 (2013): 51–60.
——, and Cynthia Farrar. "Deliberative Polling: From Experiment to Community Resource." In *The Deliberative Democracy Handbook: Strategies for Effective Civic Engagement in the Twenty-First Century,* edited by John Gastil and Peter Levine, 68–79. Jossey-Bass, 2005.
——, and Robert C. Luskin. "Broadcasts of Deliberative Polls: Aspirations and Effects." *British Journal of Political Science* 35 (2005): 184–88.
——. "Experimenting with a Democratic Ideal: Deliberative Polling and Public Opinion." *Acta Politica* 40 (2005): 284–98.
Flanagan, Francis X. "Race, Gender, and Juries: Evidence from North Carolina." *Journal of Law and Economics* 61 (2018): 189–214.
Flanigan, Bailey, et al. "Fair Algorithms for Selecting Citizens' Assemblies." *Nature* 596 (2021): 548–52.
Fleming, Thomas R. "Surrogate Endpoints and FDA's Accelerated Approval Process." *Health Affairs* 24 (2005): 67–78.
Fleischacker, Samuel. *A Third Concept of Liberty: Judgment and Freedom in Kant and Adam Smith.* Princeton University Press, 1999.
Fluit, Sam, Laura Cortés-García, and Tine von Soest. "Social Marginalization: A Scoping Review of 50 Years of Research." *Humanities and Social Sciences Communications* 11 (2024): 1665.
Font, Joan, and Ismael Blanco. "Procedural Legitimacy and Political Trust: The Case of Citizen Juries in Spain." *European Journal of Political Research* 46 (2007): 557–89.
Formisano, Ron. *American Oligarchy: The Permanent Political Class.* University of Illinois Press, 2017.
Forrester, Katrina. *In the Shadow of Justice: Postwar Liberalism and the Remaking of Political Philosophy.* Princeton University Press, 2019.
Foucault, Michel. *The Birth of Biopolitics: Lectures at the Collège de France, 1978–79.* Picador, 2010.
——. "The Ethics of the Concern of the Self as a Practice of Freedom." In *The Final Foucault,* edited by James Bernauer and David Rasmussen, 136. MIT Press, 1987.
——. *The History of Sexuality, Vol. 2: The Use of Pleasure.* Translated by Robert Hurley. Vintage Books, 1990.
——. "The Politics of Health in the Eighteenth Century." *Foucault Studies* 18 (2014): 113–27.
——. *Security, Territory, and Population.* Translated by Graham Burchell. Palgrave Macmillan, 2007.
Fournier, Patrick, et al. *When Citizens Decide: Lessons from Citizen Assemblies on Electoral Reform.* Oxford University Press, 2011.
Frank, Emma, et al. "Predictors of Intraoperative Pain During Cesarean Delivery Under Regional Anesthesia." *Proc (Baylor University Medical Center)* 14 (2022): 595–98.

Freeman, Lauren, and Heather Stewart. "The Problem of Recognition, Erasure, and Epistemic Injustice in Medicine: Harms to Transgender and Gender Non-Binary Patients—Why We Should Be Worried." In *Epistemic Injustice and the Philosophy of Recognition*, edited by Paul Giladi and Nicola McMillan, 297–325. Routledge, 2022.

Fregulia, Jeanette M. *A Rich and Tantalizing Brew: A History of How Coffee Connected the World*. University of Arkansas Press, 2019.

French, Daniel. *When They Hid the Fire: A History of Electricity and Invisible Energy in America*. University of Pittsburgh Press, 2017.

Frey, William H. "The 'Diversity Explosion' Is America's Twenty-First-Century Baby Boom." In *Our Compelling Interests: The Value of Diversity for Democracy and a Prosperous Society*, edited by Earl Lewis and Nancy Cantor, 16–35. Princeton University Press, 2016.

Fricker, Miranda. *Epistemic Injustice: Power and the Ethics of Knowing*. Oxford University Press, 2007.

———. "Epistemic Oppression and Epistemic Privilege." *Canadian Journal of Philosophy* 29 (1999): 191–210.

———. "Rational Authority and Social Power: Towards a Truly Social Epistemology." *Proceedings of the Aristotelian Society* 98 (1998): 159–77.

Friedman, Jeffrey. *Power Without Knowledge: A Critique of Technocracy*. Oxford University Press, 2019.

Friedman, Milton. *Capitalism and Freedom*. University of Chicago, 1962.

Friedrichs, Steffi, et al. "An Overview of Regulatory Approaches to Genome Editing in Agriculture." *Biotechnology Research and Innovation* 3 (2019): 208–20.

Fruh, Sharon M., et al. "Weight Bias and Stigma." *Nursing Clinics* 56 (2021): 479–93.

Fuerstein, Michael. "Democratic Consensus as an Essential Byproduct." *Journal of Political Philosophy* 22 (2014): 282–301.

Fuller, Lon L. *The Morality of Law*.Yale University Press, 1986.

Fuller, Steve, and Veronika Lipinska. *The Proactionary Imperative: A Foundation for Transhumanism*. Palgrave Macmillan, 2014.

Fung, Archon. "Minipublics: Deliberative Designs and Their Consequences." In *Deliberation, Participation and Democracy: Can the People Govern?*, edited by Shawn W. Rosenberg, 159–83. Palgrave Macmillan, 2007.

———. "Survey Article: Recipes for Public Spheres: Eight Institutional Design Choices and Their Consequences." *Journal of Political Philosophy* 11 (2003): 338–67.

———. "Varieties of Participation in Complex Governance." *Public Administration Review* 66 (2006): 66–75.

———, and Erik Olin Wright. "Thinking About Empowered Participatory Governance." In *Deepening Democracy: Institutional Innovations in Empowered Participatory Governance*, edited by Archon Fung and Erik Olin Wright. Verso, 2003.

——, Mary Graham, and David Weil. *Full Disclosure: The Perils and Promise of Transparency.* Cambridge University Press, 2007.

Furlong, Scott R., and Cornelius M. Kerwin. "Interest Group Participation in Rule Making: A Decade of Change." *Journal of Public Administration Research and Theory* 15 (2005): 353–70.

Gabric, Albert J. "The Climate Change Crisis: A Review of Its Causes and Possible Responses." *Atmosphere* 14 (2023): 1081.

Galton, Francis. *Hereditary Genius.* Macmillan, 1869.

——. *Inquiries into Human Faculty and Its Development.* Macmillan, 1883.

Gamson, William A. *Talking Politics.* Cambridge University Press, 1992.

Gao, Xue, et al. "Treatment of Autosomal Dominant Hearing Loss by In Vivo Delivery of Genome Editing Agents." *Nature* 553 (2018): 217–21.

García, José María Rodríguez. "Scientia Potestas Est—Knowledge Is Power: Francis Bacon to Michel Foucault." *Anglia: Journal of English Philology* 119 (2001): 109–21.

Garrison, Nanibaa' A. "Informed Consent and the Ethics of IRB Research: A Case Study of the Havasupai Tribe's Lawsuit Against Genetic Researchers." In *The Academic's Handbook*, edited by Lori A. Flores and Jocelyn H. Olcott, 103–8. De Gruyter, 2020.

——, et al. "Genomic Research Through an Indigenous Lens: Understanding the Expectations." *Annual Review of Genomics and Human Genetics* 20 (2019): 495–517.

Garry, John, et al. "The Perception of the Legitimacy of Citizens' Assemblies in Deeply Divided Places? Evidence of Public and Elite Opinion from Consociational Northern Ireland." *Government and Opposition* 57 (2022): 532–51.

Garvey, Todd. "A Brief Overview of Rulemaking and Judicial Review." Congressional Research Service, 2017.

Gastil, John. "The Lessons and Limitations of Experiments in Democratic Deliberation." *Annual Review of Law and Social Science* 14 (2018): 271–91.

——, and Robert Richards. "Making Direct Democracy Deliberative Through Random Assemblies." *Politics & Society* 41 (2013): 253–81.

——, and Erik Olin Wright, eds. *Legislature by Lot: Transformative Designs for Deliberative Governance.* Verso, 2019.

——, et al. "Convening a Minipublic During a Pandemic: A Case Study of the Oregon Citizens' Assembly Pilot on COVID-19 Recovery." *Digital Government: Research and Practice* 3 (2022): 1–19.

Gauckler, Charlotte. "Expert Identification for Ethics Expertise Informed by Feminist Epistemology-Using Awareness of Biases and Situated Ignorance as an Indicator of Trustworthiness." *Bioethics* 37 (2023): 523–32.

Gaus, Daniel, et al. "Defending Democracy Against Technocracy and Populism: Deliberative Democracy's Strengths and Challenges." *Constellations* 27 (2020): 335–47.

Gbashi, Sefater, et al. "Food Safety, Food Security and Genetically Modified Organisms in Africa: A Current Perspective." *Biotechnology and Genetic Engineering Reviews* 37 (2021): 30–63.

Geisler, Alexander Matthias. "Public Trust in Citizens' Juries When the People Decide on Policies: Evidence from Switzerland." *Policy Studies* 44 (2022).

Geransar, Rose. "Banking on Trust: Issues of Informed Consent in Pharmacogenetic Research." In *Emerging Technologies: From Hindsight to Foresight*, edited by Edna F. Einsiedel, 185–202. UBC Press, 2009.

Germann, Micha. *Mini-Publics, (Lack of) Representativeness, and Legitimacy Beliefs*. Cambridge University Press, 2025.

——, Sofie Marinen, and Lala Muradova. "Scaling Up? Unpacking the Effect of Deliberative Mini-Publics on Legitimacy Perceptions." *Political Studies* 72 (2022): 677–700.

Gibson, James L., et al. "Measuring Attitudes Toward the United States Supreme Court." *American Journal of Political Science* 47 (2003): 354–67.

Gillborn, David. "Softly, Softly: Genetics, Intelligence and the Hidden Racism of the New Geneism." *Journal of Education Policy* 31 (2016): 365–88.

Gimenez, Martha E. *Marx, Women, and Capitalist Social Reproduction: Marxist-Feminist Essays*. Brill, 2018.

Giraudet, Louis-Gaëtan, et al. "'Co-construction' in Deliberative Democracy: Lessons from the French Citizens' Convention for Climate." *Humanities and Social Sciences Communications* 9 (October 2022): 207.

Globus, Rea, and Udi Qimron. "A Technological and Regulatory Outlook on CRISPR Crop Editing." *Journal of Cellular Biochemistry* 119 (2018): 1291–98.

Glover, Jonathan. *Choosing Children: Genes, Disability, and Design*. Oxford University Press, 2006.

Golden, Marissa Martino. "Interest Groups in the Rule-Making Process: Who Participates? Whose Voices Get Heard?" *Journal of Public Administration Research and Theory* 8 (1998): 245–70.

Goldman, Alvin I. "Experts: Which Ones Should You Trust?" In *Social Epistemology: Essential Readings*, edited by Alvin I. Goldman and Dennis Whitcomb, 109–36. Oxford University Press, 2011.

Goodin, Robert E. "If Deliberation Is Everything, Maybe It's Nothing." In *The Oxford Handbook of Deliberative Democracy*, edited by André Bächtiger et al. Oxford University Press, 2018.

——, and John S. Dryzek. "Deliberative Impacts: The Macro-Political Uptake of Mini-Publics." *Politics and Society* 34 (2006): 219–44.

Goodin, Robert E., and Kai Spiekermann. *An Epistemic Theory of Democracy*. Oxford University Press, 2018.

Gordon, Aubrey. *What We Don't Talk About When We Talk About Fat*. Beacon Press, 2020.

——. *"You Just Need to Lose Weight": And 19 Other Myths About Fat People*. Beacon Press, 2023.

Gordon, Sanford C., and Steven D. Rashin. "Stakeholder Participation in Policy Making: Evidence from Medicare Fee Schedule Revisions." *Journal of Politics* 83 (2021): 409–14.
Graham, Ian, and Peter Llewelyn Jones. *Expert Systems, Knowledge, Uncertainty and Decision.* Chapman and Hall Computing, 1988.
Graham, Janice E. "Ambiguous Capture: Collaborative Capitalism and the Meningitis Vaccine Project." *Medical Anthropology* 35 (2016): 419–32.
Grant, Ruth, and Robert O. Keohane. "Accountability and Abuses of Power in World Politics." *American Political Science Review* 99 (2005): 29–43.
Grasswick, Heidi E. "Scientific and Lay Communities: Earning Epistemic Trust Through Knowledge Sharing." *Synthese* 177 (2010): 387–409.
Green, Cameron, and Naomi Scheinerman. "Against Genetic Determinism of Welfare and Behavior." *American Journal of Bioethics* 24 (2024): 34–36.
Green, Eric. "Sequencing the Human Genome: Elucidating Our Genetic Blueprint." In *The Genetic Revolution: Unveiling the Unity of Life*, edited by Bernard D. Davis, 35–47. Joseph Henry Press, 2002.
Green, William. *Contraceptive Risk: The FDA, Depo-Provera, and the Politics of Experimental Medicine.* New York University Press, 2017.
Grimmelikhuijsen, Stephan, et al. "Can Decision Transparency Increase Citizen Trust in Regulatory Agencies? Evidence from a Representative Survey Experiment." *Regulation & Governance* 15 (2021): 17–31.
Grönlund, Kimmo, Maija Setälä, and Kaisa Herne. "Deliberation and Civic Virtue: Lessons from a Citizen Deliberation Experiment." *European Political Science Review* 2 (2010): 95–117.
——. "Does Enclave Deliberation Polarize Opinions?" *Political Behavior* 37 (2015): 995–1020.
Grossman, Lewis A. *Choose Your Medicine: Freedom of Therapeutic Choice in America.* Oxford University Press, 2021.
——. "FDA and the Rise of the Empowered Consumer." *Administrative Law Review* 66 (2014): 627–78.
Grundmann, Reiner. "The Problem of Expertise in Knowledge Societies." *Minerva* 55 (2016): 25–48.
Guidone, Heather C. "Under the Influence: Pharmaceutical Relationships and Their Impact on Endometriosis Care." In *The Palgrave Handbook of Critical Menstruation Studies*, edited by Chris Bobel et al., 66–89. Palgrave Macmillan, 2020.
Gül, Volkan. "Representation in Minipublics." *Representation* 55 (2019): 31–45.
Gunn, Paul. "Getting Democratic Priorities Straight: Pragmatism, Diversity, and the Role of Beliefs." *Critical Review* 72 (2015): 146–73.
Gusmano, Michael K., et al. "Public Deliberation in the Wild." *Hastings Center Report* 51 (2021): S2–S10.
Gustafson, Sandra M. *Imagining Deliberative Democracy in the Early American Republic.* University of Chicago Press, 2011.
Gutmann, Amy, and Dennis Thompson. *The Spirit of Compromise: Why Governing Demands It and Campaigning Undermines It.* Princeton University Press, 2012.

——. *Why Deliberative Democracy?* Princeton University Press, 2004.
Guttmacher, Alan E., et al. "Educating Health-Care Professionals About Genetics and Genomics." *Nature Review Genetics* 8 (2007): 151–57.
Gwynne, Kylie, et al. "Genomics and Inclusion of Indigenous Peoples in High Income Countries." *Human Genetics* 142 (2023): 1407–16.
Gyngell, Christopher, et al. "The Ethics of Germline Gene Editing." *Journal of Applied Philosophy* 34 (2016): 498–513.
Habermas, Jürgen. *Between Facts and Norms.* MIT Press, 1996.
——. "Political Communication in Media Society: Does Democracy Still Enjoy an Epistemic Dimension? The Impact of Normative Theory on Empirical Research." *Communication Theory* 16 (2006): 411–26.
Haelermans, Carla, et al. "Sharp Increase in Inequality in Education in Times of the COVID-19-Pandemic." *PLoS ONE* 17 (2022): e0261114.
Haider, Najmul, et al. "Global Landmark: 2023 Marks the Worst Year for Dengue Cases with Millions Infected and Thousands of Deaths Reported." *IJID Regions* 13 (September 26, 2024): 100459.
Haidt, Jonathan. *The Righteous Mind: Why Good People Are Divided by Politics and Religion.* Pantheon Books, 2012.
Haldar, Arunima, et al. "Tokenism or Realism? Gender Inclusion in Corporate Boards." *Equality, Diversity and Inclusion* 39 (2020): 707–25.
Hall, David D. "Deliberative Democracy in the Context of Town Meetings in Seventeenth-Century New England." *Journal of Public Deliberation* 15 (2019).
Halvorsen, Kathleen E. "Assessing the Effects of Public Participation." *Public Administration Review* 63 (2003): 535–43.
Hamburg, Margaret A., and Francis S. Collins. "The Path to Personalized Medicine." *New England Journal of Medicine* 363 (2010): 301–4.
Hamel, Liz, et al. "Race, Health, and COVID-19: The Views and Experiences of Black Americans." *KFF* 2020.
Hamilton, Alexander. "Federalist 81." In *The Federalist Papers*, edited by Clinton Rossiter, 480–90. Signet Classics, 1961.
Hamilton, Lawrence C., Joel Hartter, and Kei Saito. "Trust in Scientists on Climate Change and Vaccines." *SAGE* (2015): 1–13.
Hänel, Hilkje C. "#MeToo and Testimonial Injustice: An Investigation of Moral and Conceptual Knowledge." *Philosophy and Social Criticism* 48 (2021): 1–27.
Hans, Valerie P., John Gastil, and Traci F. Brown. "Deliberative Democracy and the American Civil Jury." *Cornell Law Faculty Publications*, 2014.
Hansen, Randall, and Desmond King. *Sterilized by the State: Eugenics, Race, and the Population Scare in Twentieth-Century North America.* Cambridge University Press, 2013.
Haque, Effi, et al. "Application of CRISPR/Cas9 Genome Editing Technology for the Improvement of Crops Cultivated in Tropical Climates: Recent Progress, Prospects, and Challenges." *Frontiers in Plant Science* 9 (2018): 1–12.
Haraway, Donna. "A Cyborg Manifesto." In *Simians, Cyborgs and Women: The Reinvention of Nature.* Routledge, 1991.

Harden, Kathryn Paige. *The Genetic Lottery: Why DNA Matters for Social Equality.* Princeton University Press, 2021.

Hardin, Russell. "Do We Want Trust in Government?" In *Democracy and Trust,* edited by Mark E. Warren, 22–41. Cambridge University Press, 1999.

Harding, Alexander. "Rehabilitation or Enhancement? FDA & the Gene Therapies of Tomorrow." *Harvard Journal of Law and Technology* 33 (2020): 639–59.

Hardwig, John. "The Role of Trust in Knowledge." *Journal of Philosophy* 88 (December 1991): 693–708.

Harré, Rom. "Trust and Its Surrogates: Psychological Foundations of Political Process." In *Democracy and Trust,* edited by Mark E. Warren, 249–72. Cambridge University Press, 1999.

Harris, Gardiner. *No More Tears: The Dark Secrets of Johnson & Johnson.* Random House, 2025.

Harris, John. "Moral Enhancement and Freedom." *Bioethics Forum* 25 (2011): 102–11.

Harris, Victoria. "Sex on the Margins: New Directions in the Historiography of Sexuality and Gender." *Historical Journal* 53 (2010): 1085–104.

Hartz-Karp, Janette, and Lyn Carson. "Putting the People into Politics: The Australia Citizens' Parliament." *International Journal of Public Participation* 3 (2009): 9–31.

Harvey, R. D. "Pioneers of Genetics: A Comparison of the Attitudes of William Bateson and Erwin Baur to Eugenics." *Notes and Records of the Royal Society of London* 49 (1995): 105–17.

Hausman, Bernice L. *Anti/Vax: Reframing the Vaccination Controversy.* Cornell University Press, 2019.

Hayek, Friederick August von. *Economic Freedom.* Blackwell, 1991.

———. *The Fatal Conceit: The Errors of Socialism.* University of Chicago Press, 1988.

Hayward, Clarissa Rile. *De-Facing Power.* Cambridge University Press, 2000.

Haywood, Carlton, et al. "The Association of Provider Communication with Trust Among Adults with Sickle Cell Disease." *Journal of General Internal Medicine* 35 (2010): 543–48.

He, Weiwei, et al. "Update on Gene Therapy in the Treatment of Hereditary Hearing Loss." *Sensory Neuroscience* (2025): e70004.

"Healthcare Spending Drives Our Debt." *Congressional Record: House Section* 168 (2022). www.congress.gov/congressional-record/volume-168/issue-119/house-section/article/H6869-5.

Hébert, Martin. "Indigenous Spheres of Deliberation." In *The Oxford Handbook of Deliberative Democracy,* edited by André Bächtiger et al. Oxford University Press, 2018.

Heidari, Raheleh, et al. "CRISPR and the Rebirth of Synthetic Biology." *Science Engineering Ethics* 23 (2017): 351–63.

Heier, Jorma. "Disabling Constraints on Democratic Participation: Challenges for a Political Theory of Care." *Ethics and Social Welfare* 9 (2015): 201–8.

Heller, Patrick, and Vijayendra Rao, eds. *Deliberation and Development: Rethinking the Role of Voice and Collective Action in Unequal Societies.* World Bank Publications, 2015.

Hendriks, Carolyn M. "Coupling Citizens and Elites in Deliberative Systems: The Role of Institutional Design." *European Journal of Political Research* 55 (2016): 43–60.

———. *The Politics of Public Deliberation: Citizen Engagement and Interest Advocacy.* Palgrave Macmillan, 2011.

Hennessy, Marita, Tom Fahey, and James Larkin. "Commercial Influences on Patient and Public Involvement: A Renewed Call for Research and Action." *Health Promotion International* 39 (2024): daae188.

Herai, Roberto H. "Avoiding the Off-Target Effects of CRISPR/Cas9 System Is Still a Challenging Accomplishment for Genetic Transformation." *Gene* 700 (2019): 176–78.

Herath, R. B. *Real Power to the People: A Novel Approach to Electoral Reform in British Columbia.* University Press of America, 2007.

Herd, Pamela, et al. "Reconstructing Sociogenomics Research: Dismantling Biological Race and Genetic Essentialism Narratives." *Journal of Health and Social Behavior* 62 (2021): 419–35.

Herrnstein, Richard J., and Charles Murray. *The Bell Curve: Intelligence and Class Structure in American Life.* Free Press, 1994.

Herz, Michael. "~~Fraudulent~~ Malattributed Comments in Agency Rulemaking." *Cardozo Law Review* 42 (2020–21): 1–68.

Hew, C. L., et al. "Transgenic Salmon: Tailoring the Genome for Food Production." *Journal of Fish Biology* 47 (1995): 1–19.

Heydari Fard, Sahar. "The Transformative Power of Social Movements." *Philosophy Compass* 19 (2024): e12951.

Hobbes, Thomas. *Leviathan.* Hackett, 1668.

Hoffman, Kelly M., et al. "Racial Bias in Pain Assessment and Treatment Recommendations, and False Beliefs About Biological Differences between Blacks and Whites." *Protocols of the National Academies of Science* 113 (2016): 4296–301.

Hoffman, Neil E. "Revisions to USDA Biotechnology Regulations: The SECURE Rule." *PNAS* 118 (2021): e1118.

Hoffman-Andrews, Lily, et al. "Attitudes of People with Inherited Retinal Conditions Toward Gene Editing Technology." *Molecular Genetics and Genomic Medicine* 7 (2019): e00803.

Hollister, Brittany M., et al. "Perspectives of Sickle Cell Disease Stakeholders on Heritable Genome Editing." *CRISPR Journal* 2 (2019): 441–49.

Holst, Cathrine, and Anders Molander. "Epistemic Democracy and the Role of Experts." *Contemporary Political Theory* 18 (2019): 441–49.

Hong, Lu, and Scott E. Page. "Groups of Diverse Problem Solvers Can Outperform Groups of High-Ability Problem Solvers." *Proceedings of the National Academies of Science* 101, no. 46 (2004): 16385–89.

Hont, Istvan. *Politics in Commercial Society: Jean-Jacques Rousseau and Adam.* Harvard University Press, 2015.
Horie, Takahiro, and Koh Ono. "VERVE-101: A Promising CRISPR-Based Gene Editing Therapy That Reduces LDL-C and PCSK9 Levels in HeFH Patients." *European Heart Journal CardioVascular Pharmacotherapy* 10 (2024): 89–90.
Hornblum, Allen M. *Acres of Skin: Human Experiments at Holmesburg Prison.* Routledge, 1998.
Hou, Xin, et al. "CRISPR/Cas Genome Editing System and Its Application in Potato." *Frontiers in Genetics* 14 (2023).
Howard, Dana. "The Political Value of Letting Hopes Die." *Critical Review of International Social and Political Philosophy* 27 (2024): 928–51.
Howlett, Michael, and Sarah Giest. "The Policy-Making Process." In *Routledge Handbook of Public Policy*, edited by Eduardo Araral et al., 17–28. Routledge, 2012.
Hu, Xumeng, et al. "The Application and Progression of CRISPR/Cas9 Technology in Ophthalmological Diseases." *Eye* 37 (2023): 607–17.
Hume, David. *A Treatise of Human Nature.* Edited by L. A. Selby-Bigge. Revised by P. H. Nidditch. Clarendon Press, 1978.
Hunter, Rosemary. "More Than Just a Different Face? Judicial Diversity and Decision-Making." *Current Legal Problems* 68 (2015): 119–41.
Hurlbut, J. Benjamin. "Limits of Responsibility: Genome Editing, Asilomar, and the Politics of Deliberation." *Hastings Center Report* 45 (2015): 11–14.
Hussein, Mouraya, et al. "A CRISPR-Cas Cure for HIV/AIDS." *International Journal of Molecular Sciences* 24 (2023): 1563.
Hutt, Peter Barton. "Historical Themes and Developments at FDA over the Past Fifty Years." In *FDA in the 21st Century: The Challenges of Regulating Drugs and New Technologies*, edited by Holly Fernandez Lynch and I. Glenn Cohen, 17–38. Columbia University Press, 2015.
Huxley, Aldous. *Brave New World.* Harper & Brothers, 2006. Originally published in 1932.
Hysse, Tim. "Consensus and Power in Deliberative Democracy." *Inquiry: An Interdisciplinary Journal of Philosophy* 49 (2006): 265–89.
Imran, Hamayun, and Preethi Marri. "Gene Therapy for Sickle Cell Disease: A Breakthrough Treatment for a Child." *Sickle Cell Today* 23 (2025).
"Investigation." *American Psychological Association* 37 (2018): 307–15.
Isaacson, Walter. *The Code Breaker: Jennifer Doudna, Gene Editing, and the Future of the Human Race.* Simon & Schuster, 2021.
Ishii, Tetsuya. "Germline Genome-Editing Research and Its Socioethical Implications." *Trends in Molecular Medicine* 21 (2015): 473–81.
Jackson, Jonathan D. "A Methodological Framework for Investigating Diversity in Clinical Research." *Alzheimer's and Dementia* 8 (2022): e062321.
Jacobs, Bette, et al. "Bridging the Divide Between Genomic Science and Indigenous Peoples." *Journal of Law, Medicine, and Ethics* 38 (2010): 684–96.

Jacobs, Daan, and Wesley Kaufmann. "The Right Kind of Participation? The Effect of a Deliberative Mini-Public on the Perceived Legitimacy of Public Decision-Making." *Public Management Review* 23 (2021): 91–111.

Jacques, Jacqueline. "'Me Too': Epistemic Injustice and the Struggle for Recognition." *Feminist Philosophy Quarterly* 4 (2018).

Jacquet, Vincent, and Ramon van der Does. "The Consequences of Deliberative Minipublics: Systematic Overview, Conceptual Gaps, and New Directions." *Representation* 57 (2020): 131–41.

Jamal, Amaney, and Irfan Nooruddin. "The Democratic Utility of Trust: A Cross-National Analysis." *Journal of Politics* 72 (2010): 45–59.

James, Michael Rabinder, Hilary Pearse, and Mark E. Warren. "Descriptive Representation in the British Citizens' Assembly." In *Designing Deliberative Democracy: The British Columbia Citizens' Assembly*, edited by Mark E. Warren and Hilary Pearse, 106–26. Cambridge University Press, 2008.

Janack, Marianne. "Standpoint Epistemology Without the "Standpoint"?: An Examination of Epistemic Privilege and Epistemic Authority." *Hypatia* 12 (1997): 125–39.

Jasanoff, Sheila. *The Ethics of Invention: Technology and the Human Future*. W. W. Norton, 2016.

——. *The Fifth Branch: Science Advisers as Policymakers*. Harvard University Press, 1990.

——. *States of Knowledge: The Co-Production of Science and Social Order*. Routledge, 2004.

Jayachandran, Muthukumaran, et al. "Genetic Advancements in Obesity Management and CRISPR–Cas9-Based Gene Editing System." *Molecular and Cellular Biochemistry* 478 (2023): 491–501.

Jennings, Bruce. "Public Health and Civic Republicanism: Toward an Alternative Framework for Public Health Ethics." In *Ethics, Prevention, and Public Health*, edited by Angus Dawson and Marcel Verweij, 30–58. Clarendon Press, 2007.

Jensen, Christian B., and Robert J. McGrath. "Making Rules About Rulemaking: A Comparison of Presidential and Parliamentary Systems." *Political Research Quarterly* 64 (2011): 656–67.

Johnson, Genevieve Fuji. *Deliberative Democracy for the Future: The Case of Nuclear Waste Management in Canada*. University of Toronto Press, 2008.

——. *Democratic Illusion: Deliberative Democracy in Canadian Public Policy*. University of Toronto Press, 2015.

Johnston, Josephine, and Eric T. Juengst. "Are Parents Really Obligated to Learn as Much as Possible About Their Children's Genomes?" *Hastings Center Report* 48 (2018): S14–15.

Jones, James H. *Bad Blood: The Tuskegee Syphilis Experiment*. 1993 ed. Free Press, 1981.

Joss, Simon, and John Durant. *Public Participation in Science: The Role of Consensus Conferences in Europe*. Science Museum, 1995.

Juengst, Eric T. "What's Taxonomy Got to Do with It? 'Species Integrity,' Human Rights, and Science Policy." In *Human Enhancement*, edited by Julian Savulescu and Nick Bostrom, 43–58. Oxford University Press, 2009.

Junius, Nicolas. "What About Elite Manipulation in Deliberative Mini-Publics? Examining Threats and Resilience in the Ostbelgien Model." *European Political Science Review* 17 (2025): 415–31.

Kadlec, Alison, and Will Friedman. "Deliberative Democracy and the Problem of Power." *Journal of Public Deliberation* 3 (2007): 1–26.

Kaebnick, Gregory E. *Humans in Nature: The World as We Find It and the World as We Create It.* Oxford University Press, 2013.

Kaebnick, Gregory E., and Bruce Jennings. "De-Extinction and Conservation." *Hastings Center Report* 47 (2017): S2–S3.

Kahneman, Daniel. *Thinking, Fast and Slow.* Farrar, Straus and Giroux, 2011.

Kamm, Frances. "What Is and Is Not Wrong with Enhancement?" In *Human Enhancement*, edited by Julian Savulescu and Nick Bostrom, 91–130. Oxford University Press, 2009.

Kanaya, Tomoe, and Stephen J. Ceci. "Are All IQ Scores Created Equal? The Differential Costs of IQ Cutoff Scores for At-Risk Children." *Child Development Perspectives* 1 (March 2007): 52–56.

Karavolioas, Nicholas G., et al. "CRISPR/Cas9 Knockout of EPFL10 Reduces Stomatal Density While Maintaining Photosynthesis and Enhancing Water Conservation in Rice." *BioRxiv* (2021).

Karlsson, Martin, Joachim Åström, and Magnus Adenskog. "Democratic Innovation in Times of Crisis: Exploring Changes in Social and Political Trust." *Policy and Internet* 13 (2020): 113–33.

Karpin, Isabel A. "Choosing Disability: Preimplantation Genetic Diagnosis and Negative Enhancement." *Journal of Law and Medicine* 5 (2007): 89–103.

Karpowitz, Christopher F., and Jane Mansbridge. "Disagreement and Consensus: The Need for Dynamic Updating in Public Deliberation." *Journal of Public Deliberation* 1 (2005): 348–64.

Kasiske, B. L., et al. "Report of a Consensus Conference on Transplant Program Quality and Surveillance." *American Journal of Transplantation* 12 (2012): 1988–96.

Kasperson, Roger E., and Susanne C. Moser. *Risk Conundrums in a Fast and Complex World.* Routledge, 2017.

Keefe, Patrick Radden. *Empire of Pain: The Secret History of the Sackler Dynasty.* Vintage, 2022.

"Keep Off-Target Effects in Focus." *Nature Medicine* 24 (2018): 1081.

Kellner, Douglas. "Power, Autocracy, and the Crisis of Democracy in Donald Trump." In *Power, Politics, and Influence: Exercising Followership, Leadership, and Practicing Politics*, edited by Adebowale Akande, 383–409. Springer, 2025.

Kellough, Edward J. *The Fragility of Merit: Presidential Power and the Civil Service Under Trump.* Routledge, 2025.

Kerwin, Cornelius M., and Scott R. Furlong. *Rulemaking: How Government Agencies Write Law and Make Policy.* CQ Press, 2019.
Kettle, Nathan P. "Knowledge Co-Production in Contested Spaces: An Evaluation of the North Slope Borough–Shell Baseline Studies Program." *Arctic* 72 (2019): 43–57.
Kevles, Daniel J. *In the Name of Eugenics: Genetics and the Uses of Human Heredity.* Alfred A. Knopf, 1985.
Kidd, Ian James, and Havi Carel. "Epistemic Injustice and Illness." *Journal of Applied Philosophy* 34 (May 2017): 172–90.
King, Cheryl Simrell, et al. "The Question of Participation: Toward Authentic Public Participation in Public Administration." In *The Age of Direct Citizen Participation*, edited by Nancy C. Roberts, 383–400. Routledge, 2008.
Kirk, James. "Landscape of Distrust." *Sage Journals* 53 (2024): 28–35.
Kistler, K. E., et al. "Genome Engineering with CRISPR-Cas9 in the Mosquito Aedes Aegypti." *Cell Reports* 11 (2015): 51–60.
Kitcher, Philip. *Science, Truth, and Democracy.* Oxford University Press, 2001.
Klein, Sabra L., et al. "Opinion: Sex Inclusion in Basic Research Drives Discovery." *Protocols of the National Academies of Science* 112 (April 2015): 5257–58.
Klinke, Andreas, and Ortwin Renn. "Risk Governance: Contemporary and Future Challenges." In *Regulating Chemical Risks: European and Global Challenges*, edited by Johan Eriksson et al., 9–27. Springer, 2010.
Knight, Amber. "Gene Editing Technologies, Utopianism, and Disability Politics." *Journal of Philosophy of Disability* (2023).
Knight, Jack, and James Johnson. *The Priority of Democracy: Political Consequences of Pragmatism.* Princeton University Press and the Russell Sage Foundation, 2011.
Knight Foundation. *Crisis in Democracy: Renewing Trust in America.* Aspen Institute, 2019.
Knoppers, Bartha Maria, and Erika Kleiderman. "Heritable Genome Editing: Who Speaks for 'Future' Children?" *CRISPR Journal* 2 (2019): 285–92.
Knox, Jean. "Epistemic Mistrust: A Crucial Aspect of Mentalization in People with a History of Abuse?" *British Journal of Psychotherapy* 32 (2016): 226–36.
Koenig, Melissa A., and Paul L. Harris. "The Basis of Epistemic Trust: Reliable Testimony or Reliable Sources?" *Episteme* 4 (2007): 264–84.
Kofler, Natalie, and Colleen M. Grogan. "Giving Voice to the Voiceless in Environmental Gene Editing." *Hastings Center Report* 51 (2021): S66–S73.
Kozubek, Jim. *Modern Prometheus: Editing the Human Genome with Crispr-Cas9.* Cambridge University Press, 2018.
Kramer, Roderick, M., and Dana A. Gavrieli. "Power, Uncertainty, and the Amplification of Doubt: An Archival Study of Suspicion Inside the Oval Office." In *Trust and Distrust in Organizations: Dilemmas and Approaches*, edited by M. Roderick Kramer and Karen S. Cook, 342–70. Russell Sage Foundation, 2004.
Kübler, Daniel, et al. "Who Supports Deliberative Mini-Publics in a Context of Direct Democracy?" *International Review of Administrative Sciences* 91 (2025): 427–47.

Kuhl, Stefan. *The Nazi Connection: Eugenics, American Racism, and German National Socialism*. Oxford University Press, 2002.

Kuhn, Nicole S., et al. "Indigenous Research Ethics Requirements: An Examination of Six Tribal Institutional Review Board Applications and Processes in the United States." *Journal of Empirical Research on Human Research Ethics* 15 (2020): 279–91.

Kuo, Yu-Ying. "Public Policy in Taiwan: Harnessing Technology and Deliberative Participation." In *Handbook of Public Policy in Asia*, edited by Ora-Orn Poocharoen et al., 305–18. Edward Elgar Publishing, 2025.

Kuyper, Jonathan W., and Fabio Wolkenstein. "Complementing and Correcting Representative Institutions: When and How to Use Mini-Publics." *European Journal of Political Research* 58 (2019): 656–75.

Kuzma, Jennifer. "Policy: Reboot the Debate on Genetic Engineering." *Nature* 531 (2016): 165–67.

Kweon, Yesola, and ByeongHwa Choi. "Fueling Conspiracy Beliefs: Political Conservatism and the Backlash Against COVID-19 Containment Policies." *Governance: An International Journal of Policy, Administration, and Institutions* 37 (2023): 867–86.

Kyle, Theodore K., et al. "Regarding Obesity as a Disease: Evolving Policies and Their Implications." *Endocrinology and Metabolism Clinics of North America* 45 (2016): 511–20.

Kyrou, Kyros, et al. "A CRISPR–Cas9 Gene Drive Targeting Doublesex Causes Complete Population Suppression in Caged Anopheles Gambiae Mosquitoes." *Nature* 36 (2018): 1062–66.

Laborde, Cécile, and John Maynor. "The Republican." In *Republicanism and Political Theory*, edited by Cécile Laborde and John Maynor, 1–28. Blackwell Publishing, 2008.

Labube, Marksu K., et al. "Vulnerability and the Ethics of Human Germline Genome Editing." *CRISPR Journal* 5 (2022).

Ladd-Taylor, Molly. "Saving Babies and Sterilizing Mothers: Eugenics and Welfare Politics in the Interwar United States." *Social Politics: International Studies in Gender, State & Society* 4 (1997): 136–53.

Laffont, Jean-Jacques, and Jean Tirole. "The Politics of Government Decision-Making: A Theory of Regulatory Capture." *Quarterly Journal of Economics* 106 (1991): 1089–127.

Lafont, Cristina. *Democracy Without Shortcuts: A Participatory Conception of Deliberative Democracy*. Oxford University Press, 2019.

———, and Nadia Urbinati. *The Lottocratic Mentality: Defending Democracy Against Lottocracy*. Oxford University Press, 2024.

Lagu, Tara, et al. "'I Am Not the Doctor for You': Physicians' Attitudes About Caring for People with Disabilities." *Health Affairs* 41 (2022): 1387–95.

Lalonde, Dianne. "Regret, Shame, and Denials of Women's Voluntary Sterilization." *Bioethics* 32 (2018): 281–88.

Lam, Carla. *New Reproductive Technologies and Disembodiment: Feminist and Material Resolutions*. Ashgate Publishing Limited, 2015.

Landemore, Hélène. "Beyond the Fact of Disagreement? The Epistemic Turn in Deliberative Democracy." *Social Epistemology* 31 (2017): 277–95.

——. "Collective Wisdom: Old and New." In *Collective Wisdom: Principles and Mechanisms*, edited by Hélène Landemore and Jon Elster, 1–20. Cambridge University Press, 2012.

——. "Deliberation, Cognitive Diversity, and Democratic Inclusiveness: An Epistemic Argument for the Random Selection of Representatives." *Synthese* 190 (2013): 1209–31.

——. "Democratic Reason: The Mechanisms of Collective Intelligence." In *Collective Wisdom: Principles and Mechanisms*, edited by Hélène Landemore and Jon Elster, 251–89. Cambridge University Press, 2012.

——. "Inclusive Constitution-Making: The Icelandic Experiment." *Journal of Political Philosophy* 23 (2015): 166–91.

——. "Yes, We Can (Make It Up on Volume): Answers to Critics." *Critical Review* 26 (2014): 184–237.

Lander, Eric S., et al. "Adopt a Moratorium on Heritable Genome Editing." *Nature* 567 (2019): 165–68.

Landor, Cecilia. "Right to Informed Consent, Right to a Doula: An Evidence-Based Solution to the Black Maternal Mortality Crisis in the United States." *Michigan Journal of Gender & Law* 30 (2023): 61–106.

Landwehr, Claudia. "Facilitating Deliberation: The Role of Impartial Intermediaries in Deliberative Mini-Publics." In *Deliberative Mini-Publics: Involving Citizens in the Democratic Process*, edited by Kimmo Grönlund et al., 77–92. ECPR Press, 2014.

——. "How Can We Agree on Democratic Procedures? Dimensions of Conflict in the Political Process Space." *Journal of International Political Theory* 21 (2025): 149–65.

Larena, Maximilian, et al. "Multiple Migrations to the Philippines During the Last 50,000 Years." *PNAS* 118 (2021): e2026132118.

Latham, Stephen R. "Political Theory, Values and Public Health." *Public Health Ethics* 9 (2016): 139–49.

Lavazza, Andrea, and Mirko Farina. "The Role of Experts in the Covid-19 Pandemic and the Limits of Their Epistemic Authority in Democracy." *Frontiers in Public Health* 8 (2020).

Law, Marc T., and Gary D. Libecap. "Progressive Era Reform: The Pure Food and Drugs Act of 1906." In *Corruption and Reform: Lessons From America's Economic History*, edited by Edward L. Glaeser and Claudia Goldin, 319–42. University of Chicago Press, 2006.

Le Dantec, Christopher A., and Carl DiSalvo. "Infrastructuring and the Formation of Publics in Participatory Design." *Social Studies of Science* 43 (2013): 241–64.

Lee, Charlotte, et al. "Hurdles to Herd Immunity: Distrust of Government and Vaccine Refusal in the US, 2002–2003." *Vaccine* 34 (2016): 3972–78.

Lee, Seungil K., et al. "Misinformation of COVID-19 Vaccines and Vaccine Hesitancy." *Scientific Reports* 12 (2022): 13681.

Leino, Mikko, et al. "Expert Hearings in Mini-Publics: How Does the Field of Expertise Influence Deliberation and Its Outcomes?" *Policy Sciences* 55 (2022): 429–50.
Lemke, Thomas. *Biopolitics: An Advanced Introduction*. Translated by Eric Frederick Trump. New York University Press, 2011.
Leonard, Thomas C. "American Economic Reform in the Progressive Era: Its Foundational Beliefs and Their Relation to Eugenics." *History of Political Economy* 41 (2009): 109–41.
——. *Illiberal Reformers: Race, Eugenics, and American Economics in the Progressive Era*. Princeton University Press, 2016.
Levine, Charles H. "Citizenship and Service Delivery: The Promise of Coproduction." In *The Age of Direct Citizen Participation*, edited by Nancy C. Roberts, 78–92. Routledge, 2008.
Lewandowsky, Stephan, et al. "Beyond Misinformation: Understanding and Coping with the 'Post-Truth' Era." *Journal of Applied Research in Memory and Cognition* 6 (2017): 353–69.
Lewis, Earl, and Nancy Cantor. "Introduction: The Value of Diversity for Democracy and a Prosperous Society." In *Our Compelling Interests: The Value of Diversity for Democracy and a Prosperous Society*, edited by Earl Lewis and Nancy Cantor, 1–15. Princeton University Press, 2016.
Lewis, Myrisha S. "Is Germline Gene Editing Exceptional?" *Seton Hall Law Review* 51 (2020): 735–813.
Lin, J., and K. C. Wong. "Off-Target Predictions in CRISPR-Cas9 Gene Editing Using Deep Learning." *Bioinformatics* 34 (2018): i656–i663.
Linder, Astrid, and Mats Y. Svensson. "Road Safety: The Average Male as a Norm in Vehicle Occupant Crash Safety Assessment." *Interdisciplinary Science Reviews* 44 (2019): 140–53.
Lindley, David. *Uncertainty: Einstein, Heisenberg, Bohr, and the Struggle for the Soul of Science*. Anchor Books, 2007.
List, Christian, and John S. Dryzek. "Social Choice Theory and Deliberative Democracy: A Reconciliation." *British Journal of Political Science* 33 (2003): 1–28.
List, Christian, and Robert E. Goodin. "Epistemic Democracy: Generalizing the Condorcet Jury Theorem." *Journal of Political Philosophy* 9 (2001): 277–306.
Little, Daniel. "Governments as Regulators." In *A New Social Ontology of Governments*, 145–60. Cambridge University Press, 2020.
Little, Miles. "'Better Than Numbers . . .' A Gentle Critique of Evidence-Based Medicine." *ANZ Journal of Surgery* 73 (2003): 177–82.
Liu, Feng, et al. "Cryo-Shocked Tumor Cells Deliver CRISPR-Cas9 for Lung Cancer Regression by Synthetic Lethality." *Science Advances* 10 (2024).
Liu, Katherine, and Natalie Mager. "Women's Involvement in Clinical Trials." *Pharmacy Practice* (2016).
Locke, John. *Two Treatises of Government*. Edited by Peter Laslett. Cambridge University Press, 1988.

Lofti, Malihe, et al. "Application of CRISPR-Cas9 Technology in Diabetes Research." *Diabetic Medicine* 41 (2024): e15240.

Lohse, Simon, and Stefano Canali. "Follow the Science? On the Marginal Role of the Social Sciences in the COVID-19 Pandemic." *European Journal of Philosophy of Science* 11 (2021): 99.

Lombardo, Paul A. *Three Generations, No Imbeciles: Eugenics, the Supreme Court, and Buck v. Bell.* Johns Hopkins University Press, 2008.

Longstaff, Holly, and Michael M. Burgess. "Recruiting for Representation in Public Deliberation on the Ethics of Biobanks." *Public Understanding of Science* 19 (2010): 212–24.

Longstaff, Holly, and David M. Secko. "Assessing the Quality of a Deliberative Democracy Mini-Public Event About Advanced Biofuel Production and Development in Canada." *Public Understanding of Science* 25 (2014): 252–61.

Lopes, Rui, and Megana K. Prasad. "Beyond the Promise: Evaluating and Mitigating Off-Target Effects in CRISPR Gene Editing for Safer Therapeutics." *Frontiers in Bioengineering and Biotechnology* 11 (2024).

Lovett, Frank. *A General Theory of Domination and Justice.* Oxford University Press, 2010.

———. "What Counts as Arbitrary Power?" *Journal of Political Power* 5 (2012): 137–52.

Lublin, Nancy. *Pandora's Box: Feminism Confronts Reproductive Technology.* Rowman & Littlefield Publishers, 1998.

Lucas, Peter. "Bacon's New Atlantis and the Fictional Origins of Organized Science." *Open Cultural Studies* 2 (2018): 114–21.

Luhmann, Niklas. "Familiarity, Confidence, Trust: Problems and Alternatives." In *Trust: Making and Breaking Cooperative Relations*, edited by Diego Gambetta, 94–107. Basil Blackwell Ltd., 2000.

———. *Trust and Power.* Translated by Howard Davis et al. Wiley, 1979.

Luskin, Robert C., James S. Fishkin, and Roger Jowell. "Considered Opinions: Deliberative Polls in Britain." *British Journal of Political Science* 32 (2002): 455–87.

Lynch, Holly Fernandez, et al. "Promoting Ethical Payment in Human Infection Challenge Studies." *American Journal of Bioethics* 21 (2021): 11–31.

Lyons, Daniel A. "E-Rulemaking and the Politicization of the Comment Process." *Boston College Law School Faculty Papers* (2018).

Ma, Xueying, et al. "Would Consumers Accept CRISPR Fruit Crops If the Benefit Has Health Implications? An Application to Cranberry Products." *Agricultural and Resource Economics Review* (2024): 1–23.

MacAskill, William. *What We Owe the Future.* Basic Books, 2022.

Macintosh, Kerry Lynn. "The Regulation of Human Germline Genome Modification in the United States." In *Human Germline Genome Modification and the Right to Science*, edited by Andrea Boggio et al., 103–28. Cambridge University Press, 2020.

MacKellar, Calum, and Christopher Bechtel. *The Ethics of the New Eugenics.* Berghahn Books, 2014.

MacKenzie, Michael Kenneth, and Kieran O'Doherty. "Deliberating Future Issues: Minipublics and Salmon Genomics." *Journal of Public Deliberation* 7 (2011).
MacKenzie, Michael Kenneth, and Mark E. Warren. "Two Trust-Based Uses of Minipublics in Democratic Systems." In *Deliberative Systems: Deliberative Democracy at the Large Scale*, edited by John Parkinson and Jane Mansbridge, 95–124. Cambridge University Press, 2012.
Macklin, Ruth. "'Due' and 'Undue' Inducements: On Paying Money to Research Subjects." *IRB: Ethics & Human Research* 3 (1981): 1–6.
MacNamee, Stephen. *The Meritocracy Myth*. Rowman & Littlefield, 2018.
Madison, James. "Federalist 10." In *The Federalist Papers*, edited by Clinton Rossiter, 71–79. Signet Classics, 2003.
Maggetti, Martino. "Legitimacy and Accountability of Independent Regulatory Agencies: A Critical Review." *Living Reviews in Democracy* (2010): 1–9.
Magill, M. Elizabeth. "Courts and Regulatory Capture." In *Preventing Regulatory Capture: Special Interest Influence and How to Limit It*, 397–419. Cambridge University Press, 2014.
Mailhan, Laurence, et al. "Life Satisfaction and Disability After Severe Traumatic Brain Injury." *Brain Injury* 19 (2009): 227–38.
Majone, Giandomenico. "Political Institutions and the Principle of Precaution." In *The Reality of Precaution: Comparing Risk Regulation in the United States and Europe*, edited by Jonathan B. Wiener et al., 411–33. RFF Press, 2011.
Makary, Marty. *The Price We Pay: What Broke American Health Care—and How to Fix It*. Bloomsbury Publishing, 2019.
Manchanda, Rishi, and Jean Silver-Isenbstadt. "A Prescription for a Healthier Democracy: The Role of Health Care in Civic Participation." *National Civic Review* 99 (2010): 48–53.
Manne, Kate. *Down Girl: The Logic of Misogyny*. Oxford University Press, 2017.
Mansbridge, Jane. "Conflict and Self-Interest in Deliberation." In *Deliberative Democracy and Its Discontents*, edited by Samantha Besson and José Luis Martí, 107–32. Ashgate, 2006.
———. "Deliberative Polling as the Gold Standard." *Good Society* 19 (2010): 55–62.
———. "Should Blacks Represent Blacks and Women Represent Women? A Contingent 'Yes.'" *Journal of Politics* 61 (1999): 628–57.
Már, Kristinn, and Gastil, John. "Do Voters Trust Deliberative Minipublics? Examining the Origins and Impact of Legitimacy Perceptions for the Citizens' Initiative Review." *Political Behavior* 45 (2023): 975–94.
Marglin, Stephen A. "Farmers, Seedsmen, and Scientists Systems of Agriculture and Systems of Knowledge." In *Decolonizing Knowledge: From Development to Dialogue*, 185–248. Oxford University Press, 1996.
Markovits, Daniel. *The Meritocracy Trap: How America's Foundational Myth Feeds Inequality, Dismantles the Middle Class, and Devours the Elite*. Penguin Press, 2019.
Marsh, Stephen, and Mark R. Dibben. "Trust, Untrust, Distrust and Mistrust—An Exploration of the Dark(er) Side." In *Trust Management*, edited by P. Herrmann,

V. Issarny, and S. Shiu, Lecture Notes in Computer Science, vol. 3477 (2005), 17–33. Springer, 2005.
Martin, Randy. "From the Race War to the War on Terror." In *Beyond Biopolitics: Essays on the Governance of Life and Death*, 258–74. Duke University Press, 2011.
Martinez, Eduardo J. "Realizing the Value of Public Input: Mini-Public Consultation on Agency Rulemaking." *Philosophical Issues* 31 (2021): 240–57.
Martschenko, Daphne Oluwasen. "'The Elephant in the Room': Social Responsibility in the Production of Sociogenomics Research." *BioSocieties* (2021).
Marx, Karl. *Capital: Volume 1: A Critique of Political Economy*. Penguin, 1992.
——, and Friedrich Engels. *Communist Manifesto*. Translated by Samuel Moore. Penguin Books, 1967.
Marzinke, Mark, et al. "Limited Evidence for Use of a Black Race Modifier in eGFR Calculations: A Systematic Review." *Clinical Chemistry* 68 (2022).
Maschke, Karen J. "Navigating an Ethical Patchwork—Human Gene Banks." *Nature Biotechnology* 23 (2005): 539–45.
McCammon, Christopher. "Domination: A Rethinking." *Ethics* 125 (2015): 1028–52.
McColgan, P., and S. J. Tabrizi. "Huntington's Disease: A Clinical Review." *European Journal of Neurology* 25 (2018): 24–34.
McConkey, Jane. "Knowledge and Acknowledgement: 'Epistemic Injustice' as a Problem of Recognition." *Politics* 24 (2004): 198–205.
McCoy, Matthew S., et al. "Conflicts of Interest for Patient Advocacy Organizations." *New England Journal of Medicine* 376 (2017): 880–85.
McKean, Benjamin. *Disorienting Neoliberalism: Global Justice and the Outer Limit of Freedom*. Oxford University Press, 2020.
McKee, Chephra, and Kristin Bohannon. "Exploring the Reasons Behind Parental Refusal of Vaccines." *Journal of Pediatric Pharmacology and Therapeutics* 21 (2016): 104–9.
McKinnon, Rachel. "Epistemic Injustice." *Philosophy Compass* 11 (2016): 437–46.
Mclaverty, Peter, and Darren Halpin. "Deliberative Drift: The Emergence of Deliberation in the Policy Process." *International Political Science Review* 29 (2008): 197–214.
McMenamin, Katie E., et al. "The Journal of Deaf Studies and Deaf Education." *Journal of Deaf Studies and Deaf Education* 28 (2023): 162–70.
Medina, José. *The Epistemology of Resistance: Gender and Racial Oppression, Epistemic Injustice, and Resistant Imaginations*. Oxford University Press, 2013.
——. "Hermeneutical Injustice and Polyphonic Contextualism: Social Silences and Shared Hermeneutical Responsibilities." *Social Epistemology* 26 (2012): 201–20.
Meghani, Zahra. "Regulation of Genetically Engineered (GE) Mosquitoes as a Public Health Tool: A Public Health Ethics Analysis." *Globalization and Health* 18 (2022).
——, and Jennifer Kuzma. "The 'Revolving Door' Between Regulatory Agencies and Industry: A Problem That Requires Reconceptualizing Objectivity." *Journal of Agricultural and Environmental Ethics* 24 (2011): 575–99.

Melilo, Tara R. "Gene Editing and the Rise of Designer Babies." *Vanderbilt Journal of Transnational Law* 50 (2017): 757–90.

Mendelberg, Tali, and Christopher F. Karpowitz. "How People Deliberate About Justice: Groups, Gender, and Decision Rules." In *Deliberation, Participation and Democracy: Can the People Govern?*, edited by Shawn W. Rosenberg, 101–29. Palgrave Macmillan, 2007.

Mendell, Marguerite. *Reclaiming Democracy: The Social Economy Tradition in Quebec*. Edited by Adrian Chan. Peter Lang, 2005.

Mendonça, Ricardo Fabrino, and Hans Asenbaum. "Decolonizing Deliberative Democracy." *European Journal of Social Theory* 29 (2026): 63–84.

Mendz, George L., and Michael Cook. "Transhumanist Genetic Enhancement: Creation of a 'New Man' Through Technological Innovation." *New Bioethics* 27 (2021): 105–26.

Mensch, Elaine, and Harry Mensch. *The IQ Mythology: Class, Race, Gender, and Inequality*. Southern Illinois University Press, 1991.

Menz, Jochen, et al. "Genome Edited Crops Touch the Market: A View on the Global Development and Regulatory Environment." *Frontiers in Plant Science* 11 (2020).

Mertkan, Sefika, et al. "Diversity and Equality in Academic Publishing: The Case of Educational Leadership." *Compare: A Journal of Comparative and International Education* 47 (2016): 46–61.

Meslé, Margaux M., et al. "Estimated Number of Lives Directly Saved by COVID-19 Vaccination Programmes in the WHO European Region from December, 2020, to March, 2023: A Retrospective Surveillance Study." *Lancet* 12 (2024): P714–727.

Metzl, Jamie. *Hacking Darwin: Genetic Engineering and the Future of Humanity*. Sourcebooks, 2019.

Metzl, Jonathan M. *Why "Against Health"?* New York University Press, 2010.

Meyers, Émilie M., and Jeffrey W. Jutai. "The Disability Experience: Living with a Birth Defect Resulting from Thalidomide Exposure." *Interdisciplinary Journal of Health Sciences* 2 (2012): 28–33.

Mill, John Stuart. *On Liberty*. Hackett Publishing Company, 1978.

Miller, Amy M. "Gender Bias and the Ongoing Need to Acknowledge Women's Pain." *Practical Pain Management* 18 (2018).

Miller, Arthur G., et al. "The Prediction and Perception of Obedience to Authority." *Journal of Personality* 42 (1974): 23–42.

Miller, James. *Can Democracy Work? A Short History of a Radical Idea, From Ancient Athens to Our World*. Farrar, Straus and Giroux, 2018.

Miller, Stanley L., and Antonio Lazcano. "Formation of the Building Blocks of Life." *Life's Origin: The Beginnings of Biological Evolution*, edited by J. William Schopf, 78–112. University of California Press, 2002.

Misak, Cheryl. "A Culture of Justification: The Pragmatist's Epistemic Argument for Democracy." *Episteme* 5 (2008): 94–105.

"Mitochondrial Donation Community Consultation Citizens' Panel Statement." Australian Government National Health and Medical Research Council, 2020.

Mizrahi, Moti. *Playing God with Emerging Technologies: How to Avoid the Traps of Techno-Optimism and Techno-Pessimism*. Bloomsbury, 2025.

Molhoek, Braden. "Raising the Virtuous Bar: The Underlying Issues of Genetic Moral Enhancement." *Theology and Science* 16 (2018): 279–87.

Montalmant, Keisha E., and Anna K. Ettinger. "The Racial Disparities in Maternal Mortality and Impact of Structural Racism and Implicit Racial Bias on Pregnant Black Women: A Review of the Literature." *Journal of Racial and Ethnic Health Disparities* 11 (2024): 3658–77.

Moore, Alfred. *Critical Elitism: Deliberation, Democracy, and the Problem of Expertise*. Cambridge University Press, 2017.

Moreno, Jonathan D. *The Body Politic: The Battle over Science in America*. Bellevue Literary Press, 2011.

Morrell, Michael E. "Participant Bias and Success in Deliberative Mini-Publics." In *Deliberative Mini-Publics: Involving Citizens in the Democratic Process*, edited by Kimmo Grönlund et al., 157–75. ECPR Press, 2014.

Moynihan, Ray, and Lisa Bero. "Toward a Healthier Patient Voice: More Independence, Less Industry Funding." *JAMA Internal Medicine* 117 (2017): 350–51.

Mukherjee, Siddhartha. *The Gene: An Intimate History*. Scribner, 2016.

Muradova, Lala, et al. "Climate Change Communication and Public Engagement in Interpersonal Deliberative Settings: Evidence from the Irish Citizens' Assembly." *Climate Policy* 20 (2020): 1322–35.

Murcho, Eamon O., et al. "Oral Pre-Exposure Prophylaxis (PrEP) to Prevent HIV: A Systematic Review and Meta-Analysis of Clinical Effectiveness, Safety, Adherence and Risk Compensation in All Populations." *BMJ Open* 12 (2022): e048478.

Murray, Graham K., et al. "Could Polygenic Risk Scores Be Useful in Psychiatry?" *JAMA Psychiatry* 78 (2022): 210–19.

Musunuru, Kiran, et al. "Patient-Specific In Vivo Gene Editing to Treat a Rare Genetic Disease." *New England Journal of Medicine* 392 (2025): 2235–43.

Mutz, Diana C. *Hearing the Other Side: Deliberative versus Participatory Democracy*. Cambridge University Press, 2006.

Mykhalovskiy, Eric, and Lorna Weir. "The Problem of Evidence-Based Medicine: Directions for Social Science." *Social Science and Medicine* 59 (September 2004): 1059–69.

Nabatchi, Tina. "Addressing the Citizenship and Democratic Deficits: The Potential of Deliberative Democracy for Public Administration." *American Review of Public Administration* 40 (2010): 376–99.

——, John Gastil, Matt Leighninger, and G. Michael Weiksner, eds. *Democracy in Motion: Evaluating the Impact of Deliberative Engagement*. HalOxford University Press, 2012.

Nacos, Brigitte L., and Yaeli Block-Elkon. "How Far-Right Extremism Changed American Body Politic." *Political Science Quarterly* 139 (2024): 387–406.

Nadesan, Majia Holmer. *Crisis Communication, Liberal Democracy, and Ecological Sustainability: The Threat of Financial and Energy Complexes in the Twenty-First Century*. Lexington Books, 2016.

———. *Governmentality, Biopower, and Everyday Life*. Routledge, 2008.
Nassiri, Ashley M., Donna L. Sorkin, and Matthew L. Carlson. "Current Estimates of Cochlear Implant Utilization in the United States." *Otology & Neurotology* 43 (2022): e558–e562.
Natow, Rebecca. "Negotiated Rulemaking for U.S. Higher Education Regulatory Policy." *Journal of Deliberative Democracy* 15 (2019).
Naughton, Keith, et al. "Understanding Commenter Influence During Agency Rule Development." *Journal of Policy Analysis and Management* 28 (2009): 258–77.
Naurin, Daniel, and Christine Reh. "Deliberative Negotiation." In *The Oxford Handbook of Deliberative Democracy*, edited by André Bächtiger et al. Oxford University Press, 2018.
Negri, Antoni. *Insurgencies: Constituent Power and the Modern State*. University of Minnesota Press, 1999.
Newell, Peter. "Trade and Biotechnology in Latin America: Democratization, Contestation and the Politics of Mobilization." *Journal of Agrarian Change* 8 (2008): 345–76.
New York University Stern School of Business et al. *Regulating Wall Street: The Dodd-Frank Act and the New Architecture of Global Finance*. John Wiley & Sons, 2011.
Nicol, Dianne, et al. "The Australian Citizens' Jury and Global Citizens' Assembly on Genome Editing." *American Journal of Bioethics* 23 (2023): 61–63.
———. "Genome Editing: Formulating an Australian Community Response." *Center for Law and Genetics* 12 (2022).
Niemeyer, Simon, and Julia Jennstål. "Scaling Up Deliberative Effects—Applying Lessons of Mini-Publics." In *The Oxford Handbook of Deliberative Democracy*, edited by André Bächtiger et al., 329–47. Oxford University Press, 2018.
Niessen, Christoph. "When Citizen Deliberation Enters Real Politics: How Politicians and Stakeholders Envision the Place of a Deliberative Mini-Public in Political Decision-Making." *Policy Sciences* 52 (2019): 481–503.
———, and Min Reuchamps. "Designing a Permanent Deliberative Citizens' Assembly." *The Centre for Deliberative Democracy & Global Governance* (2019).
Noble, Charleston, et al. "Current CRISPR Gene Drive Systems Are Likely to Be Highly Invasive in Wild Populations." *eLIFE* 7 (2018): e33423.
Nowlin, Christopher. "200 Years After Frankenstein." *Perspectives in Biology and Medicine* 61 (2018): 430–49.
Nunn, Drew H. "Grounded: How the 737 MAX Crashes Highlight Issues with FAA Delegation and a Potential Remedy in the Federal Tort Claims Act." *Journal of Air Law and Commerce* 85 (Fall 2020): 703–32.
Nussbaum, Martha C. *Hiding from Humanity: Disgust, Shame, and the Law*. Princeton University Press, 2006.
Nxumalo, Zandile, et al. "Patents, Ethics, Biosafety and Regulation Using CRISPR Technology." *Progress in Molecular Biology and Translational Science* 181 (2021): 345–65.

O'Doherty, Kieran, Michael Burgess, and D. M. Secko. "Sequencing the Salmon Genome: A Deliberative Public Engagement." *Genomics, Society and Policy* 6 (2010): 15–32.

O'Flynn, Ian. *Deliberative Democracy*. Polity, 2022.

Ogbu, John U. *Cultural Amplifiers of Intelligence*. Lawrence Erlbaum Associates, 2002.

O'Grady, Cathleen. "Power to the People: Nations Are Turning to Citizen Assemblies to Weigh Up Climate Policies." *Science* 370 (2020): 518–21.

Ojakangas, Mika. *On the Greek Origins of Biopolitics: A Reinterpretation of the History of Biopower*. Routledge, 2016.

Ólafsson, Jón. "The Constituent Assembly: A Study in Failure." In *Iceland's Financial Crisis: The Politics of Blame, Protest, and Reconstruction*, edited by Valur Ingimundarson et al., 252–72. Routledge, 2018.

Olszewski, Todd M. "Between Bench and Bedside: Building Clinical Consensus at the NIH, 1977–2013." *Journal of the History of Medicine and Allied Sciences* 73 (2018): 464–500.

Opel, Douglas J., Daniel A. Salmon, and Edgar K. Marcuse. "Building Trust to Achieve Confidence in COVID-19 Vaccines." *JAMA* 3 (2020): e2025672.

Opstrup, Niels, and Anders R. Villadsen. "The Right Mix? Gender Diversity in Top Management Teams and Financial Performance." *Public Administration Review* 75 (March 2015): 291–301.

Organisation for Economic Co-operation and Development. *Innovative Citizen Participation and New Democratic Institutions: Catching the Deliberative Wave*. OECD Publishing, 2020.

Origgi, Gloria. "Epistemic Injustice and Epistemic Trust." *Epistemic Trust* 26 (June 2012): 221–35.

Orlikowski, Wanda J. "The Duality of Technology: Rethinking the Concept of Technology in Organizations." *Organization Studies* 3 (1992): 301–441.

Orr, Raymond I., et al. "American Indian Genes in the Media: Representations of the Havasupai Indian Tribe in Their Case against Arizona State University." *American Indian Culture and Research Journal* 45 (2021): 65–79.

Ostrom, Elinor. *Understanding Institutional Diversity*. Princeton University Press, 2005.

Oswald, Malcolm, and Louise Laverty. "Data Sharing in a Pandemic: Three Citizens' Juries." *National Institute for Health Research* (2021).

Owen, David, and Graham Smith. "Survey Article: Deliberation, Democracy, and the Systemic Turn." *Journal of Political Philosophy* 23 (2015): 213–34.

Padden, Carol, and Jacqueline Humphriest. "Who Goes First? Deaf People and CRISPR Germline Editing." *Perspectives in Biology and Medicine* 63 (2020): 54–65.

Padmanabhan, Sandosh, et al. "Beyond Genome-Wide Scans: Advancing Hypertension Genomics Into the Future." *Hypertension* 81 (2024): 1186–89.

Page, Scott E. *The Difference: How the Power of Diversity Creates Better Groups, Firms, Schools, and Societies*. Princeton University Press, 2007.

———. *The Diversity Bonus: How Great Teams Pay Off in the Knowledge Economy*. Princeton University Press, 2017.

——. "Making the Difference: Applying a Logic of Diversity." *Academy of Management Perspectives* 21 (2007): 6–20.
Pagliaro, Stefano, et al. "Trust Predicts COVID-19 Prescribed and Discretionary Behavioral Intentions in 23 Countries." *PLOS One* 16 (2021): e0248334.
Pamuk, Zeynep. *Politics and Expertise: How to Use Science in a Democratic Society.* Princeton University Press, 2021.
Paradise, Jordan. "21st Century Citizen Pharma: The FDA & Patient-Focused Product Development." *American Journal of Law & Medicine* 44 (2018): 309–27.
Parens, Erik. *Shaping Our Selves: On Technology, Flourishing and a Habit of Thinking.* Oxford University Press, 2015.
——, and Adrienne Asch, eds. *Prenatal Testing and Disability Rights.* Georgetown University Press, 2000.
Park, So Hyun, et al. "Comprehensive Analysis and Accurate Quantification of Unintended Large Gene Modifications Induced by CRISPR-Cas9 Gene Editing." *Science Advances* 8 (January 2022): eabj3514.
Parkinson, John. *Deliberating in the Real World: Problems of Legitimacy in Deliberative Democracy.* Oxford University Press, 2006.
——. "Rickety Bridges: Using the Media in Deliberative Democracy." *British Journal of Political Science* 36 (2006): 175–83.
Pateman, Carol. *Participation and Democratic Theory.* Cambridge University Press, 1970.
Patterson, Amy M., et al. "Emerging Roles and Opportunities for Rare Disease Patient Advocacy Groups." *Therapeutic Advances in Rare Disease* 4 (2023): 1–18.
Patton, Paul. "Power and Biopower in Foucault." In *Biopower: Foucault and Beyond*, edited by Vernon W. Cisney and Nicolae Morar, 102–17. University of Chicago Press, 2016.
Paul, Narayan Chandra, et al. "Plant and Fungal Genome Editing to Enhance Plant Disease Resistance Using the CRISPR/Cas9 System." *Frontiers in Plant Science* 12 (2021): 601614.
Paulis, Emilien, et al. "The POLITICIZE Dataset: An Inventory of Deliberative Mini-Publics (DMPs) in Europe." *European Political Science* 20 (2021): 521–42.
Pausé, Cat. "Die Another Day: The Obstacles Facing Fat People in Accessing Quality Healthcare." *Narrative Inquiry in Bioethics* 4 (2014): 135–41.
Perez, Caroline Criado. *Invisible Women: Data Bias in a World Designed for Men.* Harry N. Abrams, 2019.
Persaud, Anitra, et al. "A CRISPR Focus on Attitudes and Beliefs Toward Somatic Genome Editing from Stakeholders Within the Sickle Cell Disease Community." *Genetics in Medicine* 21 (2019): 1727–34.
Persson, Ingmar, and Julian Savulescu. *Unfit for the Future: The Need for Moral Enhancement.* Oxford University Press, 2012.
"Pesticides: Exemptions of Certain Plant-Incorporated Protectants (PIPs) Derived from Newer Technologies." *Federal Register* 85 (2020): 64308–44.
Pettit, Philip. *Republicanism: A Theory of Freedom and Government.* Oxford: Clarendon Press, 1997.

Phillips, Peter W. B., and Grant Isaac. "GMO Labeling: Threat or Opportunity?" *AgBioForum* 1 (1998): 25–30.
Phillips, Shannon, et al. "Perspectives of Individuals with Sickle Cell Disease on Barriers to Care." *PLOS One* 17 (2022): e0265342.
Phillips, Trisha. "Exploitation in Payments to Research Subjects." *Bioethics* 25 (2011): 209–19.
Pilet, Jean-Benoit, et al. "Public Support for Deliberative Citizens' Assemblies Selected Through Sortition: Evidence from 15 Countries." *European Journal of Political Research* 62 (2023): 873–902.
Pimbert, Michel P., and Boukary Barry. "Let the People Decide: Citizen Deliberation on the Role of GMOs in Mali's Agriculture." *Agriculture and Human Values* 38 (2021): 1097–122.
Pimbert, Michel P., and Tom Wakeford. "Prajateerpu: A Citizens Jury / Scenario Workshop on Food and Farming Futures for Andhra Pradesh." *Economic and Political Weekly* 37 (2002): 2778–87.
Pitkin, Hanna Fenichel. *The Concept of Representation*. University of California Press, 1967.
——. *Wittgenstein and Justice: On the Significance of Ludwig Wittgenstein for Social and Political Thought*. University of California Press, 1972.
Plato. *Republic*. Translated by G. M. A. Grube, revised by C. D. C. Reeve. Hackett Publishing, 1992.
Plotkin, Stanley A. *History of Vaccine Development*. Springer Science & Business Media, 2011.
Polanyi, Michael. *The Tacit Dimension*. University of Chicago Press, 1966.
Popescu, Delia, and Matthew Loveland. "Judging Deliberation: An Assessment of the Crowdsourced Icelandic Constitutional Project." *Journal of Deliberative Democracy* 18 (2022).
Porter, Theodore M. "Measurement and Meritocracy: An Intellectual History of IQ." *Modern Intellectual History* 6 (2009): 637–44.
Posner, Eric A. *Governance of Dual-Use Technologies*. Harvard University Press, 2020.
Posner, Richard A. "The Concept of Regulatory Capture: A Short, Inglorious History." In *Preventing Regulatory Capture: Special Interest Influence and How to Limit It*, edited by Daniel Carpenter and David A. Moss, 25–48. Cambridge University Press, 2014.
Potrykus, Ingo. "Golden Rice and Beyond." *Plant Physiology* 125 (2001): 1157–61.
Pow, James. "Mini-Publics and the Wider Public: The Perceived Legitimacy of Randomly Selecting Citizen Representatives." *Representation* (2021): 13–32.
Powell, Maria, and Daniel Lee Kleinman. "Building Citizen Capacities for Participation in Nanotechnology Decision-Making: The Democratic Virtues of the Consensus Conference Model." *Public Understanding of Science* 17 (2008): 329–48.
Powell, Melanie, and David Ansic. "Gender Differences in Risk Behaviour in Financial Decision-Making: An Experimental Basis." *Journal of Economic Psychology* 18 (1997): 605–28.

Price-Haywood, Eboni G., et al. "Hospitalization and Mortality Among Black Patients and White Patients with Covid-19." *New England Journal of Medicine* 382 (2020): 2534–43.
Priti, Disha Kamboj, et al. "Nutritional Enhancement in Horticultural Crops by CRISPR/Cas9: Status and Future Prospects." In *Sustainable Agriculture in the Era of the Omics Revolution*, edited by Channa S. Prakash et al., 399–430. Springer, 2023.
Putnam, Robert D. "What Makes Democracy Work?" *National Civic Review* 82 (1993): 101–7.
PytlikZillig, Lisa M., et al. *Deliberative Public Engagement with Science: An Empirical Investigation*. Springer International, 2018.
Quashie, Kevin. *The Sovereignty of the Quiet: Beyond Resistance in Black Culture*. Rutgers University Press, 2012.
Quine, Willard. "Epistemology Naturalized." In *Ontological Relativity and Other Essays*. Columbia University Press, 1969.
Qvortup, Matt. *Mob Rule or the Wisdom of the Crowd?* Greenhave Publishing, 2020.
Räber, Michael I. *Knowing Democracy—A Pragmatist Account of the Epistemic Dimension in Democratic Politics*. Springer, 2020.
Rackley, Erika. *Women, Judging and the Judiciary: From Difference to Diversity*. Routledge-Cavendish, 2012.
Rahman, K. Sabeel. *Democracy Against Domination*. Oxford University Press, 2017.
Rahman, Saleem Ur, et al. "Improvement of Soybean: A Way Forward Transition from Genetic Engineering to New Plant Breeding Technologies." *Molecular Biotechnology* 65 (2023): 162–80.
Rakic, Johin. "Genome Editing for Involuntary Moral Enhancement." *Cambridge Quarterly of Healthcare Ethics*, vol. CRISPR-Cas System, a Possible "Savior" of Rice Threatened by Climate Change 28 (2019): 46–54.
Rapoport, Sarah. "Rosalind Franklin: Unsung Hero of the DNA Revolution." *History Teacher* 36 (2002): 116–27.
Rawls, John. *Justice as Fairness: A Restatement*. Edited by Erin Kelly. Harvard University Press, 2001.
——. *A Theory of Justice*. Harvard University Press, 1971.
Raz, Joseph. *The Authority of Law: Essays on Law and Morality*. Clarendon Press, 1979.
Reinhardt, Uwe E. *Priced Out*. Princeton University Press, 2019.
Reiss, Dorit Rubinstein. "The Benefits of Capture." *Wake Forest Law Review* 47 (2012): 569–610.
Renn, Ortwin. *Risk Governance: Coping with Uncertainty in a Complex World*. Earthscan, 2008.
——. "Stakeholder and Public Involvement in Risk Governance." *International Journal of Disaster Risk Science* 6 (2015): 8–20.
Resnik, David B. "Environmental Disease, Biomarkers, and the Precautionary Principle." In *Genomics and Environmental Regulation*, edited by Jamie Grodsky et al., 242–57. Johns Hopkins University Press, 2008.

Retel Helmrich, Isabel Rosalie Arianne, et al. “Discrepancy Between Disability and Reported Well-Being After Traumatic Brain Injury.” *Journal of Neurology, Neurosurgery & Psychiatry* 93 (2022): 785–96.

Revesz, Richard L., and Michael A. Livermore. *Retaking Rationality: How Cost-Benefit Analysis Can Better Protect the Environment and Our Health.* Oxford University Press, 2008.

Rex, Justin. “Anatomy of Agency Capture: An Organizational Typology for Diagnosing and Remedying Capture.” *Regulation and Governance* 14 (2020): 271–94.

Ribaritis, Alexandra, et al. “Genome-Edited Plants: Opportunities and Challenges for an Anticipatory Detection and Identification Framework.” *Food* 10 (2021): 430.

Richardson, Henry S. *Democratic Autonomy: Public Reasoning About the Ends of Policy.* Oxford University Press, 2002.

Richardson, Ken. “GWAS and Cognitive Abilities: Why Correlations Are Inevitable and Meaningless.” *Science & Society* 18 (2017): 1279–83.

Risse, Thomas. “Arguing and Deliberation in International Relations.” In *The Oxford Handbook of Deliberative Democracy*, edited by André Bächtiger et al., 518–34. Oxford University Press, 2018.

Roberts, Dorothy. “Can Research on the Genetics of Intelligence Be ‘Socially Neutral’?” *Hastings Center Report* 45 (2015): S50–S53.

——. “Collateral Consequences, Genetic Surveillance, and the New Biopolitics of Race.” *Howard Law Journal* 54 (2011): 567–86.

——. *Fatal Invention: How Science, Politics, and Big Business Re-Create Race in the Twenty-First Century.* New Press, 2012.

——. “Public Deliberation in an Age of Direct Citizen Participation.” *American Review of Public Administration* 34 (2004): 315–53.

——. “The Social Immorality of Health in the Gene Age: Race, Disability, and Inequality.” In *Against Health: How Health Became the New Morality*, edited by Jonathan M. Metzl and Anna Kirkland, 61–71. New York University Press, 2010.

Roberts, Jennifer, et al. “Experts and Evidence in Deliberation: Scrutinising the Role of Witnesses and Evidence in Mini-Publics, a Case Study.” *Policy Studies* 53 (2022): 3–32.

Robertson, John A. *Children of Choice: Freedom and the New Reproductive Technologies.* Princeton University Press, 1994.

Roche, Laura, Down Adams, and Megan Clark. “Research Priorities of the Autism Community: A Systematic Review of Key Stakeholder Perspectives.” *Autism* 25 (2021): 336–48.

Røe, luf Dimitri. “The High Cost of New Cancer Therapies—A Challenge of Inequality for All Countries.” *JAMA Oncology* 3 (2017): 1169–70.

Rosanvallon, Pierre. *Counter-Democracy: Politics in an Age of Distrust.* Translated by Arthur Goldhammer. Cambridge University Press, 2008.

Rose, Nikolas. “Government, Authority and Expertise in Advanced Liberalism.” *Economy and Society* 22 (1993): 283–99.

Rose-Ackerman, Susan, Stefanie Egidy, and James Fowkes. *Due Process of Lawmaking: The United States, South Africa, Germany, and the European Union.* Cambridge University Press, 2015.
Rosen, George. *A History of Public Health.* Johns Hopkins University Press, 2015.
Rosenberg, Erica. "Environmental Procedures at the FCC: A Case of Corporate Capture." *Environment: Science and Policy for Sustainable Development* 64 (2022): 18–27.
Rosenberg, Shawn W. *Deliberation, Participation and Democracy: Can the People Govern?* Palgrave Macmillan, 2007.
Rosenblatt, Adam. "Autism, Advocacy Organizations, and Past Injustice." *Disability Studies Quarterly* 38 (2018): 450–67.
Rothbard, Murray N. *The Progressive Era.* Edited by Patrick Newman with a foreword by Judge Andrew P. Napolitano. Ludwig von Mises Institute, October 6, 2017.
Rousseau, Jean-Jacques. *The Social Contract.* Translated by Maurice Cranston. Penguin Books, 1968.
Royal, Charmaine D., et al. "Living with Sickle Cell Disease: Traversing 'Race' and Identity." In *Genetics and Global Public Health: Sickle Cell and Thalassaemia*, edited by Simon Dyson and Karl Atkin, 63–78. Routledge, 2012.
Rudow, Dianne LaPointe, et al. "Consensus Conference on Best Practices in Live Kidney Donation: Recommendations to Optimize Education, Access, and Care." *American Journal of Transplantation* 15 (2015): 914–22.
Rueda, Jon, Íñigo de Miguel Beriain, and Lluis Montoliu. "Affordable Pricing of CRISPR Treatments Is a Pressing Ethical Imperative." *CRISPR Journal* 7 (2024): 220–26.
Rulli, Tina. "Reproductive CRISPR Does Not Cure Disease." *Bioethics* 33 (2019): 1072–82.
Rusike, Elijah. "Exploring Food and Farming Futures in Zimbabwe: A Citizens' Jury and Scenario Workshop Experiment." In *Science and Citizens: Globalization and the Challenge of Engagement*, edited by Melissa Leach et al., 249–55. Zed Books, 2007.
Ryan, Alyssa, et al. "The Impact of Sex on Motor Vehicle Crash Injury Outcomes." *Journal of Transportation Safety & Security* 14 (2022): 818–42.
Ryan, Matthew, and Graham Smith. "Defining Mini-Publics." In *Deliberative Mini-Publics: Involving Citizens in the Democratic Process*, edited by Kimmo Grönlund et al., 2–39. ECPR Press, 2014.
Ryan, Molly J., et al. "Technologies of Trust in Epidemic Response: Openness, Reflexivity and Accountability During the 2014–2016 Ebola Outbreak in West Africa." *BMJ Global Health* 4 (2019): e001272.
Ryfe, David M. "Does Deliberative Democracy Work?" *Annual Review of Political Science* 8 (2005): 49–71.
Saffon, Maria Paula, and Nadia Urbinati. "Procedural Democracy, the Bulwark of Equal." *Political Theory* 41 (2013): 441–81.

Sahay, Ratna, et al. *Financial Inclusion: Can It Meet Multiple Macroeconomic Goals?* International Monetary Fund, 2015.

Saini, Angela. *Superior: The Return of Race Science*. Beacon Press, 2019.

Salcedo, Jonathan, et al. "Cost-Effectiveness of a Hypothetical Cell or Gene Therapy Cure for Sickle Cell Disease." *Scientific Reports* 11 (2021): 1–10.

Salsberg, Edward, et al. "Estimation and Comparison of Current and Future Racial/Ethnic Representation in the US Health Care Workforce." *JAMA* 4 (2023): e213789.

Sanches, Jose, et al. "Pain During Cesarean Delivery: A Patient-Related Prospective Observational Study Assessing the Incidence and Risk Factors for Intraoperative Pain and Intravenous Medication Administration." *Anaesthesia Critical Care & Pain Medicine* 43 (2024): 101310.

Sandel, Michael. *The Case Against Perfection: Ethics in the Age of Genetic Engineering*. Belknap Press, 2007.

——. *The Tyranny of Merit: What's Become of the Common Good?* Farrar, Straus and Giroux, 2020.

Sanders, Lynn M. "Against Deliberation." *Political Theory* 25 (1997): 347–76.

Savulescu, Julian. "New Breeds of Humans: The Moral Obligation to Enhance." *Reproductive Biomedicine* 10 (2005): 36–39.

Saxton, Gregory D., et al. "Rules of Crowdsourcing: Models, Issues, and Systems of Control." *Information Systems Management* 30 (2013): 2–20.

Scanlon, T. M. *What We Owe to Each Other*. Harvard University Press, 1982.

Scharp, Kristina M., Cimmiaron F. Alvarez, and Brittan A. Barker. "Conceptualizing Other-Resilience: Exploring How Hearing Parents Enact Resilience for Themselves and Their Children Who Use Cochlear Implants." *Human Communication Research* 49 (2023): 345–60.

Scheffer, Edward R., and Frazer A. Tessema. "The Double-Edged Sword of Extremely High Prices for Gene Therapies in Sickle Cell Disease." *JAMA* 332 (2024): 703–5.

Scheinerman, Naomi. "Public Engagement Through Inclusive Deliberation: The Human Genome International Commission and Citizens' Juries." *American Journal of Bioethics* 12 (2023): 66–76.

——, and Jacob S. Sherkow. "Governance Choices of Genome Editing Patents." *Frontiers in Political Science* 3 (2021).

Scheman, Naomi. "Epistemology Resuscitated: Objectivity as Trustworthiness." In *Engendering Rationalities*, edited by Nancy Tuana and Sandra Morgen, 23–38. State University of New York Press, 2001.

Scheufele, Dietram A., et al. "What We Know About Effective Public Engagement on CRISPR and Beyond." *Proceedings of the National Academies of Science* 118 (2021): e2004835117.

Schkade, David, et al. "What Happened on Deliberation Day?" *California Law Review* 95 (2007): 915–44.

Schmidthuber, Lisa, et al. "Government Openness and Public Trust: The Mediating Role of Democratic Capacity." *Public Administration Review* 81 (2021): 91–109.

Schwartz, Daniel. "Preventing Capture Through Consumer Empowerment Groups." In *Preventing Regulatory Capture: Special Interest Influence and How to Limit It*, edited by Daniel Carpenter and David A. Moss, 365–96. Cambridge University Press, 2014.

Schweigert, Francis J. "Strengthening Citizenship Through Deliberative Polling." *Journal of Community Practice* 18 (2010): 19–39.

Scott, James C. *Seeing Like a State: How Certain Schemes to Improve the Human Condition Have Failed*. Yale University Press, 1998.

Scudder, Molly. *Beyond Empathy and Inclusion: The Challenge of Listening in Democratic Deliberation*. Oxford University Press, 2021.

Scuffham, Paul A., et al. "Engaging the Public in Healthcare Decision-Making: Quantifying Preferences for Healthcare Through Citizens' Juries." *British Medical Journal* 4 (2014): e005437.

Scully, Jackie Leach. *Disability Bioethics: Moral Bodies, Moral Difference*. Rowman & Littlefield, 2008.

Scully, Matthew. *Dominion: The Power of Man, the Suffering of Animals, and the Call to Mercy*. St. Martin's Press, 2002.

Segal, Judy Z. "Sex, Drugs, and Rhetoric: The Case of Flibanserin for 'Female Sexual Dysfunction.'" *Social Studies of Science* 48 (2018): 459–82.

Seifer, David B., Fady Sharara, and Tarun Jain. "The Disparities in ART (DART) Hypothesis of Racial and Ethnic Disparities in Access and Outcomes of IVF Treatment in the USA." *Reproductive Sciences* 49 (2022): 2084–88.

Selfa, Theresa, et al. "Governing Gene Editing in Agriculture and Food in the United States: Tensions, Contestations, and Realignments." *Elementa: Science of the Anthropocene* 9 (2021): 00153.

Sénit, Carole-Anne. "Transforming Our World? Discursive Representation in the Negotiations on the Sustainable Development Goals." *International Environmental Agreements: Politics, Law and Economics* 20 (2020): 411–29.

Setälä, Maija. "The Role of Deliberative Mini-Publics in Representative Democracy: Lessons from the Experience of Referendums." *Representation* 47 (2011): 201–13.

——, et al. "Beyond Polarization and Selective Trust: A Citizens' Jury as a Trusted Source of Information." *Politics* (2021): 1–17.

Seyd, Ben, et al. "'Follow the Science': The Value and Limitations of Scientific Knowledge in Democratic Governance." *Public Understanding of Science* 34 (2024): 2–18.

Shaheen, Nabeel, et al. "CRISPR-Cas System, a Possible "Savior" of Rice Threatened by Climate Change: An Updated Review." *Rice* 16 (June 2023).

Shakespeare, Tom. "The Social Model of Disability." In *The Disability Studies Reader*, edited by Leonnard J. Davis, 195–203. Routledge, 2017.

Shanmugamab, Sabarathinam, et al. "Advanced CRISPR/Cas-Based Genome Editing Tools for Microbial Biofuels Production: A Review." *Renewable Energy* 149 (2020): 1107–19.

Shansky, Rebecca M. "Are Hormones a "Female Problem" for Animal Research?" *Neuroscience* 36 (2019): 825–26.

Shapin, Steven. *A Social History of Truth: Civility and Science in Seventeenth-Century England*. University of Chicago Press, 1994.
Shapiro, Beth. *How to Clone a Mammoth: The Science of De-Extinction*. Princeton University Press, 2015.
Shapiro, Ian. "Collusion in Restraint of Democracy: Against Political Deliberation." *Daedalus* 146 (2017): 77–84.
——. *Politics Against Domination*. Harvard University Press, 2018.
——, and Casiano Hacker-Cordón. "Promises and Disappointments: Reconsidering Democracy's Value." In *Democracy's Values*, edited by Ian Shapiro and Casiano Hacker-Cordón, 1–20. Cambridge University Press, 2010.
Shapiro, Sidney A. "Law, Expertise and Rulemaking Legitimacy: Revisiting the Reformation." *Environmental Law Review* 49 (2019): 661–82.
Shapiro, Stuart. "Does the Amount of Participation Matter? Public Comments, Agency Responses and the Time to Finalize a Regulation." *Policy Sciences* 41 (2008): 33–49.
Sharp, Burt M., et al. "Inactivation of Phosphodiesterase-4B Gene in Rat Nucleus Accumbens Shell by CRISPR/Cas9 or Positive Allosteric Modulation of the Protein Affects the Motivation to Chronically Self-Administer Nicotine." *Nature* 14 (March 2023).
Sharpes, Donald K. *The Evolution of the Social Sciences*. Lexington Books, 2009.
Shelley, Mary. *Frankenstein; or, The Modern Prometheus*. Edited by J. Paul Hunter. W. W. Norton, 2012. Originally published in 1818.
Shelton, Clifford, et al. "The 'Haves' and 'Have-Nots' of Personal Protective Equipment During the COVID-19 Pandemic: The Ethics of Emerging Inequalities Amongst Healthcare Workers." *Journal of Medical Ethics* 48 (2022): 653–57.
Shi, Yajun, et al. "CRISPR/Cas9: Implication for Modeling and Therapy of Amyotrophic Lateral Sclerosis." *Frontiers in Neuroscience* 17 (2023).
Shimazono, Susumo. "Reasons Against the Selection of Life: From Japan's Experience of Prenatal Genetic Diagnosis." In *Human Enhancement*, edited by Julian Savulescu and Nick Bostrom, 291–314. Oxford University Press, 2009.
Shrader-Frechette, Kristin. *Environmental Justice: Creating Equality, Reclaiming Democracy*. Oxford University Press, 2002.
Siegrist, Michael. "The Influence of Trust and Perceptions of Risks and Benefits on the Acceptance of Gene Technology." *Risk Analysis* 20 (2000): 195–203.
——, and George Cvetkovich. "Perceptions of Hazards: The Role of Social Trust and Knowledge." *Risk Analysis* 20 (2000): 713–19.
Silver, Hilary, Alan Scott, and Yuri Kazepov. "Participation in Urban Contention and Deliberation." *International Journal of Urban and Regional Research* 34 (2010): 453–77.
Simpson, Thomas W. "Evaluating Google as an Epistemic Tool." *Metaphilosophy* 43 (2012): 426–45.
Singh, Jennifer. *Multiple Autisms: Spectrums of Advocacy and Genomic Science*. University of Minnesota Press, 2016.

Singleton, Sara. *Constructing Cooperation. The Evolution of Institutions of Comanagement.* University of Michigan Press, 1998.
———. "Co-operation or Capture? The Paradox of Co-management and Community Participation in Natural Resource Management and Environmental Policy-Making." *Environmental Politics* 9 (2000): 1–21.
Sintomer, Yves. "From Deliberative to Radial Democracy: Sortition and Politics in the Twenty-First Century." In *Legislature by Lot: Transformative Designs for Deliberative Governance*, edited by John Gastil and Erik Olin Wright, 47–74. Verso, 2019.
Siu, Alice. "Deliberation and the Challenge of Inequality." *Daedalus* 146 (2017): 119–28.
Sloman, Steven, and Philip Fernbach. *The Knowledge Illusion: Why We Never Think Alone.* Riverhead Books, 2017.
Smith, Adam. *An Inquiry into the Nature and Causes of the Wealth of Nations.* Liberty Fund, 1981.
Smith, Graham. *Democratic Innovations: Designing Institutions for Citizen Participation.* Cambridge University Press, 2009.
———, and Corinne Wales. "Citizens' Juries and Deliberative Democracy." *Political Studies* 48 (2000): 51–65.
Smith, Rogers. *Stories of Peoplehood: The Politics and Morals of Political Membership.* Cambridge University Press, 2003.
Smith, William. "Transnational and Global Deliberation." In *The Oxford Handbook of Deliberative Democracy*, edited by André Bächtiger et al., 856–68. Oxford University Press, 2018.
Solazzo, Alexa L. "Different and Not Equal: The Uneven Association of Race, Poverty, and Abortion Laws on Abortion Timing." *Social Problems* 66 (2019): 519–47.
Song, Gaoyuan. "CRISPR/Cas9: A Powerful Tool for Crop Genome Editing." *Crop Journal* 4 (2016): 75–82.
Sorensen, Lita. *Mob Rule or the Wisdom of the Crowd?* Greenhave Publishing, 2020.
Soyer, Emre, and Robin M. Hogarth. "The Illusion of Predictability: How Regression Statistics Mislead Experts." *International Journal of Forecasting* 28 (2012): 695–711.
Sparrow, Robert. "Defending Deaf Culture: The Case of Cochlear Implants." *Journal of Political Philosophy* 13 (2005): 135–52.
———. "A Not-So-New Eugenics: Harris and Savulescu on Human Enhancement." *Hastings Center Report* 41 (2011): 32–42.
Steinberger, Peter J. *The Concept of Political Judgment.* University of Chicago Press, 1993.
Steinbock, Bonnie. "Designer Babies: Choosing Our Children's Genes." *Lancet* 372 (2008): 1294–95.
Stiglitz, Joseph E. "The Role of the State in Financial Markets." *World Bank Economic Review* 7 (1993): 19–52.
Stokel-Walker, Chris. "What Do We Know About Covid Vaccines and Preventing Transmission?" *BMJ* 376 (2022).

Stone, Peter. *The Luck of the Draw: The Role of Lotteries in Decision Making.* Oxford University Press, 2011.

Strandberg, Kim, and Kimmo Grönlund. "Online Deliberation: Theory and Practice in Virtual Mini-Publics." In *Deliberative Mini-Publics: Involving Citizens in the Democratic Process*, edited by Kimmo Grönlund et al., 93–113. ECPR Press, 2014.

Stranger, Barbara E., et al. "Progress and Promise of Genome-Wide Association Studies for Human Complex Trait Genetics." *Genetics in Medicine* 187 (2011): 367–83.

Strathern, Paul. *A Brief History of Medicine.* Carroll & Graf Publishers, 2005.

Street, Jackie, et al. "The Use of Citizens' Juries in Health Policy Decision-Making: A Systematic Review." *Social Science & Medicine* 109 (2014): 1–9.

Suarez-Alvarez, A., and A. J. Lopez-Menendez. "Is COVID-19 Vaccine Inequality Undermining the Recovery from the COVID-19 Pandemic?" *Journal of Global Health* 12 (2022): 05020.

Suiter, Jane, et al. "Scaling Up Deliberation: Testing the Potential of Mini-Publics to Enhance the Deliberative Capacity of Citizens." *Swiss Political Science Review* 26 (2020): 253–72.

Sullivan, Emily, et al. "Can Real Social Epistemic Networks Deliver the Wisdom of Crowds?" In *Oxford Studies in Experimental Philosophy, Volume 3*, edited by Tania Lombrozo et al., 29–63. Oxford University Press, 2020.

Sunshine, Jason, and Tom R. Tyler. "The Role of Procedural Justice and Legitimacy in Shaping Public Support for Policy." *Law & Society Review* 37 (2003): 513–48.

Surendran, Shilpa, et al. "A Scoping Review of the Ethical Issues in Gender-Affirming Care for Transgender and Gender-Diverse Individuals." *BMC Medical Ethics* 26 (2025).

Surowiecki, James. *The Wisdom of Crowds: Why the Many Are Smarter Than the Few and How Collective Wisdom Shapes Business, Economics, Societies, and Nations.* Doubleday, 2004.

Taeihagh, Araz, et al. "Assessing the Regulatory Challenges of Emerging Disruptive Technologies." *Regulation & Governance* 15 (2021): 1009–19.

Tallie, Lindsey Smith, and Barry Popkin. "The Promise and Pitfalls of 'Make America Healthy Again.'" *Lancet Diabetes & Endocrinology* 13 (2025): 274–75.

Talpin, Julien. "When Democratic Innovations Let the People Decide: An Evaluation of Co-Governance Experiments." In *Evaluating Democratic Innovations: Curing the Democratic Malaise?*, edited by Brigitte Geissel and Kenneth Newton, 184–214. Routledge, 2012.

Taner, Timucin, et al. "Summary of a Consensus Conference on the Management of Highly Sensitized Multiorgan Transplant Candidates." *Mayo Clinic Proceedings* 100 (2025): P700–11.

Tao, Yong, et al. "Treatment of Monogenic and Digenic Dominant Genetic Hearing Loss by CRISPR-Cas9 Ribonucleoprotein Delivery In Vivo." *Nature Communications* 14 (2023): 4928.

Tapper, Melbourne. *In the Blood: Sickle Cell Anemia and the Politics of Race.* University of Pennsylvania Press, 1999.
Tasnim, Nishat. "A Foucauldian Perspective on Covid-19 Response." *International Journal of Law and Politics Studies* 3 (2021): 1–6.
Taylor, Jeremy, and Simon Denegri. "Industry Links with Patient Organisations." *British Medical Journal* 365 (2017): j1251.
"Technology Revolution Happening Around Us." *Congressional Record: House Section* 169 (2023): H6851.
Terrier, C., and J. E. Terrier. "Efficacy and Safety of Flibanserin "New Female Viagra®?": Literature Review." *Progrès en Urologie* 26 (2016): 592–99.
Tetlock, Philip E. *Expert Political Judgment: How Good Is It? How Can We Know?* Princeton University Press, 2005.
Thapar, Nandita, et al. "Application of CRISPR/Cas9 in the Management of Alzheimer's Disease and Parkinson's Disease: A Review." *Annals of Medicine & Surgery* 86 (2024): 329–35.
Thomas, Carol. "Disability and Diversity." In *Routledge International Handbook of Diversity Studies*, edited by Steven Hertovec, 43–51. Routledge, 2014.
Thomas, Craig W. "Maintaining and Restoring Public Trust in Government Agencies and Their Employees." *Administration and Society* 30 (1998): 166–93.
Thomas, Nicole, et al. "Public Communication and Outreach by Mosquito Programs in the United States." *PLOS* 4 (December 2024).
Thompson, Dennis F. "Deliberative Democratic Theory and Empirical Political Science." *Annual Review of Political Science* 11 (June 2008): 497–520.
———. "Who Should Govern Who Governs? The Role of Citizens in Reforming the Electoral System." In *Designing Deliberative Democracy: The British Columbia Citizens' Assembly*, edited by Mark E. Warren and Hilary Pearse, 20–49. Cambridge University Press, 2008.
Thompson, Morgan. "Explanations of the Gender Gap in Philosophy." *Philosophy Compass* 12 (2017): e12406.
Tienda, Marta. "Diversity as a Strategic Advantage: A Sociodemographic Perspective." In *Our Compelling Interests: The Value of Diversity for Democracy and a Prosperous Society*, edited by Earl Lewis and Nancy Cantor, 192–205. Princeton University Press, 2016.
Tilly, Charles. *Trust and Rule.* Cambridge University Press, 2005.
Tiwari, Jagesh Kumar, et al. "CRISPR/Cas Genome Editing in Tomato Improvement: Advances and Applications." *Frontiers Plant Science* 14 (2023).
Tocqueville, Alexis de. *Democracy in America.* Translated by Harvey C. Mansfield and Delba Winthrop. University of Chicago Press, 2000.
Tomes, Nancy. *The Gospel of Germs: Men, Women, and the Microbe in American Life.* Harvard University Press, 1998.
Torres, Jada Benn. "Race, Rare Genetic Variants, and the Science of Human Difference in the Post-Genomic Age." *Transforming Anthropology* 27 (2019): 37–49.
Townley, Cynthia. *A Defense of Ignorance: Its Value for Knowers and Roles in Feminist and Social Epistemologies.* Lexington Books, 2011.

Trapido, Joe. "Ebola: Public Trust, Intermediaries, and Rumour in the DR Congo." *Lancet Infectious Diseases* 19 (2019): P457-8.

Tripathi, Jaindra Nath, et al. "CRISPR/Cas9-Mediated Editing of DMR6 Orthologue in Banana (*Musa* spp.) Confers Enhanced Resistance to Bacterial Disease." *Plant Biotechnology Journal* 19 (2021): 1291–93.

Tschandl, Philipp. "Risk of Bias and Error from Data Sets Used for Dermatologic Artificial Intelligence." *JAMA Dermatology* 157 (2021): 1271–73.

Tsosiel, Rebecca. "Indigenous Peoples and Epistemic Injustice: Science, Ethics, and Human Rights." *Washington Law Review* 87 (2012): 1133–202.

Tucker, Paul. *Unelected Power: The Quest for Legitimacy in Central Banking and the Regulatory State*. Princeton University Press, 2018.

Tucker, William H. *"The Bell Curve" in Perspective: Race, Meritocracy, Inequality, and Politics*. Springer Nature, 2023.

Ubel, Peter A., et al. "Whose Quality of Life? A Commentary Exploring Discrepancies between Health State Evaluations of Patients and the General Public." *Quality of Life Research* 12 (2003): 599–607.

Urbinati, Nadia. *Democracy Disfigured: Opinion, Truth, and the People*. Harvard University Press, 2014.

———. *Mill on Democracy: From the Athenian Polis to Representative Government*. University of Chicago Press, 2002.

Uslaner, Eric M. "Democracy and Social Capital." In *Democracy and Trust*, edited by Mark E. Warren, 121–50. Cambridge University Press, 1999.

———. *Segregation and Mistrust: Diversity, Isolation, and Social Cohesion*. Cambridge University Press, 2012.

Valgarðsson V., et al. "A Crisis of Political Trust? Global Trends in Institutional Trust from 1958 to 2019." *British Journal of Political Science* 55 (2025): e15.

Van Asselt, Marjolein B. A. *Perspectives on Uncertainty and Risk: The Prima Approach to Decision Support*. Kluwer Academic Publishers, 2000.

van Beljouw, Sam P. B., et al. "RNA-Targeting CRISPR-Cas Systems." *Nature Reviews Microbiology* 21 (2023): 21–34.

Van Crombrugge, Ronald. "The Derailed Promise of a Participatory Minipublic: The Citizens' Assembly Bill in Flanders." *Journal of Deliberative Democracy* 16 (2020): 63–72.

Van Dijk, Lisa, Emma Turkenburg, and James Pow. "The Perceived Legitimacy of Deliberative Minipublics: Taking the Perspective of Polarized Citizens." *European Political Science Review* 15 (2023): 409–26.

Vandorsten, J. P., et al. "NIH Consensus Development Conference: Diagnosing Gestational Diabetes Mellitus." *NIH Consensus and State-of-the-Science Statements* 29 (2013): 1–31.

Varden, Helga. "Self-Governance and Reform in Kant's Liberal Republicanism—Ideal and Non-Ideal Theory in Kant's Doctrine of Right." *Doispontos* 13 (2016): 39–70.

Velasco, Sebastian, and Wendy Snodgrass. "The Use of Race in eGFR: Why Racial Justice Requires Accuracy." *Delaware Journal of Public Health* 134 (2022): P827–828.

Vergne, Antoine. "A Brief Survey of the Literature of Sortition: Is the Age of Sortition upon Us?" In *Sortition: Theory and Practice*, edited by Gil Delannoi and Oliver Dowlen, 69–87. Imprint Academic, 2010.

Vermeule, Adrian. *Mechanisms of Democracy: Institutional Design Writ Small.* Oxford University Press, 2007.

Vertinsky, Liza. "Pharmaceutical (Re)Capture." *Yale Journal of Health Policy, Law, and Ethics* 20 (2021): 145–224.

Viehoff, Daniel. "Authority and Expertise." *Journal of Political Philosophy* 24 (2016): 406–26.

Vigoda, Eran. "From Responsive to Collaboration: Governance, Citizens, and the Next Generation of Public Administration." In *The Age of Direct Citizen Participation*, edited by Nancy C. Roberts. Routledge, 2008.

Villaneuva, Angela G., and Mary A. Majumder. "Hashtag Who's Missing? Lessons for Genomic Databases." *Disability and Health Journal* 14 (2021): 1–5.

Vinci, Tom, and Jason Scott Robert. "Aristotle and Modern Genetics." *Journal of the History of Ideas* 66 (2005): 201–21.

Vinck, Patrick, et al. "Institutional Trust and Misinformation in the Response to the 2018–19 Ebola Outbreak in North Kivu, DR Congo: A Population-Based Survey." *Lancet* 19 (2019): 529–36.

Vlashi, Rexhina, et al. "Potential Therapeutic Strategies for Osteoarthritis via CRISPR/Cas9 Mediated Gene Editing." *Reviews in Endocrine and Metabolic Disorders* 25 (2023): 339–67.

Waggoner, Miranda R., and Anne Drapkin Lyerly. "Clinical Trials in Pregnancy and the 'Shadows of Thalidomide': Revisiting the Legacy of Frances Kelsey." *Contemporary Clinical Trials* 119 (2022): 106806.

Wagner-Coello, Helen Urpi, Maria Elena Villar, and Matthew DeGennaro. "Incorporating Citizen Science Engagement in a Vector Surveillance Undergraduate Internship." *Discover Education* 3 (2024).

Wailoo, Keith. "Introduction: Genetic Claims and the Unsettled Past." In *Genetics and the Unsettled Past*, edited by Keith Wailoo et al., 1–10. Rutgers University Press, 2012.

——. "Sickle Cell Disease—A History of Progress and Peril." *New England Journal of Medicine* 376 (2017): 805–7.

Wallis, Jacqueline Mae. "Is It Ever Morally Permissible to Select for Deafness in One's Child?" *Medicine, Health Care and Philosophy* 23 (2020): 3–15.

Walter, Emmanuel, et al. "Evaluation of the BNT162b2 Covid-19 Vaccine in Children 5 to 11 Years of Age." *New England Journal of Medicine* 386 (2022): 35–46.

Waltz, Emily. "CRISPR-Edited Crops Free to Enter Market, Skip Regulation." *Nature Biotechnology* 34 (2016): 582.

Wang, Haoyi, and Hui Yang. "Gene-Edited Babies: What Went Wrong and What Could Go Wrong." *PLOS Biology* 17 (2019): e3000224.

Wardrop, Alistair. "Medicalization and Epistemic Injustice." *Medical Health Care and Philosophy* 18 (2015): 341–52.

Warren, Mark E. "Democratic Theory and Trust." In *Democracy and Trust*, edited by Mark E. Warren, 310–45. Cambridge University Press, 1999.
——. "Introduction." *Democracy and Trust*, edited by Mark E. Warren, 1–21. Cambridge University Press, 1999.
——, and Hilary Pearse. *Designing Deliberative Democracy: The British Columbia Citizens' Assembly*. Cambridge University Press, 2008.
Wartenberg, Thomas E. *The Forms of Power: From Domination to Transformation*. Temple University Press, 1990.
Washington, Harriet A. *Medical Apartheid: The Dark History of Medical Experimentation on Black Americans from Colonial Times to the Present*. Doubleday, 2006.
Webb, Mark Owen. "The Epistemology of Trust and the Politics of Suspicion." *Pacific Philosophical Quarterly* 73 (1992): 390–400.
Weber, Elke U., and Jessica S. Ancker. "Risk Perceptions and Risk Attitudes in the United States and Europe." In *The Reality of Precaution: Comparing Risk Regulation in the United States and Europe*, edited by Jonathan B. Wiener et al., 480–91. RFF Press, 2011.
Weber, Max. "Bureaucracy." Translated by Tony Waters and Dagmar Waters. In *Weber's Rationalism and Modern Society: New Translations on Politics, Bureaucracy, and Social Stratification*, edited by Tony Waters and Dagmar Waters, 73–128. Palgrave Macmillan, 2015.
Weldon, Laurel. *When Protest Makes Policy*. University of Michigan Press, 2012.
Wenman, Mark. *Agonistic Democracy: Constituent Power in the Era of Globalisation*. Cambridge University Press, 2013.
Werner, Hannah, and Sofie Marien. "Process vs. Outcome? How to Evaluate the Effects of Participatory Processes on Legitimacy Perceptions." *British Journal of Political Science* 52 (2022): 429–36.
White-Lewis, Damani K., et al. "'Moving the Science Forward': Faculty Perceptions of Culturally Diverse Mentor Training Benefits, Challenges, and Support." *CBE-Life Sciences Education* 21 (2022).
Whiteside, Alan, and Nana K. Poku. *AIDS and Governance*. Routledge Global Health Series. Routledge, 2007.
Wiegand, Aaron A. "'It's Probably an STI Because You're Gay': A Qualitative Study of Diagnostic Error Experiences in Sexual and Gender Minority Individuals." *BMJ Quality & Safety* 33 (2024): 432–41.
Wiener, Jonathan B. "The Rhetoric of Precaution." In *The Reality of Precaution: Comparing Risk Regulation in the United States and Europe*, edited by Jonathan B. Wiener et al., 3–35. RFF Press, 2011.
Wiggleton-Little, Jada. "'Just' a Painful Period: A Philosophical Perspective Review of the Dismissal of Menstrual Pain." *Health and Nursing* 20 (2024): 1–6.
Wild, Verina, and Nikola Biller-Andorno. "Pregnant Women's Views About Participation in Clinical Research." In *Clinical Research Involving Pregnant Women*, edited by Françoise Baylis and Angela Ballantyne, 119–36. Springer, 2016.

Williams, T. N. "Sickle Cell Disease in Sub-Saharan Africa." *Hematology/Oncology Clinics of North America* 30 (2016): 343–58.

Wilson, James Q. *Bureaucracy: What Government Agencies Do and Why They Do It.* Basic Books, 1989.

Wolfensberger, Markus, and Anthony Wrigley. *Distrust, Fear, and Science-Denial in Medicine and Healthcare.* Routledge, 2025.

Wong, Chi Heem, et al. "The Estimated Annual Financial Impact of Gene Therapy in the United States." *Gene Therapy* 30 (2023): 751–73.

Wong-Rieger, Durhane. "Moving from Patient Advocacy to Partnership: A Long and Bumpy Road." *Patient* 10 (2017): 271–76.

Woodly, Deva. *Black Lives Matter and the Democratic Necessity of Social Movements.* Oxford Academic Press, 2021.

Woolfson, Adrian. *Life Without Genes.* Harper Collins, 2000.

Wortman, Paul, et al. "Do Consensus Conferences Work? A Process Evaluation of the NIH Consensus Development Program." *Journal of Health Politics Policy and Law* 13 (1988): 469–98.

Wright, Austin, et al. "Tracking Mask Mandates During the Covid-19 Pandemic." University of Chicago, 2020.

Wrong, Dennis. *Power: Its Forms, Bases and Uses.* Routledge, 1995.

Xenos, Michael A. "Citizens Make Sense of Science Issues: Supply and Demand Factors for Science News and Information in the Digital Age." In *The Oxford Handbook of the Science of Science Communication*, edited by Kathleen Hall Jamieson et al., 283–90. Oxford University Press, 2017.

Yackee, Jason Webb, and Susan Webb Yackee. "A Bias Towards Business? Assessing Interest Group Influence on the U.S. Bureaucracy." *Journal of Politics* 68 (2006): 128–39.

——. "The Politics of Rulemaking in the United States." *American Review of Political Science* 22 (2019): 37–55.

——. "Regulatory Capture's Self-Serving Application." *Public Administration Review* 81 (2022): 63–71.

——. "The 'Science' of Policy Development During Administrative Rulemaking." *Policy Studies Journal* (2019).

——. "Sweet-Talking the Fourth Branch: The Influence of Interest Group Comments on Federal Agency Rulemaking." *Journal of Public Administration Research and Theory* 6 (2006): 103–24.

Yaqub, Ohid, et al. "Attitudes to Vaccination: A Critical Review." *Social Science & Medicine* 112 (2014): 1–11.

Yearby, Ruqaiijah. "Race Based Medicine, Colorblind Disease: How Racism in Medicine Harms Us All." *American Journal of Bioethics* 21 (2021): 19–27.

Young, Iris Marion. *Justice and the Politics of Difference.* Princeton University Press, 1990.

Young, Shaun P. "Rawlsian Reasonableness: A Problematic Presumption?" *Canadian Journal of Political Science* 39 (2006): 159–80.

Yuan, Yinglin, et al. "Complement Networks in Gene-Edited Pig Xenotransplantation: Enhancing Transplant Success and Addressing Organ Shortage." *Journal of Translational Medicine* 22 (2022).

Yudell, Michael, et al. "Taking Race Out of Human Genetics." *Science* 351 (2016): 564–65.

Zahedipour, Farzaneh, et al. "Harnessing CRISPR Technology for Viral Therapeutics and Vaccines: From Preclinical Studies to Clinical Applications." *Virus Research* 341 (2024): 199314.

Zhai, Xiaomei, et al. "Lessons from the He Jiankui Incident." *Issues in Science and Technology* 35 (2019): 20–22.

Zhang, Yingxiao, et al. "The Emerging and Uncultivated Potential of CRISPR Technology in Plant Science." *Nature Plants* 5 (2019): 778–94.

Zimring, Carl A. *Clean and White: A History of Environmental Racism in the United States.* New York: New York University Press, 2015.

Zuboff, Shoshana. *The Age of Surveillance Capitalism: The Fight for a Human Future at the New Frontier of Power.* New York: Hachette Book Group, 2019.

Index

Italic page numbers refer to illustrations.

www.ingramcontent.com/pod-product-compliance
Lightning Source LLC
LaVergne TN
LVHW050911241026
850939LV00024BA/220
* 9 7 8 1 4 6 9 6 9 7 2 2 2 *